Dahlem Workshop Reports
Life Sciences Research Report 35
Biotechnology: Potentials and Limitations

The goal of this Dahlem Workshop is:
to evaluate how manipulated genes and cells can be used for industry, medicine, and agriculture

Life Sciences Research Reports
Editor: Silke Bernhard

Held and published on behalf of the
Stifterverband für die Deutsche Wissenschaft

Sponsored by:
Fonds der Chemie
in cooperation with the
Bundesministerium für
Forschung und Technologie

Biotechnology: Potentials and Limitations

S. Silver, Editor

Report of the Dahlem Workshop on
Biotechnology: Potentials and Limitations
Berlin 1985, March 24–29

Rapporteurs:
J. Collins · R.B. Flavell · M.-R. Kula
M. Smith

Program Advisory Committee:
S. Silver, Chairperson · P. Broda · A.M. Chakrabarty
J. Collins · J.E. Davies · D.A. Hopwood · C.J. Knowles
P. Starlinger · M. van Montagu

Springer-Verlag
Berlin Heidelberg New York Tokyo

Copy Editors: K. Geue, J. Lupp
Text Preparation: J. Lambertz, D. Lewis
Photographs: E.P. Thonke

With 4 photographs, 17 figures, and 21 tables

ISBN-13:987-3-642-70537-3 e-ISBN-13:987-3-642-70535-9
DOI: 10.1007/987-3-642-70535-9

CIP-Kurztitelaufnahme der Deutschen Bibliothek

Biotechnology: potentials and limitations:
report of the Dahlem Workshop on Biotechnology, Potentials and Limitations
Berlin 1985, March 24–29/S. Silver, ed. Rapporteurs: J. Collins...
[Held and publ. on behalf of the Stifterverb. für d. Dt. Wiss.
Sponsored by: Fonds d. Chemie in cooperation with the Bundesministerium für Forschung u. Technologie].
– Berlin; Heidelberg; New York; Tokyo: Springer, 1986
(Life sciences research report; 35 (Dahlem Workshop reports)
ISBN-13:987-3-642-70537-3

NE: Silver, Simon [Hrsg.]; Collins, J. [Mitverf.];
Workshop on Biotechnology, Potentials and Limitations <1985, Dahlem, Berlin>; 1. GT

2114/3020-5 4 3 2 1 0

Table of Contents

The Dahlem Konferenzen
S. Bernhard *ix*

Introduction
S. Silver *1*

Technical Approaches
Group Report
M. Smith, Rapporteur
W.M. Barnes, H. Blöcker, U. Krawinkel, H. Lehrach, R. Marquardt, M. van Montagu, P. Weglenski, G. Winter, S.J. Wodak 5

Cloning and Expressing Genes for Clinically Useful Proteins
R.M. Kay, R. Kaufman, P. Schendel, K. Turner, and R. Kamen *19*

Synthetic Genes
H. Blöcker and R. Frank *41*

Protein Engineering
G. Winter, P. Carter, H. Bedouelle, D. Lowe, R.J. Leatherbarrow, and A.R. Fersht 55

Microbiology and Industrial Products
Group Report
M.-R. Kula, Rapporteur
Y. Aharonowitz, J.D. Bu'lock, A.M. Chakrabarty, D.A. Hopwood, B. Mattiasson, J.G. Morris, O.M. Neijssel, H. Sahm, S. Silver, J. Sobieszczański, U. Stahl, K.N. Timmis *71*

Fermentation Products: Physiological and Bioenergetic Considerations
O.M. Neijssel and D.W. Tempest 83

Bioactive Microbial Secondary Metabolites
Y. Aharonowitz and G. Cohen 99

Technological Processes for Biotechnological Utilization of Microorganisms
B. Mattiasson 113

Animals and Medicines
Group Report
J. Collins, Rapporteur
H. Betz, J.E. Davies, W. Fiers, E. Paoletti, E. Pfaff, D.J. Rowlands, H. Schaller, G. Siewert, A.E. Sippel, E.F. Wagner 127

Vaccines - The Synthetic Antigen Approach
D.J. Rowlands 139

A Modern Approach to Live Vaccines: Recombinant Poxviruses
E. Paoletti, M.E. Perkus, A. Piccini, B.R. Lipinskas, and S.R. Mercer 155

The Future Role in Medicine of Proteins Made by Genetic Engineering
W. Fiers 165

Some Applications of Modern Immunological Techniques
U. Krawinkel 179

Gene Transfer into Mouse Stem Cells
E.F. Wagner, U. Ruether, and C.L. Stewart 185

Plants and Agriculture
Group Report
R.B. Flavell, Rapporteur
P.H. Abelson, W. Barz, J.E. Beringer, P. Broda, D.E. Eveleigh, K. Hahlbrock, R. Kahmann, W.J. Peacock, H. Saedler, F. Salamini, P. Starlinger, E. Weiler, M. Zenk 199

Plant Gene Engineering and Plant Agriculture
W.J. Peacock and E.S. Dennis 223

Secondary Products
K. Hahlbrock 241

Plant-Microbe Interactions
J.E. Beringer 259

Importance of the Rhizosphere in Plant-Microbe Interactions
J. Sobieszczański 275

The Production and Utilization of Lignocellulose
P. Broda 283

Social and Ethical Considerations for Biotechnology
J.E. Davies 293

List of Participants with Fields of Research 295

Subject Index 301

Author Index 313

The Dahlem Konferenzen

Founders
Recognizing the need for more effective communication between scientists, especially in the natural sciences, the Stifterverband für die Deutsche Wissenschaft*, in cooperation with the Deutsche Forschungsgemeinschaft**, founded Dahlem Konferenzen in 1974. The project is financed by the founders and the Senate of the City of Berlin.

Name
Dahlem Konferenzen was named after the district of Berlin called "Dahlem", which has a long-standing tradition and reputation in the arts and sciences.

Aim
The task of Dahlem Konferenzen is to promote international, interdisciplinary exchange of scientific information and ideas, to stimulate international cooperation in research, and to develop and test new models conducive to more effective communication between scientists.

Dahlem Workshop Model
Dahlem Konferenzen organizes four workshops per year, each with a limited number of participants. Since no type of scientific meeting proved effective enough, Dahlem Konferenzen had to create its own concept. This concept has been tested and varied over the years, and has evolved into its present form which is known as the *Dahlem Workshop Model*. This model provides the framework for the utmost possible interdisciplinary communication and cooperation between scientists in a given time period.

**The Donors Association for the Promotion of Sciences and Humanities*
***German Science Foundation*

The main work of the Dahlem Workshops is done in four interdisciplinary discussion groups. Lectures are not given. Instead, selected participants write background papers providing a review of the field rather than a report on individual work. These are circulated to all participants before the meeting to provide a basis for discussion. During the workshop, the members of the four groups prepare reports reflecting their discussions and providing suggestions for future research needs.

Topics
The topics are chosen from the fields of the Life Sciences and the Physical, Chemical, and Earth Sciences. They are of contemporary international interest, interdisciplinary in nature, and problem-oriented. Once a year, topic suggestions are submitted to a scientific board for approval.

Participants
For each workshop participants are selected exclusively by special Program Advisory Committees. Selection is based on international scientific reputation alone, although a balance between European and American scientists is attempted. Exception is made for younger German scientists.

Publication
The results of the workshops are the Dahlem Workshop Reports, reviewed by selected participants and carefully edited by the editor of each volume. The reports are multidisciplinary surveys by the most internationally distinguished scientists and are based on discussions of new data, experiments, advanced new concepts, techniques, and models. Each report also reviews areas of priority interest and indicates directions for future research on a given topic.

The Dahlem Workshop Reports are published in two series:
1) Life Sciences Research Reports (LS), and
2) Physical, Chemical, and Earth Sciences Research Reports (PC).

Director
Silke Bernhard, M.D.

Address
Dahlem Konferenzen
Wallotstrasse 19
1000 Berlin (West) 33

Biotechnology: Potentials and Limitations, ed. S. Silver, pp. 1-3. Dahlem Konferenzen 1986. Berlin, Heidelberg, New York, Tokyo: Springer-Verlag.

Introduction

S. Silver
Dept. of Biology
Washington University
St. Louis, MO 63130, USA

We need to place this workshop report into a broad context, considering various aspects of biotechnology which differentiate it from more narrow descriptions of industrial microbiology (including food technology, industrial fermentations, and the pharmaceutical industry, which combined have a history of more than 100 years). The various aspects of biotechnology include a) the introduction of recombinant DNA and cloning techniques that allow a more rational and planned (that is, engineering) approach to production problems than was possible in the past. This new logical approach is in marked contrast to earlier use of genetics to isolate high production mutants in the fermentation industry, the pharmaceutical industry, and in agricultural plant and animal husbandry. b) The new genetic engineering expands beyond the possibilities of product production in the factory. It is clearly understood that it will have an enormous impact on human medicine, on animal medicine and meat production, and on agronomy and the production of useful agricultural plants. c) We need to pay attention to more than merely the technological questions. As introduced by Blöcker and Frank (this volume), we need to be concerned about how the new products of biotechnology will affect the human community; and we must be especially concerned about how this technology will affect lesser developed nations. Most of us at the meeting were from developed, high technology advanced nations. A few of us have backgrounds or special attachments to countries where the level of technology, including biotechnology, is lower. Although a few in each of the four groups were concerned with "biotechnology in lesser developed and developing nations," this was not in the list of topics considered during the week. In some cases, for examples, in the group on TECHNICAL APPROACHES (Smith et al., this volume), there would have been little to say. High tech progress is going to be made in high tech societies. However, it was abundantly clear for the other groups (Kula et al., Collins et al., and Flavell et al., all this volume) that there will be an enormous impact on the rapidity of development and raising of living standards in developing countries if we can focus new types of biotechnology to their special needs, which are not identical with those of the more affluent industrial nations.

The Program Advisory Committee met a year before the conference to frame the goal of the workshop: "to evaluate how manipulated genes and cells can be used for industry, medicine, and agriculture." The meeting had an unusual and desirable mixture of a) people who work in industrial firms (some established traditional companies with biotechnological interests, and some with newer, specific biotechnology companies), b) people from government laboratories with responsibilities to develop national potentials, and c) a large number of people from university settings whose interests run from basic science in the laboratory to the hope of practical applications of their efforts. In each case, the predictions of how manipulated genes and cells will create new and large product markets, new industries, new medicines, and new agricultural materials elicited common understanding and remarkable consensus. The four groups were divided along the lines of where the efforts must be placed and where the opportunities are expected to occur.

Repeatedly during the meeting there was a strong consensus about the need to intensify basic research, as well as doing the applied research that characterizes biotechnology. For human protein products, progress has depended upon the rapidly developing understanding of polypeptide regulation of growth, the complex interactions between peptide hormones, and the detailed network of the components of the immunological system. Over a small number of years, those polypeptides that are understood at the current time (at the beginning of this technology) will have been made and tested. Further progress in the introduction of new products in biotechnology will require an intense interrelationship between the biotechnologists oriented toward product production and those involved in basic research in biological and biomedical sciences, oriented only toward greater understanding.

A related question was posed by M.R. Kula (who works in a Biotechnology Institute that has to bridge the gap between the basic scientists and the applied technologists), who was concerned about the meeting being heavily one of basic science and not about the technological applications. There are at least two reasons for that. The first is that I assembled a planning committee for the meeting that was heavily biased toward the basic sciences, because of a sense that this is where the hopes for the future will lie. Although we expect in biotechnology (as in previous types of technology) that the costs in money and in years of effort to bring products into useful forms will greatly exceed the costs of the basic research, the rate-limiting hurdle will be in the latter area. The technological hurdles will be formidable but will not limit what happens: once the basic ideas are available, the technology will be developed. The unique part of biotechnology will be to imagine what the possibilities are.

There was a discussion in several of the groups on the problems of introducing a novel science into a social and economic context. What biotechnologists are learning on this matter is not novel, although that does not make it any less important or difficult. People in the development of electronics and computers, in the pharmaceutical industry, and in many other types of industry that have grown from university research have had to face these problems in the past. It is the old situation of having to reinvent the wheel again and again. There is one aspect on which biotechnology seems to have handled this inherent difficulty better than some of our predecessor technologies: the people in the biotechnology companies by and large take a rather academic approach to free communication with one another at meetings such as this and open publication of many of their basic findings in the literature. This seems unique and certainly is different from the experience of the recent Silicone Valley Industry, which in other ways tries to emulate an academic environment, but not in open and free publication. Certainly there are questions of proprietary value; some ideas and market strategies were not discussed openly during the week. However, these were few and the surprise was the ease of communication back and forth. The number of papers with obvious commercial applicability appearing in the open literature from biotechnology companies is large and admirable.

The success of the workshop depended heavily on the remarkable efforts of Silke Bernhard and her staff, starting with the arrangements prior to the meeting, including importantly the facilities, support, and encouragement during the meeting itself, and then on with their tireless efforts to revise and edit the reports and papers into a presentable form. The rapporteurs carried out the remarkable task of writing group reports under the experienced eyes of the Dahlem Konferenzen staff during a single afternoon and night! That was the hardest task for the workshop, and the four rapporteurs deserve our special thanks.

Top row, left to right:
Rüdiger Marquardt, Greg Winter, Peter Weglenski, Wayne Barnes

Middle row, left to right:
Michael Smith, Shoshana Wodak, Marc van Montagu

Bottom row, left to right:
Ulrich Krawinkel, Hans Lehrach

Not shown: Helmut Blöcker

Biotechnology: Potentials and Limitations, ed. S. Silver, pp. 5-18. Dahlem Konferenzen 1986. Berlin, Heidelberg, New York, Tokyo: Springer-Verlag.

Technical Approaches
Group Report

M. Smith, Rapporteur
W.M. Barnes
H. Blöcker
U. Krawinkel
H. Lehrach
R. Marquardt
M. van Montagu
P. Weglenski
G. Winter
S.J. Wodak

INTRODUCTION

In evaluating the available technical resources and future needs for the manipulation of genes and cells for use in industry, medicine, and agriculture one has to consider the need for increased understanding of biology as well as the identification and achievement of very specific objectives. To this end, gene synthesis, mutagenesis, protein structure modeling and engineering, gene expression (transcription, translation, and post-transcriptional processing), host-vector systems, gene isolation, reverse genetics (transplacement), and applications of immunology were reviewed. In many of these areas the technology is very sophisticated (17, 21), in others it is lacking or is understood to such a limited degree that it cannot be applied in a predictable way. Most, if not all, of the technology is the result of academic investigator-initiated research. It is essential for future developments in molecular genetics that support of this non-applied research is not diminished.

DNA SYNTHESIS (4, 9)

There are two technical steps in gene synthesis, the chemical assembly of nucleotides into short (10-100 nucleotide) single-stranded DNA fragments and then their enzymatic assembly into double-strand, clonable DNA. Short DNA fragments (oligonucleotides) have important uses in addition to total or partial gene synthesis. These include their uses as probes for isolation of specific clones from cDNA (DNA derived by reverse transcription of mRNA) and genomic DNA libraries, as probes for the identification of specific mutations in genomic DNA and in recombinant DNA clones, as primers for DNA sequence determination and as site-specific mutagens for in vitro mutagenesis.

There are two highly effective chemistries for assembly of nucleotides into oligonucleotides: the phosphate triester method and the phosphite triester method. The preferred strategy of synthesis with both chemistries involves the use of a solid support (most commonly, controlled pore-glass beads) with a nucleoside attached, to which nucleoside-3' phosphates (with appropriate protecting groups) are added one at a time. This procedure can be carried out manually and also by fully automated, commercial instruments. An alternate, solid support strategy involves the use of paper discs (probably to be superceded by glass fiber discs). Here many syntheses can be carried out simultaneously by combining the discs into one of four groups (corresponding to each of the four nucleotide components of DNA) and adding the appropriate nucleotide. The discs are then recombined into four new groups determined by the next nucleotide required for each oligonucleotide, and the cycle is repeated. At present this synthetic method is carried out manually. In principle, the method can be automated, and if this is achieved together with technical improvements which increase yields to those achieved with controlled pore-glass beads (97-99%), then it will be very useful indeed.

A general, quick, one-step, batchwise oligonucleotide purification step will be useful for removing shorter oligonucleotide by-products as an alternative to high pressure liquid chromatography (HPLC) or gel electrophoresis. However, in recent experience the products have been sufficiently pure for use after ethanol precipitation. It seems very likely that with relatively minor changes in chemistry, oligonucleotides which can be used without gel electrophoresis or HPLC purification will be available routinely.

The enzymology of assembly of oligonucleotides into double-strand DNA is well established. As a general principle, longer nucleotide building blocks reduce the number of required enzymatic ligations and increase the specificity of the reaction. It is now possible to join twenty 50-nucleotide fragments into a 500 bp (base pair) DNA duplex and to clone it in one day. Longer duplexes can, in turn, be assembled from 500 bp duplexes. New strategies for DNA assembly are directed at reducing the number of nucleotides that have to be assembled chemically. One such strategy will depend on the development of the chemical synthesis of a long single strand (100-500 nucleotides) which can be used as a template for enzymatic synthesis of its complement. A variant of this uses short complementary oligonucleotides to promote the ligation of a long single strand from adjacent 50-nucleotide fragments. For these strategies to be useful, it is essential that the chemical synthesis be very reliable. At present, error frequency appears to be about one nucleotide in 500. It is not clear whether modification of the existing chemistry (shorter reaction time or milder conditions) or a change in the chemistry (new condensing agents or base-protecting groups) will be needed to increase fidelity of chemical synthesis.

The present status of chemical synthesis of genes is such that it probably offers the fastest route to a gene when its sequence is accurately known. Total gene synthesis is a very useful tool. However, synthesis applied to genes is most likely to be partial synthesis to adapt a gene from one organism so that it is expressed efficiently in a heterologous host or to construct specific mutational changes. And, as was indicated earlier, there are major applications of synthetic oligonucleotides in molecular biology which do not involve assembly into DNA.

MUTAGENESIS (9, 11, 13, 14, 20)
The focus of discussion was on the use of synthetic oligonucleotides, as heteroduplexes with wild-type DNA, to produce specific mutations. It should be noted that there have been major developments in methods and strategies for producing random point mutations and that these methods are very useful for analysis of promoters and of structural genes whose products are not molecularly defined (e.g., where the three-dimensional structure of a protein is not known).

In oligonucleotide-directed mutagenesis, the availability of oligonucleotides of more than fifteen nucleotides has increased the effectiveness, reliability, and diversity of the method. Multiple replacements (several nucleotides at one time), extensive deletions, and the substantial insertions are now possible. It is likely that long synthetic oligonucleotides, with a low level of random misincorporations, will be useful mutagenic tools. The next important advances will lie in more efficient methods for producing mutants from the heteroduplex DNA and in producing sophisticated vectors which combine the ability to switch from the production of double-strand to single-strand DNA and which allow replication and selection in two hosts and direct gene expression in at least one of the hosts.

With regard to efficiency of production of mutants, there are a number of useful developments. The objective is to bias things in favor of the production of progeny of the strand of DNA containing the synthetic oligonucleotide. One very effective strategy involves selection by joining the mutagenic oligonucleotide to a second oligonucleotide containing a selectable marker (or a DNA fragment with such a marker) which is lacking in the template DNA. Transfection into the appropriate host strongly favors the production of a desired mutant (>50% of the progeny being mutant). An alternate, also highly effective strategy creates a template DNA with a small fraction of the deoxythymidylate residues replaced by deoxyuridine. Such DNA is recognized by a host cell as being damaged. The resultant repair replaces patches of DNA in the damaged strand and coincidentally incorporates the desired mutation.

While there are effective methods for incorporating oligonucleotide-directed mutations into plasmid DNA, the attractiveness of single-strand DNA packaged as filamentous phage makes it desirable to have the best of both worlds, a clone of DNA replicating as a plasmid which can be induced to produce secreted, packaged single-strand DNA. The pEMBL (European Molecular Biology Laboratory) vectors have this feature as do improved versions, some of which are shuttle vectors. One problem with the present systems is that they require superinfection by wild-type phage to induce packaging of recombinant single-strand DNA. This process does not work efficiently with all constructions. A completely cellular inducible system would be a real benefit.

PROTEIN ENGINEERING (7, 20, 25)

The objective of protein engineering is to generate a new protein or obtain modifications of existing proteins by total or partial gene synthesis or by directed mutagenesis of a clone of an existing gene. Thus there are two technical components, gene synthesis or modification and in vivo expression (transcription and translation) of the modified gene. The first component has been discussed in the preceding sections of this report and expression will be discussed later. What is considered here is the kind of analysis and modification of protein structure that is feasible. There are three approaches: de novo design of a protein, the identification of the basic structural framework (scaffolding) of a type of protein (for example, immunoglobulin) with the mounting of new functions on that framework, and the introduction of small changes (such as replacing individual amino acids) on a protein whose structure is known in detail in order better to understand and/or modify its function(s).

The de novo design of a protein is limited by the lack of detailed understanding of the mechanisms by which proteins function and the limitations of our ability to predict the way in which a protein will fold. However, it has been possible to design a small functional RNA-binding protein. It is likely that a limited number of proteins of this type or proteins containing simple repeating sequence motifs can be obtained by de novo design and gene synthesis.

The mounting of protein structural changes on a basal skeleton is a much more attractive possibility. The potential of the strategy is naturally explicit in the diversity of immunological responses achieved by in vivo recombination of DNA segments and somatic mutation all superimposed on and incorporated in the immunoglobulin class of molecules. It is also clearly, albeit less dramatically, evident from the common structural elements which are being found in proteins whose enzymatic activities do not automatically suggest homology. In vitro mutagenesis directed at generating an enzymatic function in an immunoglobulin could be aided by the information obtained from immunization with ligands or substrates or transition state analogs. It seems certain that the attractiveness of this type of strategy for mutant construction will result in considerable research activity, especially with immunoglobulins. Clearly there is going to be a requirement for host cells which will be suitable recipients for a modified light chain gene or a truncated heavy chain gene or a complete heavy chain gene, or combinations of light and heavy chain genes.

The production of mutants where single amino acid or a very limited number of amino acids is or are replaced in a protein is already a very active field of endeavor which is bound to grow. The main emphasis is on the definition of specific functions. There is the long-term hope that it will be possible to produce proteins with predetermined new properties.

For systematic protein engineering at this level, the three-dimensional structure at appropriately high resolution is required. At present about 100 structures are known. Improvements in crystallization techniques as well as the use of area detectors and synchrotron X-ray sources have increased the ability for determining additional structures. An immediate application of in vivo mutagenesis would be the introduction of a cysteine residue to facilitate preparation of isomorphous heavy metal derivatives.

Once structural information is available, particularly for protein-ligand (where the ligand can be DNA, RNA, or another protein) or protein-substrate analog complexes, then it is possible to try to improve substrate affinity by optimizing local interactions or to try to improve reaction rate by changing the catalytic site or to create a new specificity by modifying the substrate-binding pocket. The design of experiments of this type requires accurate model building and energy calculations (see next section). There are bound to be unsuspected subtleties of protein structure and pleiotropic consequences of side chain substitutions which confound simple predictions. However, there can be no doubt that extensive studies in these areas will lead to greatly increased understanding of enzyme mechanisms and protein-ligand interactions. At the moment our ability to achieve a predicted, beneficial change in protein function is not great, but this has to increase and with it there should be a significantly useful increase in our ability to produce useful new protein functions. Already, at a more elementary structural level it has been possible to stabilize lysozyme by introducing a disulfide bond or to determine the pairing of cysteines in a disulfide bond by systematic replacement of cysteine residues and to change the specificity of a protease inhibitor.

Apart from changes which immediately affect the environment of the substrate, such as pH optima, changes which decrease a protein's susceptibility to inactivation by oxidation (replacement of cysteine or methionine residues at critical positions) and which increase or decrease lability to heat would be of industrial significance. The introduction of side chains which facilitate attachment to solid supports could be useful. Alpha-interferons of novel sequence and altered specificity have already been made. Targets of immediate potential are the modification of penicillin acylase to produce a cephalosporin C acylase and of a mouse immunoglobulin to make it structurally like a human immunoglobulin.

PROTEIN STRUCTURE MODELING (2, 3, 10, 12, 15, 26)

There are two potential classes of application, the prediction of tertiary structure from primary amino acid sequence in the absence of three-dimensional information and the prediction of the effect of change in established structures.

The present state of knowledge precludes the prediction of three-dimensional structures or even of accurate prediction of secondary structure. This is because interactions between regions which lie far apart on the polypeptide chain play important roles in established secondary and tertiary structure. Progress in prediction could result from simulation of the folding process (very likely) or from developing a computerized data base derived from raw structural and sequence data combined with higher level descriptions of folding motifs, topological features, and side chain physical chemistry. One has to hope that this latter possibility will be actively pursued. In the interim, complete prediction of tertiary structure is only possible when the structure of a homologous protein is available. Here the structure is constructed by model building using graphic display and conformational energy calculations. The quality of the resulting structure will depend on the degree of sequence homology between the established and unknown structures.

The application of molecular modeling to the engineering of an established structure involves the use of molecular dynamics and energy minimization (to a limited extent) to evaluate the effect of local changes in aminio acid sequence on the three-dimensional structure of the protein and on changes in thermal fluctuations of positions of atoms. Molecular dynamics uses calculated forces of interatomic interactions to solve Newton's equations of motion, at very short time intervals. Available computers allow calculations in the time range (picoseconds) of local structural fluctuations which occur at equilibrium. The calculations reproduce the fluctuations rather well. The simulation of biologically relevant phenomena such as enzyme substrate interaction is progressing steadily with the advent of faster, specialized computers and improved methods. Energy minimalization is used to find "stable" protein conformations near to that of an established crystal structure. The use of sufficiently detailed force fields in either molecular dynamics or energy minimization gives a good description of conformational changes of structures. The calculation of free energy changes is difficult because the methods for calculating entropy contribution are computationally prohibitive.

Molecular graphics are an integral component of any protein engineering project because it is essential to visualize and manipulate three-dimensional models in order to rationalize structure-function relationships. Present-day graphic facilities offer a variety of ways for displaying protein structures: bond connectivity, surfaces, volumes, and physical properties of side chains. Calligraphic (vector) displays offer the highest image resolution and fastest real time transformations of large structures. Raster graphic systems are undergoing rapid improvements which are beginning to be competitive.

EXPERIMENTAL METHODS (6, 16)

There are a number of experimental methods for characterization of proteins which are an essential complement to modeling since progress in prediction methods can only be achieved by confirmation by experiment.

Obviously, enzyme kinetics are required for studies with enzymes and include steady-state and fast kinetic studies. Equally obvious is crystallography of engineered proteins. Other important techniques include nuclear magnetic resonance (NMR) because of the detailed information it provides on the structure of proteins in solution. NMR can be used to derive a low-resolution, three-dimensional structure of a protein of less than 20,000 molecular weight. The particular attractions of NMR are in the characterization of changes introduced by mutation of an established structure and in studies on ligand-protein interactions. Time-resolved Raman spectroscopy could be useful for studies of conformational changes under different conditions. Microcalorimetry is used to evaluate thermodynamic parameters of mutant proteins. Optical spectroscopic and fluorescence techniques will be very useful with certain classes of proteins (e.g., heme proteins). Antibodies, monoclonal antibodies in particular, also provide an important tool for analyzing conformation properties of proteins and their mutants.

GENE EXPRESSION (1, 5, 8, 17, 24)
The focus here is on the efficient production of a protein in a heterologous host cell. Factors directly influencing this gene are copy number, transcription efficiency, mRNA maturation (in eukaryotes), translation efficiency, and protein folding and stability. If the protein is to be secreted and/or modified in any way in vivo, then an additional number of poorly defined factors are involved.

Gene copy number can be increased by increasing recombinant plasmid numbers or by chromosomal gene amplification. In E. coli the preferred route involves the use of inducible, runaway plasmids. These undergo temperature-induced, approximately 50-fold amplification of the plasmid copy number, normally present in 20-30 copies per cell. The particular application of this type of plasmid is with a gene which is expressed very poorly; gene amplification above the level provided with normal pBR322—derived plasmids is not needed for efficiently expressed genes. Gene amplification in other bacteria has not been extensively studied. However, there are interesting examples in Bacillus and Streptomyces of mutations involving substantial gene amplification. It would be useful if this could be used to amplify a gene of interest, either one normally present or a transforming gene.

In yeast it has been possible to obtain moderately high copy numbers by inserting a gene in a transposon which is amplified by conversion of other transposons. Yeast plasmids are available which provide about 100 copies per cell; it might be useful if a runaway plasmid could be developed for yeast. The system for obtaining high gene copy numbers in higher eukaryotes involves gene amplification. For example, Chinese hamster ovary (CHO) cells, defective in dihydrofolate reductase ($DHFR^-$) are transformed using your favorite gene (YFG) covalently linked to a DHFR gene. The YFG-DHFR fragment transforms by chromosomal integration. Methothrexate, in progressively increasing concentration, selects for cells of increasing drug resistance. This results in at least 20-fold amplification of the DHFR gene and, as a consequence, the co-integrant YFG is also amplified. After removal of methotrexate, the amplified genes are stable for up to fifteen passages of the CHO cells. Alternatively, nondividing cells can be "milked" of the YFG product.

Viral vectors can also be used to achieve high gene copy numbers in higher eukaryotes. Plasmids containing the polyoma virus origin replicate in a transformed cell and then can integrate many copies. Papilloma viruses can be used to produce a fixed high copy number.

Transcription efficiency in E. coli has been studied in great detail and a wide variety of very efficient promoters, both constitutive and inducible, have been isolated or constructed. Because overproduction of a protein can be toxic or harmful, inducible promoters are preferred. The most convenient of these are the lac promoter, induced by IPTG, the trp promoter (repressed by tryptophan), and the phage λP_L or P_R promoters used in concert with a temperature-sensitive repressor which is inactivated at 42°C. Up to 500-fold induction is possible. The low basal level of expression of inducible promoters remains a problem with protein products which are toxic to E. coli and require special expression strategies.

In yeast a number of efficient constitutive promoters are available, mainly those for glycolytic enzymes. The use of regulated promoters is in its infancy. While the availability of such promoters will be useful, the primary interest in gene expression has been focussed on secretion where inducible systems may not offer any advantages.

In mammalian systems the whole picture with promoters is complicated by the additional factor of promoter enhancers which are required for efficient transcription. The SV40 virus enhancer coupled to the early promoter is relatively nonspecific with regard to host cell requirement. Other enhancers show high cell type specificity. Clearly it would be useful to have much more information because of the growth in interest in use of mammalian cells in culture for gene expression. Regulated promoters are also of great interest. The metallothionein promoter is stimulated by metal cations (Zn^{2+} or Cd^{2+}). The promoter for the mouse mammary tumor virus is stimulated by glucocorticoids but requires a host cell with a glucocorticoid receptor. It would be of interest to exploit the characteristics of heat shock protein promoters.

Efficient translation is the least defined of expression processes. In E. coli it requires a ribosomal binding site sequence appropriately spaced from an initiation codon. But it is clear that mRNA sequence upstream, between, and downstream of these two elements can dramatically influence the efficiency of translation for reasons that are not understood. At the moment trial and error is the only route to efficient translation in E. coli. The translation initiation signal of the C protein of ATPase appears to be a generally good translation signal and may offer an avenue to greater understanding.

Translation initiation in eukaryotes involves the signal Pu - AUG and can be blocked by an upstream AUG, but little else is known about function.

GENE IDENTIFICATION AND ISOLATION (9 , 17, 19, 20, 23, 24)
The focus of this section is on methods for specific identification of a gene or its products, mRNA (as cDNA) or protein, and their use for monitoring gene of cDNA purification by molecular cloning. Most commonly used of all strategies is that where synthetic oligonucleotides, whose sequence is determined by applying the genetic code to a partial protein sequence, are used to recognize a specific DNA and to monitor its cloning. Two variants of this procedure are used. The first uses mixtures of relatively short oligonucleotides (14, 17, or 20 nucleotides in length) which contain (theoretically) all possible sequences which could code for the equivalent peptide. The second variant uses a longer synthetic oligonucleotide (40 or more nucleotides) which is a unique sequence (or contains a very limited number of sequences) determined by the target organism's codon preference and adjacent dinucleotide preference (low frequency of sequence CG). Both strategies work, but with less than 100% reliability. Failures with the short, complete mixture probably result from incomplete representation of all possibilities (use of the disc support method to synthesize a lot of unique oligonucleotides in parallel is a potential remedy) or from a high level of false positives (a possible remedy is the use of tetramethylammonium ion-containing buffer during probing to eliminate false positives due to GC-rich mismatches). Failure in the second strategy results from the target gene having atypical codons. It is a good idea to use two probes to identify clones which respond to both. These strategies require some protein sequence. Modern methods for peptide sequence determination can be applied to 1-10 μg of protein.

Antibody screening requires that clones be carried in an expression vector and are usually used with cDNA libraries to allow expression in E. coli. Phage lambda and plasmid-derived expression vectors are available. All involve the fusion of the cDNA with a protein coding sequence (usually β-galactosidase, or an N-terminal fragment of a β-galactosidase whose expression is determined by E. coli promoter and translation signals. The phage lambda vector (gt11) is very effective in generating a large number of clones from a small amount of cDNA, but the plasmid-derived systems are also very effective.

The foregoing strategies require at least a small sample of pure gene product (protein). This is not always available or even identifiable. Hence a number of indirect strategies have been developed.

In some cases tissues are rich in a particular mRNA and cDNA is readily identified. In other cases a particular mRNA can be segregated because it is expressed at a differentially higher level at a particular time in the life cycle of a cell, tissue, or organism. Alternatively, the presence of a specific mRNA can be the most significant difference in gene expression between two cell types (e.g., lymphocytes). In both of these categories differential hybridization is used to enrich the target mRNA (or its derived cDNA).

Genetic complementation, using genomic DNA as donor, can be used to isolate a specific gene. This is a powerful tool for isolating prokaryote and yeast genes (some of the latter can be identified by complementation of E. coli functions). The common occurrence of introns in the genes of organisms other than prokaryotes and yeast makes interspecies complementation by genomic DNA difficult or impossible. Strategies for expression of cDNA allow for a broader range of donor species. The number of CHO cell mutant strains is increasing and they offer potential for isolation of higher eukaryote genes by complementation. Even more indirect techniques are required for isolation of genes lacking a molecular phenotype. The most useful route in this case involves insertional mutagenesis by a transposon for which a molecular probe is available. This can be used to isolate a mutant gene which then provides access to the wild-type gene. Because transposons are present in eukaryotes in multiple copies, a transposon probe does not identify a mutant gene uniquely. In Drosophila melanogaster genetic mapping coupled with cytogenetics allows identification of the desired clone. In other organisms the increasing availability of separated chromosomes will partially compensate for the lack of high-resolution cytogenetics. Transposon insertion mutagenesis and cloning of developmental genes is proving to be a powerful tool in plant molecular genetics. The development of an equivalent system in the mouse would be a major boon. In this and other higher eukaryote systems, retroviral DNA insertion or integrative transformation with DNA will provide a partial answer to the need for a transposon-like system but in its current form will not be capable of the saturation of a genetic map, which is required.

A final type of system, particularly appropriate for gene isolation in humans, involves molecular chromosome walking, hopping, or jumping starting from a closely linked marker. Monitoring of progress towards the target gene involves studying patterns of restriction nuclease fragment length polymorphisms and their relationship to familial inheritance of a defect in the gene of interest. Genetic localization to subregions of chromosomes can in favorable cases allow the identification of the mutated gene by a combination of genetic and molecular approaches. It is likely that a first test and success of this strategy will occur with the isolation of the gene for Huntington's disease. In the human, substantial progress has been made towards the isolation of a set of DNA fragment markers for the complete genome (23): isolation should be completed in about two years.

VECTOR-HOST SYSTEMS (1, 8, 17, 23, 27)

Some features of vectors have been discussed in the section dealing with heterologous gene expression. The focus of discussion of vectors in this section relates to their stable maintenance and segregation in proliferating host cells. In Gram-negative bacteria, broad host-range plasmids are available and are very useful for starting a project with a new host. However, these plasmids are lost from some host cells under the nonselective conditions required by many practical applications, and it seems likely that the native plasmids will be the best sources for vectors which are inherited stably under nonselective conditions. Thus it is important to isolate native plasmids and to define their essential functions as a key step in developing new vector-host systems. Gram-positive bacteria are less well studied although they are potentially attractive host cells. Plasmids from Staphylococcus aureus have a usefully broad host range in Gram-positive bacteria, and broad host-range plasmid and phage vectors suitable for use in a wide range of actinomyces are becoming available.

In yeast, plasmid vectors containing a 2-micron plasmid origin or a chromosomal origin (ARS) are available but are unstable. These are present as 10-20 copies per cell. Some 2-micron derived vectors are present in larger numbers (100-200 copies per cell). Stability under nonselective conditions is conferred by introduction of a centromere into the plasmid, but copy numbers are reduced to about one per cell. It would be useful if plasmids could be obtained in which limitation of copy number and stable maintenance were dissociated. One useful type of yeast plasmid, from the industrial aspect, contains an essential gene which complements a mutant in the host yeast cell. This allows maintenance in inexpensive media.

The major considerations in choosing host cell systems has been their freedom from proteolytic enzymes, the absence of pyrogens, and their ability to carry out post-translational modifications of proteins and to secrete them. When large amounts of proteins are synthesized in E. coli they are produced intracellularly in insoluble inactive form. This requires that renaturation be carried out in vitro, and the latter is an unsolved problem for very large proteins. However, insolubility protects against proteolysis and facilitates purification.

A host cell which secretes the product protein is very attractive because of economy in the cost of purification. For this reason, Bacillus subtilis has received a lot of attention, but its proteolytic enzymes have so far provided insurmountable problems and there are problems with the plasmid stability. Attempts to develop a secretory system in E. coli have not resulted in a very efficient protein production. Streptomyces and Staphylococci secrete proteins and warrant further attention.

Yeast has a secretory system which can be exploited for some proteins of medical interest, e.g., smaller molecules such as some hormones and growth factors. Aspergillus species secrete a wider range of proteins and the possibility of exploiting this potential should be investigated. Mammalian cells in culture are now serious contenders for the production of proteins of medical importance. Their attraction lies in their ability to introduce post-translational modifications which are essential for function. It is not clear whether the cost of growing mammalian cells in culture can be reduced significantly. If this cost remains high, it may be worth transferring genes for key post-translation modifying enzymes to a lower eukaryote such as yeast.

REVERSE GENETICS (TRANSPLACEMENT) (17, 20)
One aspect of genetical engineering that is important in an indirect way to industry and directly to medicine and biology in general is the more precise definition of genetic functions. In many cases this requires gene transplacement, the use of genetic transformation to exchange alleles. This process involves homologous recombination between a DNA fragment and its chromosomal allele. In bacteria and in lower eukaryotes transplacement is readily achieved. However, in animals and plants, transformation predominantly involves integration of transforming DNA into chromosomes by random, nonhomologous recombination. There is a real need to develop strategies for increasing the efficiency of homologous transformation and/or identifying cells in which transformation by homologous recombination has occurred.

IMMUNOLOGY AND CELL SORTING (18, 22)
Here we are concerned with the application of these techniques to gene identification and isolation. Immunology, in combination with the fluorescence-activated cell sorter (FACS) allows the isolation of a very rare subpopulation of cells (one in 10^6) expressing a unique cell surface antigenic determinant. The FACS can separate the cell according to three characteristics: red fluorescence (rhodmaine), green fluorescence (FITC), and size (light scattering). The FACS can sort 10^8 cells per day with an efficiency of 99%. In experiments with lymphocytes, it is possible to isolate cells containing as few as 3000 stainable molecules per cell. The FACS has been used to isolate hybridoma cells exhibiting a rare phenotype, e.g., class-switching variants and idiotope variants. FACS technology is not confined to lymphocytes and, generally speaking, any stainable cell can be separated from a nonstaining population. FACS can also be used to separate chromosomes according to size. As such it is a powerful aid to gene mapping and identification and isolation of DNA fragment clones.

Chimeric antibodies, formed by combination of murine variable region genes with human heavy chain genes have potential clinical applications. Other antibody chimeras, such as a fusion between an antibody gene and an enzyme, can provide a simple route to enzyme purification by passage of the hybrid protein through a column containing a hapten recognized by the antibody.

CONCLUSION
This report illustrates the wide variety of techniques that are required by genetic engineers and which are already deployed. The combined application of this array of scientific specialities to a specific problem, the isolation of a defined gene, represents a largely unsung tour-de-force of modern science. Many of the technologies are fully developed. The most clear requirements for the future are techniques for identification and isolation of unusual genes from higher eukaryotes, particularly animals

including humans, and systems in which those genes can be expressed efficiently in functional form (or in a form which can easily be converted to the desired protein).

REFERENCES

(1) Botstein, D., and Davis, R.W. 1982. Principles and practice of recombinant DNA research with yeast. In The Molecular Biology of the Yeast Saccharomyces. Metabolism and Gene Expression, eds. J.N. Strathern, E.W. Jones, and J.R. Broach, pp. 607-636. Cold Spring Harbor: Cold Spring Harbor Laboratory.

(2) Chothia, C. 1984. Principles that determine the structures of proteins. Ann. Rev. Biochem. **53:** 537-572.

(3) Connolly, M.L. 1983. Solvent-accessible surfaces of proteins and nucleic acids. Science **221:** 709.

(4) Davies, J.E., and Gassen, H.G. 1983. Synthetic gene fragments in genetic engineering - the renaissance of chemistry in molecular biology. Angew. Chem. Int. Ed. Engl. **22:** 13-31.

(5) Elder, J.T.; Sprintz, R.A.; and Weissman, S.M. 1981. Simian virus 40 as a eukaryote cloning vehicle. Ann. Rev. Genet. **15:** 295-340.

(6) Fersht, A. 1985. Enzyme Structure and Mechanism. New York: W.H. Freeman.

(7) Fersht, A.R.; Shi, J.-P., Wilkinson, A.J.; Blow, D.M.; Carter, P.; Waye, M.M.Y.; and Winter, G.P. 1984. Analysis of enzyme structure and activity by protein engineering. Angew. Chem. Int. Ed. Engl. **23:** 467-473.

(8) Inouye, M., ed. 1983. Experimental Manipulation of Gene Expression. New York: Academic Press.

(9) Itakura, K.; Rossi, J.J.; and Wallace, R.B. 1984. Synthesis and use of synthetic oligonucleotides. Ann. Rev. Biochem. **53:** 323-356.

(10) Karplus, M., and McCammon, J.A. 1981. The internal dynamics of globular proteins. Crit. Rev. Biochem. **9:** 293.

(11) Kunkel, T.A. 1985. Rapid and efficient site-specific mutagenesis without phenotypic selection. Proc. Natl. Acad. Sci. USA **82:** 488- 492.

(12) Langridge, R.; Ferrin, R.E.; Kuntz, I.D.; and Connolly, M.L. 1981. Real-time color graphics in studies of molecular interactions. Science **211:** 661-666.

(13) Lathe, R.F.; Lecocq, J.P.; and Everett, R. 1983. DNA engineering: the use on enzymes, chemicals and oligonucleotides to restructure DNA sequences in vitro. In Genetic Engineering, ed. R. Williamson, vol. 4, pp. 1-56. New York: Academic Press.

(14) Levinson, A.; Silver, D.; and Seed, B. 1984. Minimum size plasmids containing an M13 origin for production of single-strand transducing particles. J. Molec. Appl. Genet. **2:** 507-517.

(15) Levitt, M. 1982. Protein conformation, dynamics, and folding by computer simulation. Ann. Rev. Biophys. Bioeng. **11:** 251.

(16) Murkley, J.L., and Ulrich, E.L. 1984. Detailed analysis of protein structure and function using NMR spectroscopy. Ann. Rev. Biophys. Bioeng. **13:** **493.**

(17) Setlow, J.K., and Hollaender, A., eds. 1979. Genetic Engineering. Principles and Methods, vol. 1. New York: Plenum Press (continuing annual series).

(18) Sikora, K., and Smedley, H.M. 1984. Monoclonal Antibodies. Oxford: Blackwell Scientific Publications.

(19) Smith, M. 1983. Synthetic oligodeoxyribonucleotides as probes for nucleic acids and as primers in sequence determination. **In** Methods of DNA and RNA Sequencing, ed. S.M. Weissman, pp. 23-68. New York: Praeger.

(20) Smith, M. 1986. In vitro mutagenesis. Ann. Rev. Genet. **19,** in press.

(21) Ulmer, K.M. 1983. Protein engineering. Science **219:** 666-671.

(22) Watson, J. 1984. Cell and chromosome sorting. **In** Molecular Biology and Human Disease, eds. A. MacLeod and K. Sikora, pp. 66-85. Oxford: Blackwell Scientific Publications.

(23) White, R. 1985. DNA sequence polymorphisms revitalize linkage approaches in human genetics. Trends Genet. **1:** 177-181.

(24) Williamson, R.W., ed. 1981. Genetic Engineering, vol. 1. New York: Academic Press (four-volume series).

(25) Winter, G., and Fersht, A.R. 1984. Engineering enzymes. Trends Biotechnol. **2:** 115-119.

(26) Wodak, S.; Alard, P.; Delhaise, P.; and Renneboog-Squilbin, C. 1985. Simulation of conformational changes in 2Zn Insulin. J. Molec. Biol. **181:** 317.

(27) Yelton, M.M.; Timberlake, W.E.; and van den Hondel, C.A.M.J.J. 1985. A cosmid for selecting genes by complementation in Aspergillus nidulans: selection of the developmentally regulated yA locus. Proc. Natl. Acad. Sci. USA **82:** 834-838.

Biotechnology: Potentials and Limitations, ed. S. Silver, pp. 19-40. Dahlem Konferenzen 1986. Berlin, Heidelberg, New York, Tokyo: Springer-Verlag.

Cloning and Expressing Genes for Clinically Useful Proteins

R.M. Kay, R. Kaufman, P. Schendel, K. Turner, and R. Kamen
Genetics Institute, Inc.
Cambridge, MA 02140, USA

Abstract. Clinically useful proteins come in all shapes and sizes and while some are extremely rare in natural occurrence, others are quite abundant. Some have complex modifications to their primary sequence which are necessary for their function. As a result, the methods used for the isolation of the genes encoding these proteins are diverse and in many cases expression of the proteins requires the development of novel host vector systems. This review will not try to cover all the possible alternative cloning and expression methods available but will give the reader an insight into the strategies frequently used to clone and express genes of clinical interest while selecting some examples to illustrate these approaches.

INTRODUCTION

Since the first isolation of mammalian gene sequences almost ten years ago, a wide range of technologies have been developed to isolate gene sequences from both genomic DNA and mRNA sources. Given the great diversity of structural features amongst clinically useful proteins, their widely differing sizes and post-translational modifications, one would expect that the successful cloning group would be experienced in the use of many of these techniques. Some clinically useful proteins are small with no apparent requirement for secondary modification for function, examples being growth hormone, insulin, the interferons, and interleukin-2 (IL-2). Other small polypeptides such as erythropoietin and colony-stimulating factor are extensively glycosylated. Certain extremely large proteins promise clinical utility, the prime example being the anti-hemophilic factor, clotting factor VIII:C. This is a protein comprised of 2351 amino acids and extensive glycosylation, and which is furthermore processed by post-translational cleavage. Another member of the clotting cascade, factor IX, useful in the treatment of hemophilia B, is a vitamin K-dependent protease that requires γ-carboxylation of certain glutamic acid residues for activity. In some cases, the requirement for post-translational modification is unknown at the time the gene is cloned and is only defined by attempts to express active gene products in heterologous host cell systems. Proteins of clinical interest are frequently stabilized by disulfide bridges; correct folding of these extensively cross-linked proteins appears to require membrane translocation during their synthesis.

The natural sources of clinically useful proteins are also diverse, coming from plasma to urine, tissues to cell culture conditioned medium, and each represents its own challenges to the researcher. Furthermore, human proteins are almost exclusively preferred to avoid immunogenic reaction.

As we will see below, no one cloning or expression strategy can be depended on to yield the desired fully active product and normally several approaches are attempted simultaneously. Perhaps with time more universal cloning strategies will be developed and the ground rules for efficient gene expression will be defined. In the meantime, as the diversity of clinically useful proteins expands, so the range of techniques used to deal with them becomes more complex. In this brief review we will describe the salient features of the cloning and expression strategies which we have found to be most successful.

GENE CLONING

Construction of Gene Libraries

Gene libraries can be prepared from either fragmented genomic DNA (genomic libraries) which will contain the genetic information of the organism or from complementary DNA (cDNA libraries) synthesized from cellular mRNA. These latter libraries will contain only gene sequences expressed in one cell type or tissue. The choice of which library to screen is essentially based on whether the tissue or cell source of a protein is known. If a cell line produces readily detectable amounts of the protein or biological activity, a cDNA library derived from mRNA isolated from the cell line is the best line of attack. When the source of the protein is either unknown or is suspected to be synthesized by a small subset of the cells in a tissue, genomic library screening is the best course of action. A variety of screening procedures are now available for cDNA libraries and, for genomic libraries, both hybridization probe and gene expression screening have been used. In any event, genomic clones must be considered an intermediate step towards the final isolation of a cDNA clone since it is this cDNA sequence which will be used for the eventual large-scale expression of the encoded protein.

Preparation of genomic DNA libraries

Methods for the preparation of genomic DNA libraries are well established and, in fact, libraries derived from many commonly used animal species are available within the scientific community. Genomic libraries are normally constructed in bacteriophage λ cloning vectors in order to obtain enough independent recombinants to be confident of having a complete genome representation in the library (8.1 x 10^5 recombinants yields a 99% probability of the presence of any one sequence in a mammalian genomic library of 17 Kb fragments (31)). In outline, the method involves insertion of genomic DNA fragments into a restriction site in the vector, replacing a segment of vector DNA. The recombinants are packaged into phage particles in vitro and then transfected into E. coli. Each type of cloning vector only accommodates a narrow range of inserted DNA fragment size, and thus it is best to prepare a library from either sheared DNA or partially restriction enzyme-digested DNA (using a frequent cutter) in order to avoid inconvenient restriction sites which either may be within a probe sequence or may produce fragments of inappropriate size. (For detailed descriptions of genomic cloning methods using bacteriophage λ, see (31)).

An alternate system for cloning genomic DNA fragments uses cosmid vectors (7). These vectors contain the COS sites of bacteriophage λ necessary for packaging and plasmid sequences for selection and replication in E. coli. DNA fragments are inserted into these vectors and the vectors are packaged into phage particles in vitro as for phage λ cloning. Transfection into E. coli thus yields very high transformation efficiencies, but the cosmids replicate as extrachromosomal plasmid elements and clones are screened as bacterial colonies instead of phage plaques. The main advantage of this vector system is in the very large genomic DNA fragments that can be cloned (30-40 Kb).

Preparation of cDNA libraries

cDNA cloning involves the preparation of a double-stranded DNA copy (dscDNA) of mRNA and its insertion into a plasmid or bacteriophage vector for cloning. This first step of dscDNA synthesis remains the most difficult aspect of this cloning process. In the most careful hands with optimized reactions and highly purified enzymes, large cDNA sequences can be synthesized and converted to double-stranded molecules.

Two basic methods are widely used to obtain dscDNA. Both involve priming first-strand synthesis with reverse transcriptase and oligo dT. In order to promote completion of first-strand synthesis it is common to denature the mRNA with methyl mercuric hydroxide (2) in the hope of eliminating secondary structure barriers to reverse transcriptase transfer along the mRNA. Addition of RNase inhibitors such as the placental inhibitor (8) can limit degradation of the mRNA during the reaction. It is possible to synthesize with good efficiency single-stranded cDNA sequences many kilobases in length.

Second-strand synthesis is accomplished in one of two ways. Efstradiatis et al. (11) showed that cDNA sequences formed hairpin structures at their 3' ends (probably in most cases fortuitously) which can act as primers for the second-strand synthesis by DNA polymerase I (reverse transcriptase has also been used). Thus, the second-strand synthesis reaction yields a hairpin structure which can be cleaved at the single-stranded terminal loop by S1 nuclease to produce a blunt-ended, double-stranded cDNA. Two major problems are presented by this method. The S1 nuclease reaction produces secondary cleavages in the cDNA other than at the loop. This results in reduced yields of dscDNA, reduced cloning efficiency, and loss of cDNA length. More recently a considerably improved second-strand synthesis method has been developed (14, 37) and can be considered the method of choice. The hybrid mRNA/cDNA product of first-strand synthesis is incubated with RNase H and DNA polymerase I such that single-strand nicks in the mRNA, produced by RNase H action, are used as primers for DNA polymerase I activity in a repair synthesis reaction. E. coli DNA ligase appears unnecessary for completion of a contiguous second strand (14).

Insertion of cDNA sequences into plasmid vectors for propagation and cloning is accomplished by either complementary oligonucleotide tailing of the cDNA and vector or addition of synthetic restriction site-containing linkers to the cDNA. Since cleavage of the linkers is necessary to expose the restriction site and the presence of restriction sites within the cDNA identical to the linkers will be unknown, restriction site methylation of the cDNA is carried out before linker cleavage.

The vector system chosen for cDNA cloning depends on the screening method to be applied and on the expected abundance of the mRNA of interest. For extremely rare mRNAs, perhaps derived from a subpopulation of cells within a tissue, the much higher cloning efficiencies of bacteriophage λ (10^8/µg DNA) are preferable for obtaining large libraries of independent clones. Where hybridization probe screening is contemplated, the large numbers of phage plaques which can be screened offers a further advantage in time and effort. On the other hand, if pooled DNA preparations are to be used for some form of expression cloning or where only moderate-size clone banks are required (20,000 to 50,000 independent clones), cloning in plasmid vectors in bacteria can offer advantages in manipulation of the cloned DNA. Plasmid vectors constructed specifically for particular screening approaches will be noted in the appropriate sections. The reader is referred to Maniatis et al. (31) for a detailed description of cDNA cloning methods.

Isolation of Gene Sequences from Libraries

The most vexing problem faced by all investigators wishing to isolate their favorite gene is choosing the best library screening method. No one method can give assurance of successful gene isolation in a reasonable period of time since all have potential pitfalls and limitations. The choice can only be made on the basis of assessing the quality of the resources at hand and the nature of the protein being stalked. The frequently voiced opinion that "cloning is the easy part" should be interpreted as reflecting not the ease of cloning but rather the multitude of other technical hurdles along the way to producing large amounts of a recombinant protein.

Protein sequence-based oligonucleotide probe screening

Protein microsequencing techniques using gas phase reactions (18) can provide extensive amino acid sequence information from even microgram quantities of protein. Armed with this information it is possible to predict the nucleotide sequence of the gene encoding these amino acid sequences, albeit with some uncertainty arising from redundancy in the genetic code. If one is fortunate, a region of this predicted sequence (perhaps 14-20 bases long) will have few ambiguities and all possible combinations of nucleotide sequence can be synthesized. This mixture of oligonucleotides can then be end-labelled with ^{32}P and used for in situ hybridization to phage plaques or bacterial colonies. Many reports of the successful use of such low-redundancy oligonucleotides exist, and in a few cases highly redundant sequences (576X) have been used (49). The GC content of the individual oligonucleotides within a group can vary widely, presenting some difficulty in choosing hybridization conditions that are both stringent enough to ensure specificity of hybridization and relaxed enough to ensure hybridization of the lowest GC content oligonucleotide sequence. In some instances it is possible to synthesize the oligonucleotides in small groups and select a group on the basis of specific sequence priming on mRNA or by hybridization to mRNA sequences of the expected size. Alternatively, a recently developed hybridization solution containing tetramethyl ammonium chloride can be used which considerably reduces the difference in stability between G:C and A:T base pairing, allowing simultaneous hybridization of oligonucleotides with widely varying G/C content.

More often than not, however, the amino acid sequences obtained do not offer the opportunity for low-redundancy oligonucleotide synthesis. Thus, while the protein purification and sequencing groups go back to work to obtain more information, the impatient cloning groups make the best of a difficult job. Several amino acid sequence regions may offer the opportunity for high-redundancy probes. Thus by screening with some or all of these, clones which hybridize to several independent, mixed oligonucleotide pools can be considered highly likely to contain the desired gene. An alternative approach involves the use of oligonucleotide sequence approximation (commonly known as "guessmers") in which the sequence complexity of the probe is reduced by choosing, at positions of ambiguity, the most probable or least "detrimental" nucleotide at that position, the choice being based on a) the frequency of the specific codon usage in the species, b) the low frequency of CpG in coding regions, and c) the relative stability of G:T versus A:C mismatches. Examples of various types of oligonucleotide probes are described by Jacobs et al. (22) and Toole et al. (45).

The wide popularity of the oligonucleotide probe approach to clone isolation can in part be attributed to the high level of confidence that, with due diligence and perseverance, the desired cloned sequence will eventually turn up. Furthermore, one has little choice but to use this approach when the site of synthesis of the protein is unknown or a producing cell line is unavailable and one has to resort to screening genomic libraries.

Establishing that the isolated clones do in fact encode the protein of interest is usually accomplished by sequencing of the cloned DNA in the region of oligonucleotide hybridization. This sequence should encode the determined amino acid sequence provided of course that, when dealing with genomic clones, introns are not present in the sequenced segment.

Cloning by expression of biological activity

Frequently one is faced with the problem that a useful biological activity has been extensively studied and is believed to be of great interest but little is known of its molecular structure and it has not been purified even in the minute quantities now required for high-sensitivity protein sequencing. Good examples of such proteins are the immune and hemopoietic regulators. Many of these lymphokines and monokines are synthesized in very small amounts and are assayed by in vitro or in vivo induction of cellular differentiation or maturation. An effective approach to the cloning of genes encoding these proteins is to express the cloned gene sequences in a heterologous system and assay directly for the production of bioactivity, thus circumventing the need to have sequence data or antibodies to the protein.

Normally, when attempting cloning by these methods, one knows little about the level of post-translational processing or how important this is for functional activity. This caveat can largely be eliminated by first demonstrating the translation of bioactivity from mRNA injected into Xenopus oocytes. Furthermore, prior size fractionation of the mRNA can show whether two or more distinct mRNA species are needed to produce the bioactivity. Four quite different variations on this clone selection theme are presented below.

Transient expression in mammalian cells. In this approach double-stranded cDNA is inserted into a mammalian cell expression vector plasmid downstream of a potent mammalian cell transcriptional promoter (23, 25, 26, 48). An example of such a vector is shown in Fig. 1. The vector contains a transcription unit consisting of the adenovirus major late promoter and tripartite leader sequence followed by an intron, a cloning site (EcoRI), a 3' non-coding region, and finally, a poly-A addition sequence from SV40. The presence of an SV40 origin of replication on the vector allows replication in monkey COS cells which support the replication of this extrachromosomal DNA by constitutively producing the SV40 large T antigen. In order to enhance the efficiency of translation of the transcripts, the adenovirus VA gene has been included in the vector. VA RNA has been shown to stimulate translation of exogenous mRNA up to 10-fold, although the mechanism of action of VA RNA is unclear. One possibility is that double-stranded RNA generated by the transfecting plasmid suppresses host translation by inducing phosphorylation of the translation factor EIF2 and that the function of VA RNA is to suppress this effect.

An outline of the expression cloning approach is shown in Fig. 2 (48). Using this vector, groups of up to 500 cDNA clones can be pooled and plasmid DNA isolated. This DNA is then used directly to transform monkey COS cells and, following a suitable incubation period, conditioned medium (in the case of screening for a secreted protein) or cell lysates are assayed for the presence of the bioactivity of interest. Any pool of recombinant plasmids shown to express the gene of interest is further screened using subpools or individual clones in separate transformations. In this way the link between

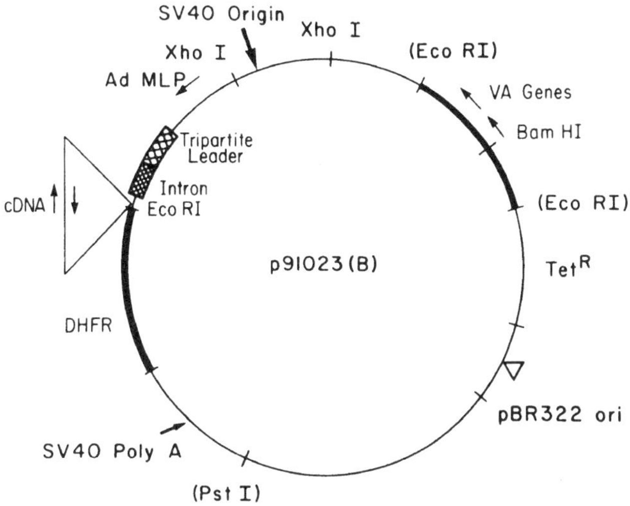

FIG. 1 - Structure of the expression vector p91023(B). A detailed description of this vector has been presented by Wong et al. (48) (see text).

nucleic acid sequence and biological activity can be bridged in a single procedure. We should note that this procedure is not universally applicable and a variety of criteria need to be met for successful clone identification. Clearly one cannot screen the very large libraries required for isolation of a clone from a tissue source where only a small percentage of the cells actually synthesize the protein of interest. Thus, a suitable mRNA source for this method would be a cloned cell line known to produce the protein. A prerequisite for the expression system to work is the presence of complete coding segments within the cDNA. As we have already discussed, the synthesis of large cDNA sequences with high efficiency remains a major problem. As a result, only small proteins (50,000 D) can reasonably be expected to be detected by these methods. Since the available transient expression systems are relatively inefficient but a large number of clones need to be pooled in each transformation in order to cover a large library, the assay for the protein of interest needs to be sensitive and relatively easy to perform in large numbers. An almost unavoidable limitation of this method would occur where the chimeric transcript was unstable or subject to aberrant processing. Thus, this method offers several clear advantages but the risk of failure is higher than that of some other methods.

One of the first novel genes to be successfully isolated using this approach was the gene encoding human granulocyte macrophage colony stimulating factor (GM-CSF) obtained from a T cell-derived cDNA library (48). Two hundred pools of clones each containing 200-500 clones were screened, from which six positive pools were identified. Three were further selected and all shown to contain a cDNA encoding the same GM-CSF sequences. Impressively, the abundance of the GM-CSF mRNA was estimated to be only 0.01% of total mRNA in the T cell line.

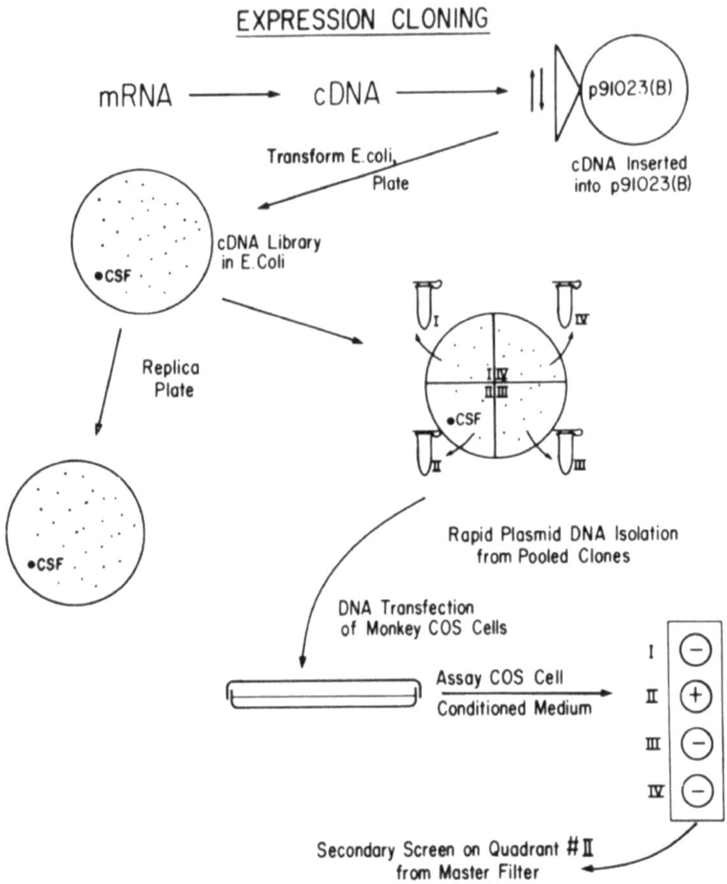

FIG. 2 - Outline of the expression cloning methodology used by Wong et al. (49). cDNA libraries are prepared using the vector p91023(B)1 to transform E. coli to tetracycline resistance. Approximately 1000-2000 transformants are plated on nitrocellulose filters and replicates of the filters prepared. One replica is divided into quadrants and the colonies on each separate quadrant are collected and used to prepare a pooled plasmid DNA preparation. This plasmid DNA is subsequently used to transform monkey COS cells and, 72 hours later, conditioned medium from the cultures is used to assay for the synthesis of bioactivity. Quadrants which produced a positive assay response by this analysis are then re-searched by analysis of single colonies picked from the quadrant.

Hybrid selection screening. Having a tissue or preferably a cell line source of mRNA, a reproducible, sensitive assay, and an in vitro translation system yielding an assayable product might lead one to choose hybrid selection of mRNA to screen clone banks. In outline, plasmid DNA from pools of cDNA clones are filter hybridized to mRNA and the specifically hybridized mRNA is eluted and translated. A suitably sensitive translation system is the microinjection of Xenopus oocytes where only sub-nanogram amounts of specific mRNA/oocyte are normally required to obtain detectable translation of biological activity. The small amounts of mRNA hybridized can thus be analyzed directly for their ability to direct the synthesis of the protein of interest. Once a positive pool of cDNA clones is found, further rounds of selection will yield the required cDNA. Although this approach avoids the requirement for structural information on the protein and is thus similar in principle to the transient mammalian cell expression cloning method, there is no requirement for a complete cDNA sequence. One has the added confidence that one can demonstrate in advance that the translation system can produce detectable activity, an inherent uncertainty with the expression cloning system. On the other hand, substantial amounts of mRNA are needed for the hybridizations. Several important clones have been isolated in this way, including interferon and IL-2 (8a, 36a, 44a).

Translation of in vitro generated transcripts. Recently an in vitro transcription system was developed utilizing the SP6 phage promoter and SP6 RNA polymerase to generate substantial amounts of specific RNAs (33). If a cDNA sequence is inserted into this transcription unit and the RNA product is capped using guanylytransferase, also in vitro, then the RNA can be translated by microinjection into Xenopus oocytes (27). Because of the substantial amounts of mRNA that can be obtained and the efficiency of oocyte translation, it should be possible to pool cloned plasmids for transcription and subsequently translate the mixed mRNA by microinjection. The limit on the number of plasmids in a pool will depend essentially on the sensitivity of the translation assay. We know of no report of the successful application of this method at present but consider it to be an eminently reasonable alternative to hybrid selection or transient expression in mammalian cells.

Cloning by expression of antigenic determinants. The insertion of cDNA sequences in phase within the coding region of a transcription unit should result in the synthesis of a fusion protein containing some of the antigenic determinants of the protein encoded by the cDNA. Screening of clones prepared in this way with specific antibodies should identify a clone encoding the protein of interest. Several groups have reported on methods of antigenic screening of expressed proteins in bacterial or phage systems. Recently Young and Davis (50) reported the construction of an effective λ phage expression system using a phage construction designated λgt11. This phage contains a β-galactosidase gene with a unique EcoRI restriction site upstream of the termination codon. cDNA sequences are inserted into this site and the recombinant phage used to transfect a high-frequency lysogenic strain of E. coli. Cultures are plated on nitrocellulose and subsequently prophage production and β-galactosidase gene expression are induced by temperature shift in the presence of isopropylthiogalactoside. Accumulated intracellular fusion protein is exposed by in situ lysis of the bacteria.

In principle, screening of the phage plaques for expression could be performed either with a monoclonal or polyclonal antibody with a second radiolabelled or colorimetric marker of antibody presence. However, polyclonal antisera is preferable because, in using a monoclonal, one is dependent on the expression of a single epitope which may not be formed by the fusion protein. Nonspecific backgrounds produced by contaminating antibodies (usually against coliform proteins) can present a further problem in identifying clones, and as a result this method has not been widely preferred. In fact, in considering one's cloning strategy, this method would only be used if there were unique problems associated with purification of the protein whereby amino acid sequence was not obtainable.

cDNA probe enrichment by subtractive hybridization

Much of the sequence complexity in the mRNA of a cell is common to all cells in an organism because it encodes essential housekeeping functions. Furthermore, closely related cell types such as B and T lymphocytes have more than 97% of their mRNA sequences in common. Frequently, the protein one wishes to clone is synthesized by one cell line and not by a closely related cell line such that mRNA from the related cell type can be used to remove all common sequences by subtractive hybridization. Such an approach has been used to identify mRNA sequences unique to T lymphocytes and has allowed the isolation of the T cell antigen receptor (17).

Radiolabelled cDNA is synthesized from mRNA derived from the cell line of interest. This cDNA is then hybridized to mRNA from a closely related cell line. Unhybridized cDNA, presumably specific to the cell line of interest, is selected by hydroxyapatite chromatography and then hybridized back to the original mRNA. This hybridized cDNA is then used as a cell line-specific hybridization probe. Highly specific probes can be generated by this procedure, identifying a limited set of clones which can then be screened by other methods if necessary.

GENE EXPRESSION

Systems for the large-scale synthesis of gene products have been developed using both prokaryotic and eukaryotic cell hosts. In general outline, all these systems involve the in vitro construction of a DNA sequence containing a promoter of transcription upstream of the desired coding sequence, a terminator of transcription downstream, and a translation initiator either provided by the coding sequence or inserted during the construction process. This transcription unit is inserted in an appropriate DNA vector which will allow either extrachromosomal replication of the DNA or integration of the DNA into the host chromosome. Selectable marker genes are included to allow isolation of transformed cells and maintain the introduced DNA. Many factors affect the decision as to which system will be most applicable for a particular protein. In general, high-volume products such as insulin must be made in easily fermentable microorganisms such as yeasts or bacteria, and low-volume proteins can be made in higher eukaryotic systems such as mammalian cells. Perhaps the most important factor affecting the choice of host expression system is the complexity of the protein to be produced. Most extracellular proteins are held in their active conformation by disulfide bonds. In small proteins only one or two such bonds may exist, but in large proteins as many as twenty or more can occur. Although the principle that the polypeptide chain alone can define its active conformation and will thus self-assemble has been established (1), where multiple disulfide bond formation is required, the efficient assembly of the protein may only be possible during the sequential process of protein secretion into the oxidizing environment either of the rough endoplasmic reticulum lumen (eukaryotes) or the periplasmic space or medium (prokaryotes). At present, efficient prokaryotic secretion systems for producing heterologous products are not well developed (or at least are not in the public domain), and one is dependent on eukaryotic secretion systems such as yeast or mammalian cells to obtain active proteins of this type. Since most clinically useful proteins are in fact secreted and thus also contain an N-terminal leader sequence required for directing their secretion, mammalian cell expression systems are often preferred. A good example of this type of protein is presented by the serine protease tissue plasminogen activator (t-PA), a 68,000 dalton protein with 17 potential disulfide bonds. Intracellular synthesis of this protein in bacteria can be achieved with good yield but the protein is essentially inactive. Attempts to denature this protein and allow it to refold slowly in an oxidizing environment yield largely misfolded protein. Secretion of t-PA protein from mammalian cells, on the other hand, produces a high specific activity, fully functional protein with correct N-terminal processing (24).

Another structural consideration is the functional importance of secondary modifications such as the presence of oligosaccharide side chains, γ-carboxylated glutamic acid residues, or β-OH-aspartic acids. These modifications can critically affect the in vivo pharmacokinetics of the product and its functionality. Mammalian cell systems glycosylate heterologous proteins efficiently and probably correctly, a process foreign to bacteria. Thus, where an active product can be made in sufficient quantities in either system, only a critical assessment of the in vivo properties of the product will provide information leading to the best choice of expression system.

A further consideration to be addressed with many proteins is the requirement for cleavage of the primary translation product either to activate the product or to remove N-terminal signal sequences. Insulin is particularly interesting in this regard requiring the removal of an internal domain (the C peptide) for activation. The large amounts of insulin needed have required the use of bacterial expression systems, and various approaches have been devised to synthesize the active product. The two active polypeptides can be synthesized independently in bacteria, purified, and then renatured under controlled conditions. Alternatively, the polypeptide chain can be engineered to build in sites for chemical or enzymatic removal of the C peptide.

Finally, it is worth noting that, with proteins which are expected to be administered over extended periods of time such as growth hormone or Factor VIII:C, minor conformational differences between the native and recombinant protein can induce an immune response. The extent of this problem is unclear, but where possible it is probably preferable to synthesize the protein in a higher eukaryotic cell in a manner similar to its natural synthetic route.

We will deal here with the three major expression host systems now widely in use, those being various mammalian cell systems, yeast systems, and E. coli. Many others are under development and will not be dealt with here.

Bacterial Expression Systems

Intracellular expression systems

Although the fermentation characteristics of E. coli were not as well understood as those of some other species, this bacterium is the host of choice for intracellular bacterial expression largely because its genetics have been so extensively studied. Several examples of the very successful production of clinically useful proteins have been reported, including the interferons, growth hormone, and insulin. The expression system is usually designed such that the transcription unit is inducible and the organism can be grown to high density in the absence of product formation. Induction then rapidly produces large amounts of product (up to 40% of total cell protein can be achieved) which accumulates in dense inclusion bodies, essentially containing a precipitate of the product. This effect has the pleasing advantage of simplifying the early steps in purification of the product.

No universally efficient promoter system has been identified, but most work has been done with the Trp promoter, inducible by tryptophan depletion, the P_L promoter from bacteriophage λ using a temperature-sensitive repressor so that induction is achieved by temperature shift, and the Tac promoter, a fusion of the Trp promoter sequence with the Lac operator so that induction is achieved by adding isopropylthiogalactoside (IPTG). The Trp promoter is generally viewed as the best available system because of its promoter strength and the physiological conditions under which it is induced. In fact, induction will occur spontaneously in late log-phase growth as tryptophan is depleted from the fermentation broth. The P_L system does work well in some cases, but the temperature shift needed to induce this system can co-induce several proteases and cause product instability.

Since the primary translation product of this type of expression is intended to be the final form of the protein, some engineering of the encoding sequence is required to remove 5' non-coding sequences and any secretory leader signal sequences. The first codon of the mature protein is thus ligated to the initiator AUG codon of the bacterial promoter to provide a bacterial 5' non-coding segment with a ribosome binding site. A major problem encountered here is the removal of an N-terminal methionine residue in the recombinant protein, not present in the natural protein, which can potentially induce immunological reactions in clinical use if allowed to remain.

Since high rates of translation are necessary to drive product formation during the small time interval of late log-phase growth, secondary structures in the hybrid mRNA sequence which can slow down translation initiation by sequestering initiator codons into double-stranded structures or slow ribosome translocation along the mRNA can present a problem. It has also been noted that, since codon usage frequencies (or, more properly, specific t-RNA abundancies) differ between bacterial and mammalian systems, the presence of a series of consecutive "rare" codons in the hybrid mRNA can slow translation rates. These problems are difficult to identify prior to direct DNA sequencing but are somewhat more straightforward to eliminate by modification of the coding sequence using site-directed mutagenesis.

Extracellular expression systems

As we have mentioned previously, intracellular synthesis of the product does not promote disulfide bond formation, and yet many clinically useful proteins such as the serine proteases of the coagulation and fibrinolytic pathways contain multiple disulfide bonds critical to their activity. Much attention is now being given to the secretion of products from bacterial systems. In these systems the coding sequence of the mature protein is ligated to an N-terminal secretory leader segment. Examples of useful leaders include β-lactamase, lamB, ompF, phoA,and lpp, all of which can direct the secretion of the product into the periplasmic space. High specific activities have been found for the secreted products, implying correct folding of the polypeptide, but very low levels of expression are at present reported. This may be due in part to the fusion of the secretory leader to the mature product with the result that the correct processing site for proteolytic removal of the leader is either destroyed or inefficiently utilized.

Some bacterial species such as Bacillus and Pseudomonas secrete substantial amounts of protein into the medium. Harnessing of these secretory mechanisms would be an attractive future development. At present little data is available either on the levels of secretion possible with heterologous proteins or on the efficiency of folding of the secreted product, but it can be expected that such systems will in due course be developed.

Yeast Expression Systems

Within the last several years, the importance of yeast as an alternative system for expression of foreign proteins has emerged. There are several advantages for selecting yeast as a host. One attractive advantage is the ability to culture high-density, large-scale yeast fermentations for which the growth characteristics have been extensively investigated and are well understood. Since yeast secretes proteins by a mechanism very similar to that observed in mammalian cells, harnessing these mechanisms for the production of foreign proteins should yield large quantities of highly active products. Yeast cells also glycosylate secreted proteins using the same signal sequences in the polypeptide backbone as mammalian cells.

Another advantage is that yeast secretes a relatively small percentage of its total cellular protein into the medium, facilitating separation and purification of the desired product. Moreover, yeast cells do not produce endotoxins and are nonpathogenic to man, which should simplify the acceptance of human pharmaceutical products made in yeast. As a result of the increasing interest in yeast as a host for production of foreign proteins, many types of expression and secretion vectors for the introduction and maintenance of genes in yeast have been developed. The following is a brief outline of some of these systems.

Vector systems

Expression vectors for yeast can either integrate into the genome or remain in a stable extrachromosomal configuration. Since yeast expression plasmids can be maintained at a relatively high copy number, 20-200 copies per cell, they have become the most frequently used expression vectors. A typical yeast/bacterial shuttle vector consists of a yeast origin of replication, which is usually the DNA segment encoding the origin of the yeast 2µ circle, or a chromosomal replication origin (e.g., ARS1), and a yeast selectable marker such as the genes coding for URA3, LEU2, or TRYP1. These vectors are maintained in the host by complementation of an auxotrophic requirement of the host cell. For manipulation and characterization of the plasmid in E. coli, the shuttle vectors also contain a bacterial origin of DNA replication and a bacterial selectable marker gene such as Amp^R or Tet^R.

Expression systems

In most systems, two types of yeast transcriptional promoters have been used to express high levels of foreign proteins in yeast. Where accumulation of the foreign protein is expected to be toxic to cell growth, inducible promoters have been employed. Two examples are the inorganic phosphate-repressible acid phosphatase and the glucose-repressible invertase promoters which are induced when inorganic phosphate or glucose, respectively, are depleted from the medium. The other class of promoters are active constitutively in the cell and include the glycolytic promoters, glyceraldehyde 3-phosphate dehydrogenase (GPD4), triose phosphate isomerase (TP1), and 3-phosphoglycerate kinase (PGK). Some other useful promoters have been obtained from ornithine carbamoyl transferase (ARG3), alcohol dehydrogenase (ADH1), and galactokinase (GAL1) genes. Although the level of expression of foreign proteins varies depending on the protein, the use of these promoters has resulted in expression levels between 0.1-10% of total cellular protein. Examples of proteins successfully expressed in yeast include interferon (20), α1-antitrypsin (6), proinsulin (44), prorennin (13), and hepatitis B antigen (19, 47).

In addition to the need for a transcriptionally strong promoter, it has also been reported that efficient transcription termination is important for high level expression. For example, mutations affecting transcription termination of the cycl-512 gene resulted in a dramatic decrease in transcript levels (51). However, introduction of a yeast termination sequence from the invertase SVC2 gene at the 3' end of the transcriptional unit had no effect on the expression of calf pro-chymosin in yeast. Another factor which may be important for high level expression of foreign proteins in yeast and human cells which might affect translational efficiency is the difference in codon usage between yeast and mammalian systems (13). Codon biases have been noted for abundant yeast proteins (3).

Secretion systems
Although intracellular expression of heterologous proteins in yeast has been quite successful in several cases (most notably the hepatitis B surface antigen), some proteins have been synthesized intracellularly as inactive products. One explanation for the observed lack of activity of the cytoplasmically expressed proteins is that they require processing via the secretory pathway to assume the correct active conformation. Hence, a lot of effort has been directed towards development of secretion vectors using a variety of different leader sequences. In several cases such as the interferons (20), the natural human signal sequence has been found to be efficiently recognized and processed by yeast cells. However, for other proteins, such as α1-antitrypsin (16), the natural leader was not efficiently recognized by the yeast cell, resulting in the accumulation of unprocessed pre-protein.

As an alternative to the use of the natural leader sequence, several groups have substituted yeast-derived secretory leaders, resulting in increased secretion efficiency and higher yields of secreted products. For example, fusion of the invertase leader to the coding sequence of mature human leukocyte interferon increased the level of secretion of the protein by severalfold. The yeast mating α-factor and α-factor pre-pro regions have also been employed successfully to direct foreign proteins into the medium. Some examples using the α-factor pre-pro leader include human epidermal growth factor which was observed to accumulate in the medium at approximately 5-10 μg/ml culture (5), human β-endorphin, and human α-interferon (4). In each example, the biologically active protein was recovered from the medium and shown to be correctly processed at the natural N-terminus.

As the technology for developing high level expression and secretion systems for yeast becomes increasingly available, the promise of yeast as an alternative host for production of foreign proteins is being met with the development of a yeast-derived hepatitis β-vaccine (43): it is expected that within a short period of time the production of other foreign proteins made in yeast will become commercially available.

Mammalian Cell Expression Systems
The complex nature of many proteins of potential therapeutic benefit has required their expression in mammalian cell systems. Advantages of producing a protein in a mammalian cell system include a) the expressed

proteins are glycosylated similarly to their natural counterparts, b) the expressed proteins are secreted into the medium to facilitate purification; c) the proteins expressed in mammalian cells yield products which are properly folded and which contain the proper disulfide bond formation; and d) very large proteins (greater than 100 Kd) have not been successfully produced in microbial systems, whereas success has been obtained from mammalian cell expression systems. The primary problem with the production of pharmaceuticals in mammalian cells is the high cost of mammalian cell culture. Thus, expression in mammalian cell systems has been limited to proteins needed in small quantities but which are of high value. However, as the need to grow mammalian cells in large volume grows, so engineering advances are being made to reduce dramatically the cost of growing mammalian cells. Eventually the cost may become very competitive with costs of preparing pharmaceuticals from microbial fermentations. Several systems have been utilized to express proteins at high levels in mammalian cells. For the purpose of this review, focus will be limited to three general systems: 1) lytic viral vectors, 2) non-lytic viral vectors, and 3) DNA cotransformation with subsequent selection for gene amplification.

Lytic viral vector systems

The most extensively studied lytic viral vector system is the SV40 virus. This is primarily a result of the wealth of information which exists on the replication and expression of SV40 in different host cells. Foreign DNA may be substituted into the early or late region of SV40 and the virus propagated in the presence of a helper virus (12, 15, 36). More recently, the availability of a monkey kidney cell line which expresses the early region of SV40 (COS cells) has allowed the generation of helper-free recombinant viruses (21). Although SV40-based viral vectors have proved useful for significant expression of some proteins (for example, hepatitis surface antigen (35) and human growth hormone (39)), several severe limitations exist. First, the SV40 chromosome is limited in size, so genes larger than 2.5 kilobases become very difficult to propagate efficiently. Second, most efficient viruses result from mixed viral infections. In many cases, the helper virus will predominantly replicate over the recombinant during continued virus passage. Third, these viruses lyse the host cells such that harvesting of the product from the medium must be carried out before cell lysis occurs. Finally, the large-scale production of products from recombinant SV40 viruses also requires the large-scale production of high titer recombinant viruses. In many cases this is cost-prohibitive. Another lytic virus system which has limited success is human adenovirus. This expression system also suffers the same limitations as the SV40 recombinant viruses. Although adenoviruses may be constructed to transduce 7-8 Kb of foreign DNA, their use is more limited due to the short lytic cycle. Adenovirus-infected cells lyse 20-24 h post-infection, and hence very little time is available to accumulate product before cell lysis.

Non-lytic viral vectors

There are primarily two classes of non-lytic viral vector systems, based on either RNA or DNA viruses, which have been used to express heterologous genes. RNA tumor viruses (retroviruses) have been used to transfer DNA into heterologous cells. Particular advantages of retroviral vectors include a) the ability to transduce genes in a variety of cell types and of a variety of species, b) since the retrovirus integrates into the host chromosome, the DNA is very stable, c) the efficiency of transfer into the host cell can approach 100%, and d) high-titer recombinant virus stocks can be prepared. However, expression of protein from retroviral-based vectors has been low and problems with RNA splicing have hampered the usefulness of this vector system. Furthermore, packaging constraints again limit the size of the inserted segments to approximately 6-7 kilobases.

The prototype of DNA-based non-lytic viral vectors, bovine papilloma virus (BPV), is a small DNA virus that morphologically transforms a variety of cells. Vectors containing the entire BPV genome or a 5.5 Kb subgenomic transforming fragment are in many cases stable as multicopy (20-100 copies/cell) extrachromosomal elements in transformed cells (9, 29, 40, 52). In other cases, the vector sequences are maintained as oligomeric plasmids integrated into the host chromosome in a head-to-tail tandem array (10). The multicopy nature of the DNA in BPV-transformed cells is in part responsible for efficient expression of the foreign gene. Derivatives of BPV vectors which contain selectable markers have been produced, so it is not necessary to rely on morphological transformation (30).

Many secreted proteins, such as β-interferon (34, 53) and human growth hormone (38), have been expressed in BPV-based vectors. Although there has been considerable success with BPV expression systems, the biology of BPV expression from its genome and requirements for its plasmid maintenance and regulation of copy number are poorly understood. Results have been variable. Potential problems include obtaining proper transcription of the inserted gene, the instability of plasmid maintenance, and DNA rearrangements which occur upon propagation. As more features of the biology of BPV become understood, this will probably become a more generally useful system for expression of a wide variety of heterologous genes.

Co-transformation with subsequent DNA amplification

A particularly useful system for obtaining permanent cell lines expressing high levels of heterologous proteins is to co-transfect and subsequently co-amplify the copy number of the integrated DNA. Although a number of amplifiable markers are potentially available, the most characterized system is the amplification of the dihydrofolate reductase (DHFR) gene which occurs upon selection of cultured cells in stepwise increasing concentrations of methotrexate (MTX). Since MTX binds to and inhibits DHFR stoichiometrically, the amplification of the DHFR gene, with a resultant increase in intracellular DHFR, is responsible for the cells' ability to grow in higher concentrations of MTX. Upon selection at very high concentrations of MTX in the culture medium, cells are selected which can contain up to 10,000 copies of the DHFR gene/cell (for review see (42)).

The MTX selection and amplification system has proven particularly successful since Chinese hamster ovary (CHO) cells which are deficient in DHFR (46) can be transformed with a DHFR gene along with any particular gene of interest. $DHFR^+$ transformants can be selected on the basis of growth in the absence of nucleotides. Subsequent selection in increasing MTX results in amplification of the transfected DNA. DNA which has been co-amplified in this manner is also co-expressed (23, 28). This approach has been used to obtain cell lines which constitutively express and secrete high levels of β-interferon (32) and γ-interferon (6, 41). More recently high levels of human tissue plasminogen activator have been produced by CHO cells transformed and selected in this manner (24).

CONCLUSIONS

It is our opinion that we are only in the first of many phases of the growing use of recombinant DNA research to produce clinically useful products. Our technology is at present most readily applied to the isolation and expression of proteins which can be directly used in the treatment of disorders. As we learn more about the structure/function relationships in proteins and the mechanisms of their interaction with biological systems, it is to be expected that the true meaning of the term "genetic engineering" will be realized with the design of new proteins which can be applied in more convenient forms or can perform new tasks not carried out by natural proteins.

In the meantime, the startling accomplishments of many groups, who have isolated and characterized proteins and genes whose presence was previously known only by inference, have mitigated many of the fears that the biotechnology industry would not live up to the high expectations of the medical community and the public in general.

REFERENCES

(1) Anfinsen, C.B. 1973. Principles that govern the folding of protein chains. Science **181:** 223-230.

(2) Bailey, J.M., and Davidson, N. 1976. Methylmercury as a reversible denaturing agent for agarose gel electrophoresis. Anal. Biochem. **70:** 75-85.

(3) Bennetzen, J.L., and Hall, B.D. 1982. Codon selection in yeast. J. Biol. Chem. **257:** 3026-3031.

(4) Bitter, G.A.; Chen, K.K.; Banks, A.R.; and Lai, P.-H. 1984. Secretion of foreign proteins from Saccharomyces cerevisiae directed by α-factor gene fusions. Proc. Natl. Acad. Sci. USA **81:** 5330-5334.

(5) Brake, A.J.; Merryweather, J.P.; Colt, D.G.; Heberlein, U.A.; Masiarz, F.R.; Mullinback, G.T.; Urdea, M.S.; Valenzuela, P.; and Barr, P.J. 1984. α-Factor-directed synthesis and secretion of mature foreign proteins in Saccharomyces cerevisiae. Proc. Natl. Acad. Sci. USA **81:** 4642-4646.

(6) Cabezon, T.; DeWilde, M.; Herion, P.; Loriau, R.; and Bollen, A. 1984. Expression of human α_1-antitrypsin cDNA in the yeast Saccharomyces cerevisiae. Proc. Natl. Acad. Sci. USA **81:** 6594-6598.

(7) Collins, J., and Hohn, B. 1978. Cosmids: A type of plasmid gene-cloning vector that is packageable in vitro in bacteriophage λ heads. Proc. Natl. Acad. Sci. USA **75:** 4242-4246.

(8) de Martynoff, G.; Pays, E.; and Vassart, G. 1980. Synthesis of a full length DNA complementary to thyroglobulin 33 S messenger RNA. Biochem. Biophys. Res. Commun. **93:** 645-653.

(8a) Derynck, R.; Content, J.; DeClercq, E.; Volckaert, G.; Tavernier, J.; Devos, R.; and Fiers, W. 1980. Isolation and structure of a human fibroblast interferon gene. Nature **285:** 542-547.

(9) DiMao, D.; Corbin, V.; Sibley, E.; and Maniatis, T. 1984. High-level expression of a cloned HLA heavy chain gene introduced into mouse cells on a bovine papillomavirus vector. Molec. Cell. Biol. **4:** 340-350.

(10) DiMao, D.; Triesman, R.; and Maniatis, T. 1982. Bovine papillomavirus vector that propagates as a plasmid in both mouse and bacterial cells. Proc. Natl. Acad. Sci. USA **79:** 4030-4034.

(11) Efstradiatis, A.; Kafastos, F.C.; Maxam, A.M.; and Maniatis, T. 1976. Enzymatic in vitro synthesis of globin genes. Cell **7:** 279-288.

(12) Gething, M.-J., and Sambrook, J. 1982. Construction of influenza haemagglutinin genes that code for intracellular and secreted forms of the protein. Nature **300:** 598-603.

(13) Goff, C.G.; Moir, D.T.; Kohno, T.; Gravins, T.C.; Smith, R.A.; Yamasaki, E.; and Taunton-Rigby, A. 1984. Expression of calf prochymosin in Saccharomyces cerevisiae. Gene **27:** 35-46.

(14) Gubler, U., and Hoffman, B.J. 1983. A simple and very efficient method for generating cDNA libraries. Gene **25:** 263-269.

(15) Hamer, D.H., and Leder, P. 1979. SV40 recombinants carrying a functional RNA splice junction and polyadenylation site from the chromosomal mouse β major globin gene. Cell **17:** 737-747.

(16) Haynes, J., and Weissman, C. 1983. Constitutive, long-term production of human interferons by hamster cells containing multiple copies of a cloned interferon gene. Nucl. Acids Res. **11:** 687-706.

(17) Hendrick, S.M.; Cohen, D.I.; Nielsen, E.A.; and David, M.M. 1984. Isolation of cDNA clones encoding T cell-specific membrane-associated proteins. Nature **308:** 149-153.

(18) Hewick, R.M.; Hunkapiller, M.E.; Hood, L.E.; and Dryer, W.J. 1982. A gas-liquid solid phase peptide and protein sequenator. J. Biol. Chem. **256:** 7990-7997.

(19) Hitzeman, R.A.; Chen, C.Y.; Hagie, F.E.; Patzer, E.J.; Liu, C.C.; Estell, D.A.; Miller, J.V.; Yaffe, A.; Kleid, D.G.; Levinson, A.D.; and Oppermann, H. 1983. Expression of hepatitis B virus surface antigen in yeast. Nucl. Acids Res. **11:** 2754-2763.

(20) Hitzeman, R.A.; Leung, D.W.; Perry, L.J.; Kohr, W.J.; Levine, H.L.; and Goeddel, D.V. 1983. Secretion of human interferons by yeast. Science **219:** 620-625.

(21) Horowitz, M.; Cepko, C.L.; and Sharp, P.A. 1983. Expression of chimeric genes in the early region of SV40. J. Molec. Appl. Genet. **2:** 147-159.

(22) Jacobs, K.; Shoemaker, C.; Rudersdorf, R.; Neill, S.D.; Kaufman, R.J.; Mufson, A.; Seehra, J.; Jones, S.S.; Hewick, R.; Fritsch, E.F.; Kawakita, M.; Shimizu, T.; and Miyake, T. 1985. Isolation and characterization of genomic and cDNA clones of human erythropoietin. Nature **313:** 806-810.

(23) Kaufman, R., and Sharp, P. 1982. Amplification and expression of sequences cotransfected with a modular dihydrofolate reductase complementary DNA gene. J. Molec. Biol. **159:** 601-621.

(24) Kaufman, R.; Wasley, L.; Spiliotes, L.; Gossels, S.; Latt, S.; Larsen, G.; and Kay, R.M. 1985. Coamplification and coexpression of human tissue-type plasminogen activator and murine dihydrofolate reductase sequences in Chinese hamster ovary cells. Molec. Cell. Biol. **5:** 1750-1759.

(25) Kaufman, R.J. 1985. Identification of components necessary for adenovirus translational control and their utilization in cDNA expression vectors. Proc. Natl. Acad. Sci. USA **82:** 689-693.

(26) Kaufman, R.J., and Sharp, P.A. 1982. Construction of a modular dihydrofolate reductase cDNA gene: analysis of signals utilized for efficient expression. Molec. Cell. Biol. **2:** 1304-1319.

(27) Krieg, P., and Melton, D. 1984. Functional messenger RNAs are produced by SP6 in vitro transcription of cloned cDNAs. Nucl. Acids Res. **12:** 7057-7071.

(28) Lau, Y.-F.; Lin, C.C.; and Kan, Y.W. 1984. Amplification and expression of human α-globin genes in Chinese hamster ovary cells. Molec. Cell. Biol. **4:** 1469-1475.

(29) Law, M.-F.; Lowy, D.R.; Dvoretzky, I.; and Howley, P.M. 1981. Mouse cells transformed by bovine papillomavirus contain only extrachromosomal viral DNA sequences. Proc. Natl. Acad. Sci. USA **78:** 2727-2731.

(30) Lusky, M., and Botchan, M.R. 1984. Characterization of the bovine papilloma virus plasmid maintenance sequences. Cell **36:** 391-401.

(31) Maniatis, T.; Fritsch, E.F.; and Sambrook, J. 1982. Molecular Cloning. A Laboratory Manual. New York: Cold Spring Harbor Laboratory.

(32) McCormick, F.; Trahey, M.; Innis, M.; Dieckmann, B.; and Ringold, G. 1984. Inducible expression of amplified human β interferon genes in CHO cells. Molec. Cell. Biol. **4:** 166-172.

(33) Melton, D.; Krieg, P.; Rebagliati, M.; Maniatis, T.; Zinn, K.; and Green, M. 1984. Efficient in vitro synthesis of biologically active RNA and RNA hybridization probes from plasmids containing a bacteriophage SP6 promoter. Nucl. Acids Res. **12:** 7035-7056.

(34) Mitrani-Rosenbaum, S.; Maroteaux, L.; Mory, Y.; Revel, M.; and Howley, P.M. 1983. Inducible expression of the human interferon β_1 gene linked to a bovine papilloma virus DNA vector and

maintained extrachromosomally in mouse cells. Molec. Cell. Biol. **3:** 233-240.

(35) Moriarty, A.M.; Hoyer, B.H.; Shih, J.W.-K.; Gerin, J.L.; and Hamer, D.H. 1981. Expression of the hepatitis B virus surface antigen gene in cell culture by using a simian virus 40 vector. Proc. Natl. Acad. Sci. USA **78:** 2606-2610.

(36) Mulligan, R.C.; Howard, B.H.; and Berg, P. 1979. Synthesis of rabbit β-globin in cultured monkey kidney cells following infection with a SV40 β-globin recombinant genome. Nature **277:** 108-114.

(36a) Nagata, S.; Taira, H.; Hall, A.; Johnsrud, L.; Streali, M.; Escodi, J.; Boll, W.; Cantell, K.; and Weissmann, C. 1980. Synthesis in E. coli of a polypeptide with human leukocyte interferon activity. Nature **284:** 316-320.

(37) Okayama, H., and Berg, P. 1982. High-efficiency cloning of full-length cDNA. Molec. Cell. Biol. **2:** 161-170.

(38) Pavlakis, G.N., and Hamer, D.H. 1983. Regulation of a metallothionein-growth hormone hybrid gene in bovine papilloma virus. Proc. Natl. Acad. Sci. USA **80:** 397-401.

(39) Pavlakis, G.N.; Hizuka, N.; Gorden, P.; Seeburg, P.; and Hamer, D.H. 1981. Expression of two human growth hormone genes in monkey cells infected by simian virus 40 recombinants. Proc. Natl. Acad. Sci. USA **78:** 7398-7402.

(40) Sarver, N.; Gruss, P.; Law, M.-F.; Khoury, G.; and Howley, P.M. 1981. Bovine papilloma virus deoxyribonucleic acid: a novel eucaryotic cloning vector. Molec. Cell. Biol. **1:** 486-496.

(41) Scahill, S.J.; Devos, R.; Van der Heyden, J.; and Fiers, W. 1983. Expression and characterization of the product of a human immune interferon cDNA gene in Chinese hamster ovary cells. Proc. Natl. Acad. Sci. USA **80:** 4654-4658.

(42) Schmike, R.T. 1984. Gene amplification in cultured animal cells. Cell **30:** 705-713.

(43) Scholnick, E.M.; McLean, A.A.; West, D.J.; McAleer, W.J.; Miller, W.J.; and Buynak, E.B. 1984. Clinical evaluation in healthy adults of a hepatitis B vaccine made by recombinant DNA. JAMA **251:** 2812-2815.

(44) Stepien, P.P.; Broussean, R.; Wu, R.; Narang, S.; and Thomas, D.Y. 1983. Synthesis of a human insulin gene. VI. Expression of the synthetic proinsulin gene in yeast. Gene **24:** 289-297.

(44a) Taniguchi, T.; Matsui, H.; Fujita, T.; Takaoka, C.; Kashima, N.; Yoshimoto, R.; and Hamuro, J. 1983. Structure and expression of a cloned cDNA for human interleukin-2. Nature **302:** 305-310.

(45) Toole, J.J.; Knopf, J.L.; Wozney, J.M.; Sultzman, L.A.; Buecker, T.L.; Pittman, D.D.; Kaufman, R.J.; Brown, E.; Shoemaker, C.; Orr, E.C.; Amphlett, G.W.; Foster, W.B.; Coe, M.L.; Knutson, G.J.; Fass, D.N.; and Hewick, R.M. 1984. Molecular cloning of a cDNA encoding human antihaemophilic factor. Nature **312:** 342-347.

(46) Urlaub, G., and Chasin, L.A. 1980. Isolation of Chinese hamster cell mutants deficient in dihydrofolate reductase activity. Proc. Natl. Sci. USA **77:** 4216-4220.

(47) Valenzuela, P.; Medina, A.; Rutter, W.J.; Ammerer, G.; and Hall, B.D. 1982. Synthesis and assembly of hepatitis B virus surface antigen particles in yeast. Nature **298:** 347-350.

(48) Wong, G.G.; Witek, J.S.; Temple, P.A.; Wilkens, K.M.; Leary, A.C.; Luxenberg, D.P.; Jones, S.S.; Brown, E.L.; Kay, R.M.; Orr, E.C.; Shoemaker, C.; Golde, D.W.; Kaufman, R.J.; Hewick, R.M.; Wang, E.A.; and Clark, S.A. 1985. Human GM-CSF: Molecular cloning of the complementary DNA and purification of the natural and recombinant proteins. Science **228:** 810-815.

(49) Woods, D.E.; Edge, M.D.; and Colten, H.R. 1984. Isolation of a complementary DNA clone for the human complement protein C2 and its use in the identification of a restriction fragment length polymorphism. J. Clin. Invest. **74:** 634-638.

(50) Young, R.A., and Davis, R.W. 1983. Efficient isolation of genes by using antibody probes. Proc. Natl. Acad. Sci. USA **80:** 1194-1198.

(51) Zaret, K.S., and Sherman, F. 1982. DNA sequence required for efficient transcription termination in yeast. Cell **28:** 563-573.

(52) Zinn, K.; DiMao, D.; and Maniatis, T. 1983. Identification of two distinct regulatory regions adjacent to the human β-interferon gene. Cell **34:** 865-879.

(53) Zinn, K.; Mellon, P.; Ptashne, M.; and Maniatis, T. 1982. Regulated expression of an extrachromosomal human β-interferon gene in mouse cells. Proc. Natl. Acad. Sci. USA **79:** 4897-4901.(31)

Biotechnology: Potentials and Limitations, ed. S. Silver, pp. 41-54. Dahlem Konferenzen 1986. Berlin, Heidelberg, New York, Tokyo: Springer-Verlag.

Synthetic Genes

H. Blöcker and R. Frank
Gesellschaft für Biotechnologische Forschung mbH (GBF)
3300 Braunschweig, F.R. Germany

Abstract. Chemically synthesized DNA fragments are versatile tools in modern bioscience. They are nowadays routinely used not only in recombinant DNA projects as probes, primers, linkers, and adaptors, but also in the construction of complete genes. The present technology for total gene synthesis is described. We believe that further improvements in the methodology will make the present instrumentation for DNA synthesis obsolete, enable us to circumvent, when desired, techniques such as site-directed mutagenesis and open up better possibilities, e.g., in the design of proteins. In addition, we consider social responsibilities of DNA chemists.

INTRODUCTION

Biotechnology has been understood and practiced as the technical use of biological systems, and most people think of biotechnology and its potential for pure scientists, lawyers, instrument manufacturers, pharmaceutical industries, politicians, and - hopefully - also for the sick and the starving, when they consider the rest of the century as the era of the biologists. However, it is obvious that the biotechnology of today is an alloy of many disciplines and has been influenced by many parent sciences.

One of the major achievements of synthetic chemists for biotechnology has been to elaborate reliable procedures for the chemical synthesis of DNA fragments which may be used in the construction of long double-stranded DNA and of genes in particular.

Some thirty years ago when Michelson (15) chemically synthesized for the first time a simple dinucleotide in the laboratory of Lord Todd, probably nobody could foresee that basically this type of reaction would be repeated hundredfold every day all around the world with the aim of making completely synthetic genes. This changed, to some extent, when in the late sixties and early seventies the pioneering work of Khorana and co-workers (2, 12, 13) culminated in the total synthesis of two t-RNA genes. However, following Khorana's phosphodiester approach to DNA synthesis was too laborious for most interested groups, and only when the modified phosphotriester approach by the groups of Letsinger, Reese, Cramer, Narang, and many others as well as (more recently) the phosphoramidite approach by Caruthers and co-workers replaced the diester method (for a historical review see (1, 8, 9, 16, 17, 21)) did the chemical synthesis of DNA become more widespread. Numerous genes have been synthesized and successfully expressed since then.

Though rational methodologies for chemical and enzymatical syntheses of long double-stranded DNA are at hand, it still seems to be under discussion whether chemical gene synthesis is really efficient for realistic pursuit of biotechnological projects. Only recently Davies and Gassen stated that the synthesis of structural genes was not a job for DNA chemists, because intact genes could more efficiently be excised from the genome or from plasmids (4).

PRESENT STRATEGIES

Chemical Work

Numerous strategies for the chemical synthesis of oligodeoxyribonucleotides have been developed and are still being improved substantially (1, 8, 9, 16, 17, 21). Some basic principles are summarized in Table 1. The researcher finds the best way to synthesize DNA in this table by a sensible combination of the options from left to right. However, there is no general answer to the question of how to obtain synthetic DNA. One must first analyze the specific situation: what are the budget, available know-how, and annual demand? Apparently, most of the DNA synthesis performed these days is done via the phosphotriester or the phosphoramidite routes in 3' to 5' direction, and for the special requirements of genetic engineering and gene synthesis in particular (small amounts), people are working now on solid supports with the equipment depending on their budget and annual demand.

The synthesis of less than ca. fifty oligonucleotides per annum could quite conveniently be done with conventional manual procedures (for example, beads as solid support in a syringe or a manual synthesizer) if the oligonucleotides were not all needed at a certain point of time. For higher demands one may consider buying one of the reliable autosynthesizers which are sometimes incorrectly called "gene machines." Since these machines produce, however, only a crude mixture of short, single-stranded DNA that is chemically modified and biologically dead, they should be classified as "synthesis machines," "synthesis automats," "autosynthesizers," or just "synthesizers." All these conventional systems reported so far use solid supports in a granulated form (beads, fibers, etc.) placed into a reaction vessel, preferably a column-type one. Automation of the synthesizer is then achieved by a suitable solvent and reagent delivery system. However, all oligonucleotides are still synthesized separately, either one after the other or parallel in different reaction vessels.

TABLE 1 - How to make (obtain) synthetic DNA fragments.

chemistry	synthesis proceeds	reaction vessel/ equipment	phase	solid phases	other sources
diester	5' - 3'	flask	homogeneous (in solution)	beads	friends
triester	3' - 5'	syringe	heterogeneous	fibers	companies
phosphite (incl. phos- phoramidite)	bidirectional	funnel manual synthesizer (column) autosynthesizer (column)	(on polymers)	segmental supports - filters - distinct column segments or membranes filled with beads or fibers	

Further progress in the DNA synthesis field has commonly, until recently, been expected from better (shorter cycle times) and cheaper machines. Contrary to these expectations, we have introduced a new route to mass production of oligonucleotides: the principle of segmental supports (6). This technology allows one to assemble simultaneously hundreds of oligonucleotide chains within a few days. Syntheses by this method can be performed manually with very simple and inexpensive equipment. However, automation is also possible. The basic idea of this new principle is as follows: one uses non-interchangeable entities as polymer supports. These are either clearly marked, paper-like flat materials (for example, cellulose disks) or conventional-type materials (beads or fibers) sealed in membranes or kept in distinct column segments. Each of the segments in a "multisynthesis" refers to one oligonucleotide to be synthesized.

For each of the elongation steps one combines the segments in not more than four groups, if monomer additions are planned. The four groups refer to the four natural bases A, C, G, and T. One combines the segments in a way that allows the treatment of all segments of one group simultaneously with the same solvents, reagents, and the elongation mix. The sequential steps of an elongation cycle can be done simultaneously for all the four groups. For those steps that are not sequence-specific, one can even combine all the segments in one vessel or flush them in series. After an elongation cycle is over, one looks at the list of oligonucleotide sequences. According to which type of elongation (A-, C-, G-, or T-addition) is needed next for the individual oligonucleotide, the corresponding segments are recombined - again in the four A-, C-, G-, and T-groups. This is repeated until all chains are completed.

The segmental support approach, which has been successively practiced for the construction of various long genes and for other purposes, saves a lot of labor time. Since in this methodology the number of elongation cycles practically no longer depends on the number of oligonucleotides to be synthesized, one can calculate that in the synthesis of (say) an interferon gene one can reduce the number of cycles by about 90% (ca. 900 cycles vs. ca. 60 cycles).

One should, however, never forget that assembling a protected oligonucleotide is, whatever strategy one follows, only 10-20% of the work of building a DNA fragment which can be used in biological experiments.

Even if one bought the building blocks and other starting materials, isolating and characterizing the finished oligonucleotide would account for more than half of the labor. The scheme in Fig. 1 perhaps does not completely reflect how much manual work is required, how many small operations such as evaporations, desaltings, measurements, etc., contribute to the high expenditure of work, or how much equipment has to be used. This unpleasant situation has recently encouraged people to reinvestigate this part of the methodology and suggest alternative procedures to achieve less work load and/or higher reliability.

The introduction of β-cyanoethyl phosphoramidites and the use of additional base-protecting groups on T and G in the phosphotriester approach allow one to use ammonia as the only alkaline deprotection reagent. Chain scission by nucleophilic attack of OH^- on phosphorus is apparently no longer a practical problem, at least for shorter oligonucleotides of up to twenty-five bases (25mers).

Purifications according to lipophilicity (based on the high lipophilicity of the 5' protecting group which by the end of the synthesis should only be present on the target sequence if an efficient capping step has been included after each coupling) can be performed simultaneously for 5-10 oligonucleotides on disposable cartridges rather than consecutively on hplc.

Some synthesis groups restrict themselves by first deblocking completely and then purifying their oligonucleotides only by polyacrylamide-gel electrophoresis. On the one hand, this is certainly a good procedure if only small quantities must be purified, and of course, many samples can be worked up simultaneously on a slab gel. On the other hand, some sequence isomers, base-modified sequences, and other impurities which may be formed during the chemical syntheses and which will later cause higher rates of false sequences after cloning cannot be sorted out by this time-saving procedure. Therefore, to be on the safe side, many people still rely on a two-step purification based on two different separation principles (for example, reverse phase and ion exchange chromatography).

Unfortunately, unlike all other organic chemists, oligonucleotide chemists do not carefully analyze their products. This is mainly due to three reasons:

1 - Classical techniques such as elemental analysis and nuclear magnetic resonance (NMR) do not reveal much and cannot be usefully applied.

First alkaline treatment
(detaches DNA from
support, removes
phosphate-protecting groups)

↓

Second alkaline treatment
(removes base-protecting groups)

↓

Purification according to
lipophilicity
(on reversed phase
material; purifies from
truncated sequences and
removed protecting
groups)

↓

Acid treatment
(removes lipophilic 5'-O-protecting group)

↓

Purification according to
chain length
(Paired ion/ion exchange
hplc or gel electrophoresis;
removes traces of modified material
and shorter sequences)

↓

Sequence analysis
(wandering spot, Maxam/Gilbert,
FAB mass spectrometry)

FIG. 1 - Flow chart of the commonly followed procedures for the isolation and characterization of synthetic DNA fragments.

2 - Other techniques such as the two-dimensional wandering spot or the Maxam/Gilbert-type chemical degradation sequencing methods are sometimes considered too laborious, whereas the very fast sequencing by the "Fast-Atom-Bombardment"-mass spectrometry (FAB-MS) is still limited to oligonucleotides of thirteen bases in length and requires up to one optical density (O.D.) unit per analysis (10). Furthermore, suitable instrumentation for the latter method is currently available only in a few places.

3 - Some producers of DNA fragments do not collect analytical data of their products at all. They prefer to sequence their DNA only after cloning.

When making a decision in this matter, one should also take into account how reliable an experimenter can be and how reliable the chosen chemistry is. On the other hand, one can imagine how desperate the situation of a chemist may be if a biologist pressures him by requesting still more synthetic DNA to put forward his biotechnology projects. Nevertheless, one should avoid taking chances (that means not to increase the work load on the level of the cloned DNA) and gather at least a few analytical data. This can easily be done by proper sizing on denaturing polyacrylamide gels after electrophoresis. Since the mobility of an oligonucleotide in such gels depends not only on the chain length but also greatly on the base composition, one can calculate this mobility relative to homo-oligomeric chain length markers with a simple formula (for example, dT_{3-60} and dA_{3-60}) (7). By co-electrophoresis of oligonucleotides and chain length markers one can thus evaluate whether the product is most likely the target molecule. Along with these data one obtains a nice picture of the homogeneity of the oligonucleotide preparation.

Enzymatic Work

The enzymatic part of gene construction can be described more easily than the chemical part. There are nowadays basically two approaches to the construction of long double-stranded DNA (Fig. 2).

After the discovery of DNA ligases, Khorana and co-workers (11) established a procedure which is basically still valid for the construction of genes (Fig. 2, Ia). On planning the construction, one cuts the double-strand in such a way that it looks like two rows of bricks in a wall. The experiment proceeds as follows: the oligonucleotides which are designed to form the double-strand by complete overlaps and which usually bear free hydroxyls on either end after their chemical synthesis are 5'-phosphorylated with ATP by polynucleotide-kinase. They are then heated to high temperatures (90°-100°C) to destroy any unwanted hybridization and slowly cooled down to allow wanted hybridization (annealing). The DNA is then partially present as a double-strand with nicks (not gaps) at the ligation sites. After addition of DNA-ligase these nicks are sealed to give the target double-stranded DNA. The overlaps of two complementary fragments in this strategy are about half of the fragment length (that is, longer than five bases), and the length of the fragments usually ranges between twelve and twenty nucleotide units. The products are (nowadays) isolated from preparative slab gels after electrophoresis.

There are a few "secrets" as to how one best achieves high yields of intact synthetic genes.

A clean chemistry will provide oligonucleotides with very little base-modified or (n-1) or (n+1) contaminants. These contaminants cause most of the mutations observed in synthetic genes.

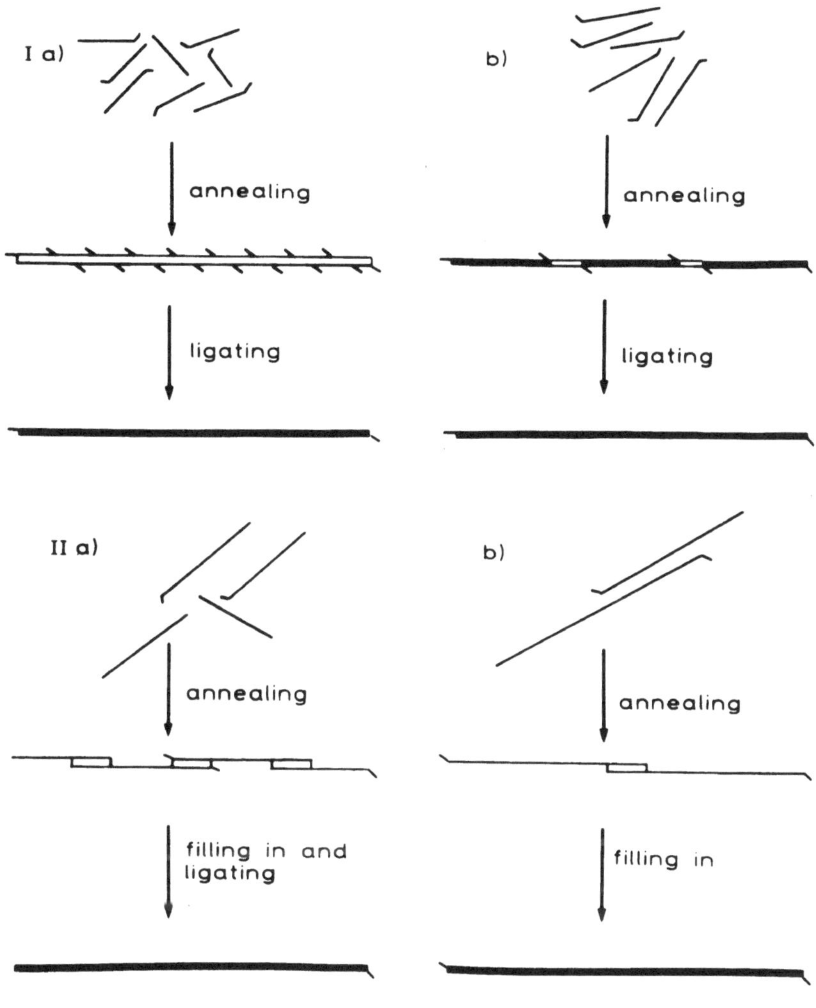

FIG. 2 - Scheme of strategies for the enzymatic construction of long, double-stranded DNA. Lines representing DNA single-strands, bars represent DNA double-strands. 5'-phosphorylated DNA carries an oblique line on one end.

Complete phosphorylation of the 5'-hydroxyls must be achieved. Otherwise, after annealing, two hydroxyls may be adjacent at the nicks and the ligase will by no means be able to seal the nicks. Though very rapid tests for phosphorylation yields are described in the literature, according to our knowledge, quantitative phosphorylation by the very popular kinasing reaction is a problem in many laboratories.

Computer-aided design of the fragments that form the double-strand will reduce or practically exclude mishybridizations during annealing. These competing reactions are a real danger and can sometimes prevent the formation of the desired double-strand. For longer genes, planning by visual inspection cannot be done in a reasonable time.

Ligation at the highest possible reaction temperature is another safety barrier against unwanted hybridizations and wrong ligation products.

Following these rules and taking these precautions, we found that relatively long DNA double-strands (up to 300 base pairs) can now be formed in one-step ligations (for an example see Fig. 3). Furthermore, the yields are surprisingly high (about 90%, above 30% after isolation from gel) and the number of mutations detected after cloning is low (in our hands now less than one in 500 base pairs). We are convinced that without computer planning we would have isolated much less material, if any at all.

Besides this Khorana-type route to enzymatical gene construction (11), Fig. 2 describes a few variations and alternative routes.

Procedure Ib has been used by some scientists since the chemical synthesis of longer oligonucleotides has becomes possible. In this variation of the Khorana approach, the length of the oligonucleotides usually varies between thirty and fifty bases. For the most part, the ligation sites are no longer placed opposite the center of the complementary fragment, thus generating a series of double-stranded fragments which overlap specifically with their neighboring fragments on the same strand by only a few bases. The double-stranded fragments have high stacking energies and will therefore be formed at relatively high temperatures. In this experiment one may first form all the short double-strands in separate tubes, then combine them in one tube, and finally ligate them like short restriction fragments.

Procedures II, basically proposed by Itakura and co-workers (19), go beyond the use of longer oligonucleotides. They are based on a useful property of another enzyme: the template-directed and primer-dependent 5'-3' synthesis capability of DNA polymerase I (Klenow fragment). After annealing of the long oligonucleotides the polymerase uses deoxynucleoside-triphosphates to fill in the gaps as shown in Fig. 2. Any nick in the resulting, completely double-stranded DNA is sealed by DNA-ligase. In theory this approach is very elegant. As in procedure Ib, only longer oligonucleotides (about 40mers) are used. This reduces the number of oligonucleotides that have to go through laborious workup procedures as mentioned earlier. An additional reduction of some 40% is gained by not completely overlapping the oligonucleotides and by filling in the gaps enzymatically. However, in the laboratory procedure IIa currently results in the formation of too many mutants to be of practical value.

Based on these negative results and on the fact that very recently the synthesis of rather long oligonucleotides (up to and above 100 bases) has become possible, a variation of procedure IIa has been used as outlined under IIb in Fig. 2 (5). Two oligonucleotides overlap only with their 3' end regions. This construct is completed to a full-length double-strand by filling in with polymerase. Experimentally this procedure results in a smear on the gel. Nevertheless, it has been shown that the target DNA can be sorted out after cloning with a reasonable effort in time. Though this variation does not require DNA-ligase for sealing any nicks, it does require (like procedure IIa) cutting at the ends with an appropriate restriction endonuclease if single-stranded overhangs are needed for subsequent cloning steps.

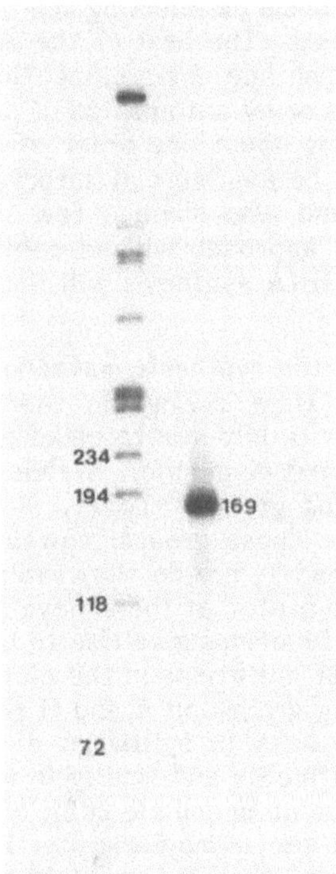

FIG. 3 - End product of a one-step ligation of 22 oligonucleotide fragments to give a 169 base pair, DNA double-strand analyzed on a non-denaturing 10% polyacrylamide gel.

FUTURE DEVELOPMENTS AND CONCLUSIONS

No biotech company would ever spend any money on a DNA synthesis project if it would not contribute to higher profits. Since many companies do have DNA synthesis capabilities available, these capabilities must open up new possibilities or more effective routes. To get an idea of what DNA synthesis can offer, let us consider the total synthesis of a gene of about 500 base pairs and assume that we would apply the best available technology. Provided that all starting materials and instrumentation were available, we could expect that the project would proceed as follows:

1 -	computer planning	less than 1 day
2 -	chemical synthesis	less than 1 week
3 -	isolation and characterization	2-5 weeks, depending on fragment length/number of fragments
4 -	ligation	3 days, including analysis on gel
5 -	cloning and sequencing	2 weeks

Clearly, the chemical synthesis (assembling the oligonucleotide chains) is no longer the rate-limiting step. The best of the currently available, conventional autosynthesizers finish one chain elongation cycle every six minutes. By multi-column switching or by automation of a segmental support method one can reduce this time by about one order of magnitude. Since machines with such capabilities will be available at surprisingly low prices within the next one or two years, and also since a few more improvements of the manual segmental support approach will be achieved, one can predict that the present machines for DNA synthesis will disappear from the market in the near future.

Further time reduction in the synthesis methodology can no longer significantly contribute to any time savings in the pursuit of gene synthesis projects. Therefore, one will only aim to make chemical synthesis of oligonucleotides more reliable and economic. A redesign of the chemistry must include new base-protecting groups. Today, most DNA chemists still use the Khorana-type groups. These groups, however, do not protect derivatizable O-functions on G and T, nor do they replace (because they are simple acyl groups) the second proton of the exocyclic amino functions of A, G, and C. These unprotected functions give rise to base-modification reactions and hence account for most mutations in the cloned material. Furthermore, the conventional protecting groups on A and G reduce the stabilities of the glycosidic bonds. This results in splittings of the latter bonds (depurinations) during acidic treatments and finally in less and less pure material. It would be very useful from an economic point of view to have reliable procedures for the workup of the monomeric starting material because these compounds, which are usually applied in tenfold to twentyfold excesses, are nowadays discarded after one cycle.

A sensible redesign of the oligonucleotide chemistry will greatly influence the ease and the quality of the workup and characterization procedures. And this part is obviously the bottleneck of the present methodology for gene synthesis. However, the workup will in the future be speeded up considerably. In one of our last publications (14) we introduced an automatable system for the workup nucleotides. After completion of a chain synthesis and the alkaline treatment, the solution containing the oligonucleotide is directly washed on a small disposable column. The column filling has hydrophobic as well as ion exchange properties. By a suitable washing procedure one can not only separate the product from the phosphate and base-protecting groups and from truncated sequences, but also split off the final, acid-labile protecting group. Though we still struggle with a few capacity problems, we believe that the future autosynthesizers will have such an automated workup procedure built in as an on-line option.

At present, some people still argue that the simplest way to reduce the effort for the workup of oligonucleotides and hence for gene synthesis is to synthesize fewer but longer instead of more but shorter oligonucleotides. However, shorter oligonucleotides are not only more easily purified, but also less specialized with respect to the genetic information they carry. Since we can synthesize hundreds and thousands of oligonucleotides per annum and laboratory, now that we have banks of oligonucleotide sequences and proper retrieval programs in our computers, reusability of oligonucleotides has become a point to think about.

Over the years we have heard about and also experienced a few things in gene technology projects which at first glance appeared to be rather mysterious. In quite a few cases it came down to the level of synthetic DNA when the reason for the mystery was finally found: mostly spoiled chemicals, wrong labels on bottles, and other human errors. In none of these cases had the synthetic DNA been sequenced prior to use. We therefore think that a reliable and rapid sequencing method will be commonly accepted and applied. Such a procedure could evolve from the Maxam/Gilbert-type chemical degradation sequencing on solid support as recently published from groups in the USSR (3) and the German Democratic Republic (18). Another choice will certainly be the FAB mass spectrometry sequencing. Stronger magnets will allow us within the next few months to sequence bidirectionally oligonucleotides longer than 20mers if the oligomers are to be sequenced completely from either end. If we only want to look for a few overlaps in the middle of the oligonucleotides, the stronger magnet will take us beyond 40mers. A computer-driven "hopping technique" (only relevant mass regions are looked at), multi-scan facilities, and automated data acquisition and data processing will reduce by a factor of ten the amount of DNA required to run the sequencing and make this ultrafast method even more convenient. Since FAB mass spectrometry can also be used as the method of choice for the identification and partial sequencing of recombinant DNA gene products (proteins), it will find widespread application for DNA sequencing as detailed above.

Putting together all these hopes, expectations, and predictions, how might the situation be for the synthesis of a 500 base-pair fragment? Provided that all starting material and instrumentation were available, some six to eight days after computer planning a high-quality, ligated 500 base-pair fragment would be ready for cloning, and the situation for a 1 or 2 Kb fragment would not differ much from this. With these potentials of the methodology in prospect, one can give some reasons why one should synthesize genes or other long DNA double-strands.

Whenever the primary structure of a gene or a gene product is known, one will prefer to synthesize the gene rather than to obtain it through cDNA cloning. Recombinant DNA techniques represent for many people the most fascinating part of a biotechnology project. However, large-scale fermentation of over-producing strains, protein isolation, restoring the activity of the protein, and establishing the product in industrial processes or for medicinal treatments is by far the longer way to go. Any company can therefore catch up with its competitor by gene synthesis if the project is still on the cloning level - with the inherent danger that nobody will continue to publish gene sequence or protein sequence data.

Protein design, systematical bettering or creation de novo of proteins, is commonly regarded as one of the most challenging fields in biotechnology for the future. If one wants to create proteins de novo, one can synthesize them either by peptide synthesis or via gene synthesis. Since it is hard to produce large quantities of longer peptides (longer than about thirty amino acids) of high quality by chemical synthesis, one will prefer to make longer peptides in protein design projects via gene synthesis. When bettering proteins, one has an additional alternative on the DNA level.

Site-directed mutagenesis techniques rely to date on short synthetic primers designed to introduce the mutations in either single-stranded or gapped-duplex DNA carrying the DNA stretch that has to be mutagenized specifically (mismatch primer mutagenesis). Although substantial progress has been made in this field, the yield of mutants is sometimes low and the procedure laborious. Since, on the other hand, the synthesis of long DNA double-strands is going to be a matter of days, one should consider excising the DNA fragment to be mutagenized from the vector, adding to the cutting mix in large excess a synthetic DNA which represents the mutated fragment, then religating and cloning the DNA (cassette mutagenesis). Such a procedure would work on any vector DNA, especially on double-stranded vectors.

STATEMENT ON SOCIAL RESPONSIBILITIES

Our world is threatened by many things, amongst others, hunger, sickness, natural catastrophes, but also hate, racism, environmental problems, and last but not least, something the experts call "taylorization" (20). One feature of this is a pronounced specialization of workers and a higher qualification of a few managers. This is, to some extent, also true for those who run biotechnology projects. The bricklayer (DNA chemist) does not really know what the tailor (cloner) does, and an upper manager pulls the strings (and sells the project). Some politicians, especially in a crisis, show a tendency to jump on new technologies that appear to be in fashion, because they think this to be the easiest way to get rid of most of their problems. It seems as if at present biotechnology is such a technology. Indeed, we will work out a better understanding of biological systems, we will design new and better proteins for various purposes, we will have high-score techniques for germ-line manipulations. However, on the other hand, we have already experienced a certain eager desire expressed by institutes run by armed forces and also from other unpleasant places. Like other specialists, the DNA chemist will play a key role in biotechnology. He had better think twice before he gives away his synthetic genes.

Acknowledgements. We thank D.N. Lincoln for critically reading the manuscript and B. Seeger-Kunth for typing the text.

REFERENCES

(1) Blöcker, H.; Frank, R.; and Fritz, H.-J., eds. 1986. Chemical Synthesis and Biological Application of DNA Fragments. Weinheim: VCH Verlagsgesellschaft (in preparation).

(2) Brown, E.L.; Ramamoorthy, B.; Ryan, M.J.; and Khorana, H.G. 1979. Chemical synthesis and cloning of a tyrosine tRNA gene. Methods Enzymol. **68:** 109-151.

(3) Chuvpilo, S.A., and Kravchenko, V.V. 1985. A simple and rapid method for sequencing DNA. FEBS Lett. **179:** 34-36.

(4) Davies, J.E., and Gassen, H.G. 1983. Synthetic gene fragments in gene technology - renaissance of chemistry in molecular biology. Angew. Chem., Intl. Ed. Eng. **22:** 13-31.

(5) Engels, J.; Leineweber, M.; and Uhlmann, E. 1985. Chemical and enzymatic synthesis of genes. Proceedings of the Fifth International Conference on Organic Synthesis, Freiburg, F.R. Germany, August 27-30, 1984, in press.

(6) Frank, R.; Heikens, W.; Heisterberg-Moutsis, G.; and Blöcker, H. 1983. A new general approach for the simultaneous chemical synthesis of large numbers of oligonucleotides: segmental solid supports. Nucl. Acids Res. **11:** 4365-4377.

(7) Frank, R., and Köster, H. 1979. DNA chain length markers and the influence of base composition on electrophoretic mobility of oligodeoxyribonucleotides in polyacrylamide-gels. Nucl. Acids Res. **6:** 2069-2087.

(8) Gait, M.J., ed. 1984. Oligonucleotide Synthesis. Eynsham: IRL Press.

(9) Gassen, H.G., and Lang, A., eds. 1982. Chemical and Enzymatic Synthesis of Gene Fragments: A Laboratory Manual. Weinheim: Verlag Chemie.

(10) Grotjahn, L.; Frank, R.; and Blöcker, H. 1982. Ultrafast sequencing of oligonucleotides by FAB mass spectrometry. Nucl. Acids Res. **10:** 4671-4678.

(11) Gupta, N.K.; Ohtsuka, E.; Sgaramella, V.; Büchi, H.; Kumar, A.; Weber, H.; and Khorana, H.G. 1968. Studies on polynucleotides LXXXVIII. Enzymatic joining of chemically synthesized segments corresponding to the gene for alanine-tRNA. Proc. Natl. Acad. Sci. USA **60:** 1338-1344.

(12) Khorana, H.G.; Agarwal, K.L.; Besmer, P.; Büchi, H.; Caruthers, M.H.; Cashion, P.J.; Fridkin, M.; Jay, E.; Kleppe, K.; Kleppe, R.; Kumar, A.; Loewen, P.C.; Miller, R.C.; Minamoto, K.; Panet, A.; RajBhandary, U.L.; Ramamoorthy, B.; Sekiya, T.; Takeya, T.; and van de Sande, J.H. 1976. Total synthesis of the structural gene for the precursor of a tyrosine suppressor transfer RNA from Escherichia coli. I. General introduction. J. Biol. Chem. **251:** 565-570.

(13) Khorana, H.G.; Agarwal, K.L.; Büchi, H.; Caruthers, M.H.; Gupta, N.K.; Kleppe, K.; Kumar, A.; Ohtsuka, E.; RajBhandary, U.L.; van de Sande, J.H.; Sgaramella, V.; Terao, T.; Weber, H.; and Yamada, T. 1972. Studies on polynucleotides CIII. Total synthesis of the structural gene for an alanine transfer ribonucleic acid from yeast. J. Molec. Biol. **72:** 209-217.

(14) Meyerhans, A.; Heisterberg-Moutsis, G.; Kurth, G.; Blöcker, H.; and Frank, R. 1985. Rapid purification of chemically synthesized oligodeoxyribonucleotides. Nucleosides Nucleotides **4:** 245 (Abstract).

(15) Michelson, A.M., and Todd, A.R. 1955. Synthesis of a dithymidine dinucleotide containing a 3':5'-internucleotide linkage. J. Chem. Soc. **Part III:** 2632-2638.

(16) Ohtsuka, E.; Ikehara, M.; and Söll, D. 1982. Recent developments in chemical synthesis of polynucleotides. Nucl. Acids Res. **10:** 6553-6570.

(17) Reese, C.B., ed. 1984. Recent Aspects of the Chemistry of Nucleosides, Nucleotides and Nucleic Acids. Tetrahedron Symposia-In-Print Number 13. Tetrahedron **40:** 1-163.

(18) Rosenthal, A.; Schwertner, S.; Hahn, V.; and Hunger, H.-D. 1985. Solid-phase methods for sequencing of nucleic acids I. Simultaneous sequencing of different oligodeoxyribonucleotides using a new, mechanically stable anion-exchange paper. Nucl. Acids Res. **13:** 1173-1184.

(19) Rossi, J.J.; Kierzek, R.; Huang, T.; Walker, P.A.; and Itakura, K. 1982. An alternate method for synthesis of double-stranded DNA segments. J. Biol. Chem. **257:** 9226-9229.

(20) Taylor, F.W. 1911. The Principles of Scientific Management. London, New York: Harper and Bros.

(21) Winnacker, E.-L. 1985. From Genes to Clones. Weinheim: VCH Verlagsgesellschaft.

Biotechnology: Potentials and Limitations, ed. S. Silver, pp.55-69. Dahlem Konferenzen 1986. Berlin, Heidelberg, New York, Tokyo: Springer-Verlag.

Protein Engineering

G. Winter*, P. Carter*, H. Bedouelle*, D. Lowe**, R.J. Leatherbarrow*, and A.R. Fersht**
*Laboratory of Molecular Biology, Cambridge CB2 2QH
**Dept. of Chemistry, Imperial College, London SW7 2AY, England

Abstract. Site-directed mutagenesis of a gene allows the sequence of the encoded protein to be altered, and thereby its functions. The technology of making mutants and engineering new properties in proteins will be reviewed.

INTRODUCTION

It is now possible to alter the nucleotide sequence of a gene at will, and therefore the sequence of the encoded protein. This not only permits an analysis of the roles of individual residues in a protein (25), but the construction of protein molecules with new properties (3, 4, 18, 23). However, although the amino acid sequence of a protein dictates its three-dimensional structure and function, the rules by which sequence determines structure or structure determines function are not known. So while sequence can be changed to order, changing the three-dimensional structure or function of a protein to order is much more difficult. One strategy towards defining the rules governing sequences and structure, and structure and function, is to make limited changes in a protein of known three-dimensional structure. By inspection of the mutant structures and/or their new properties it may be possible to deduce some of the rules. For example, the X-ray crystallography of variant hemaglobins with differing physiological properties has provided a detailed insight into the structural mechanism of hemaglobin. Likewise the energetics of hydrogen bonding at the active site of the enzyme tyrosyl tRNA synthetase has been analyzed by protein engineering (7). Ideally the crystallographic structure of each mutant should be solved (at least a difference map for isomorphous derivatives). However, some mutations are unlikely to create major structural changes, as in (7), and it may even be possible to use a second mutation to probe the extent of structural change introduced by the first (2). Furthermore, computer graphic model building, perhaps incorporating energy calculations (as in the minimal perturbation method), may help predict the orientation of the new side chain and the likelihood of structural change elsewhere. Thus for the hundred or so proteins for which there is a crystallographic structure, interpreting the effects of carefully chosen mutations might not strictly require the X-ray crystallographic structure of each mutant. For these proteins, mutants could be used, for example, to analyze the forces between enzymes and substrates. Knowing the energies of the multiple contacts between an enzyme and substrate, it might then be possible to alter the contacts to improve substrate affinity or catalytic rate (23).

For many industrial enzymes, the crystallographic structure is not available. What can protein engineering offer? First, if suitable crystals can be obtained, with improving technologies such as synchrotron sources of X-rays (suitable for small crystals) and area detectors for data collection, it may be possible to solve the crystallographic structure. Introducing new surface cysteine residues might then help in the crystallization of isomorphous heavy atom derivatives, which are required in addition to the native crystals to solve the structure. Second, if there is some sequence homology with another enzyme of known structure, it might then be possible to build a model using computer graphics techniques. Alternatively, other physical techniques such as nuclear magnetic resonance or circular dichroism and/or protein structure prediction rules (such as Lim or Chou-Fasman rules) may enable a tentative structure prediction. Protein engineering might then be used to refine the tentative structure by making mutants which distinguish between alternative models or which test critical features of one model. Thus if substrate binding were lost in mutants which clustered together in a surface patch in the predicted structure, this would tend to confirm the structure and localize the binding site. Such an approach could, however, at best provide only a low-resolution model of the protein.

Gene structure can be altered either by total synthesis of a new gene segment or by site-directed mutagenesis of the gene using synthetic oligonucleotides. Although the total synthesis approach is suitable for construction of many mutations in a small gene, it is long-winded for the construction of single mutants. Thus site-directed mutagenesis has emerged as the paramount technology in protein engineering. We therefore review this technique, and how it has been used to deduce the roles of individual side chains in proteins and to change protein function.

SITE-DIRECTED MUTAGENESIS IN M13 VECTORS

M13 vectors have emerged as the vectors of choice for constructing mutations with oligonucleotides, because they are a good source of single-stranded template. This facilitates both mutant construction and subsequent sequencing of the mutant DNA to check the lesion. Expression of the encoded protein may well then require recloning into a suitable plasmid expression vector, although it has proved possible to express some proteins well in M13 vectors. The frequency of mutants in M13 vectors can be variable (ranging from 0.1%-50%), although mutant clones can be readily identified by dot blot hybridization to the mutagenic primer. However, if mutants could be routinely obtained at a frequency of >70%, they could be simply screened by sequencing a couple of plaques. Hence efforts have been made to improve the yield of mutants with dramatic results.

In Vitro Priming

First, a single-stranded copy of the gene is prepared by cloning the gene into the replicative form of the bacteriophage M13. Second, a short (15-20 bases) oligonucleotide is synthesized which is complementary to the region to be changed, apart from a limited internal mismatch which directs the mutation. At its simplest, "single priming," the oligonucleotide is hybridized to the single-stranded template and is extended with the Klenow subfragment of DNA polymerase in the presence of T4 DNA ligase. It is thought that the mismatch can be edited out via 5'-3' exonucleases (for example, DNA polymerase) in vivo, and therefore in the following strategies the 5' end is protected by ligation. In the "all the way round" method, the oligonucleotide is extended right round the template and the closed circular DNA purified on an alkaline sucrose gradient (27). In the "double primer" method, a second primer to the 5' side of the mutagenic oligonucleotide is extended concurrently to join up with the 5' end of the mutagenic oligonucleotide (28). In the "gapped duplex" method, the oligonucleotide is extended within a "window" of exposed single-strand: the gapped duplex is made by hybridizing M13 recombinant template with the linearized replicative form of the parent vector (11).

Transfection/Asymmetric Replication/Mismatch Repair

The mismatched heteroduplex is then transfected into a male E. coli strain made competent for DNA uptake. In the process of replication both plus (template) and minus (in vitro synthesized) strands of the heteroduplex make a contribution to the progeny phage, but the contribution from the minus strand is the major one. Each plaque may be mixed (containing wild-type and mutant phage), although one type usually predominates (6, 9). The frequency of mutant plaques (assuming perfect in vitro synthesis of the heteroduplex) thus depends on the relative contribution of the original plus and minus (mutagenized) strands to phage progeny and on the extent of mismatch repair in the cell. Certain mismatches are repaired very efficiently, others not at all (9). Transfection into cells deficient in mismatch repair, such as mutL or mutS strains of E. coli, has given dramatically enhanced frequencies of mutants for some point mismatches (1, 10).

Strand Selection

To improve mutant frequencies further, it is possible to eliminate the contribution of the plus strand to progeny phage by constructing a selectable marker on the minus strand. In (10) the gapped duplex method was

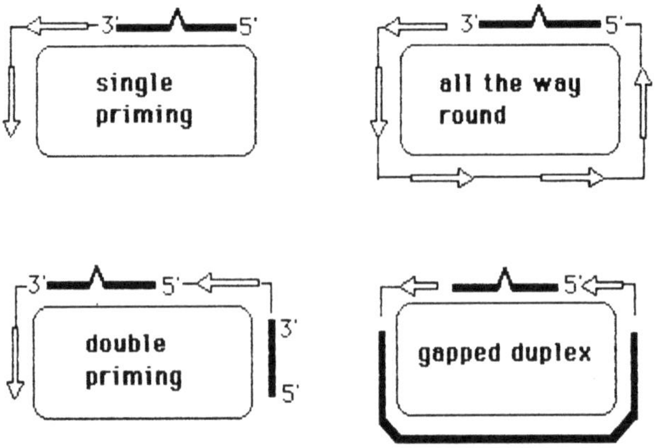

FIG. 1 - Four approaches to constructing mutations in M13 single-strand DNA.

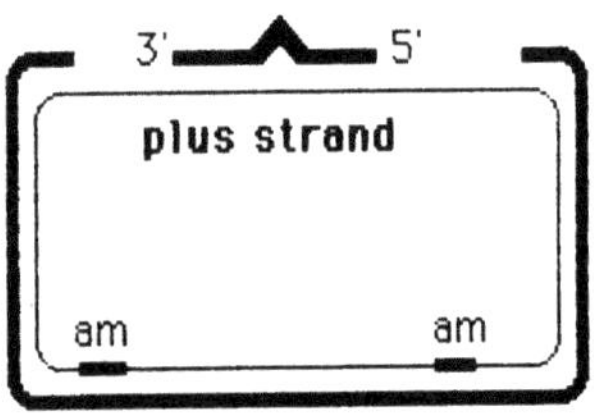

FIG. 2 - Coupled priming by the gapped duplex method.

utilized with two amber mutations on the M13 plus strand and the minus strand supplied by wild-type replicative form. Transfection into non-suppressor repair-deficient cells allows selection against progeny derived from replication of the plus strand (Fig. 2).

Much the same approach has been used with a variant of the "double priming" technique: an amber mutation was first constructed in M13 gene IV by site-directed mutagenesis, converting M13mp18 and M13mp19 to am IV versions. By priming on the single-stranded template with the mutagenic primer in the insert and with another primer (the selection oligonucleotide) which removes the amber mutation in gene IV, a partial duplex is constructed as in the "double priming" approach. Transfection into repair-deficient non-suppressor cells then selects against those phage progeny derived from plus-strand replication (1). This approach has been called "coupled priming" (Fig. 3).

Similarly, a cassette containing an EcoK or EcoB site has been introduced into M13 vectors as a strand marker for "coupled priming": the host restriction system (EcoK or EcoB) can recognize such sites and destroy the

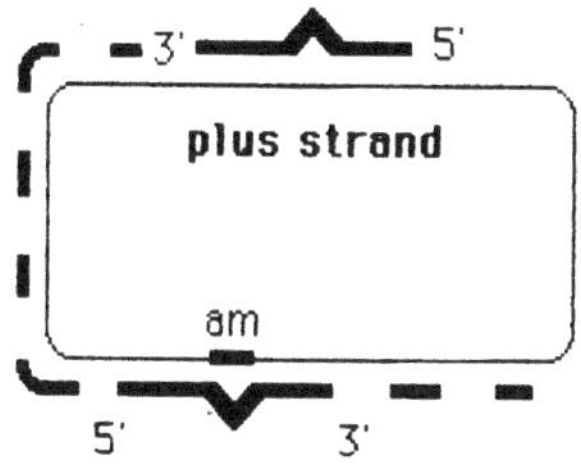

FIG. 3 - Coupled priming with a selection primer.

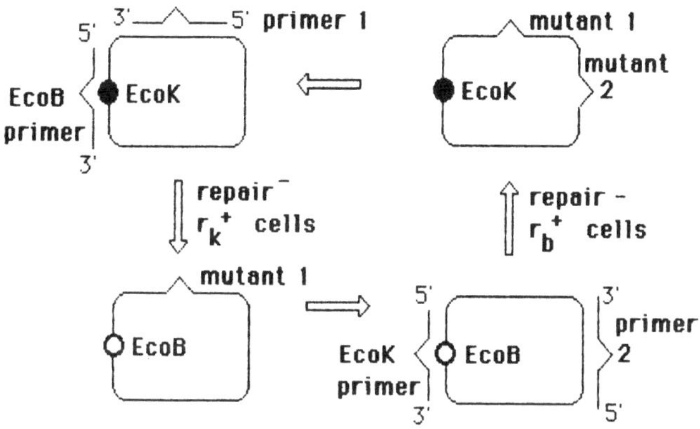

FIG. 4 - Cyclic selection mutagenesis using a reciprocating EcoK/EcoB marker.

vector. Thus a "selection primer" is used to remove the EcoK site in the vector, and by transfecting into repair$^-$rk$^+$ cells it is possible to select against progeny derived from plus-strand replication. At the same time that the EcoK site is removed, an EcoB site may be created, allowing further rounds of mutagenesis under "coupled priming" conditions (Fig. 4). Thus in the second round a "selection primer" is used which reconverts the EcoB site into an EcoK site, and by transfecting into repair$^-$rb$^+$ cells it is again possible to select against progeny derived from the plus strand. Compared with selection against the amber mutation in gene IV where the marker is lost after one round, this reciprocating genetic marker can allow multiple rounds of mutation, first under EcoK selection and then under EcoB selection (1).

There is an alternative approach in which the plus-strand contribution can be eliminated without coupling the silent mutation to a genetic marker. In (12), the M13 phage was grown in a <u>dut</u> <u>ung</u> mutant so as to incorporate uridine residues into the template strand: the plus strand was then directly eliminated in the heteroduplex by treating with uracil glycosylase and alkali. This approach seems promising especially as improved mutant frequencies can be obtained by simply growing the template in the dut ung mutant.

Multiple Mutations
Point mutations are readily constructed, but also mutants with multiple changes using a single oligonucleotide. For example, three adjacent nucleotides can easily be changed at the same time, and it is therefore possible to convert any amino acid in a protein. Given a long enough oligonucleotide, it is even possible to change several amino acids, and this can offer an attractive way of inserting a small region of one protein into the structure of another. The sequence in an α-helix of the tyrosyl tRNA synthetase from Bacillus stearothermophilus was converted into the sequence of the closely related E. coli enzyme (using a 38mer to direct six amino acid changes), and the loop preceding the helix into that of the distantly related methionyl tRNA synthetase from E. coli (using a 44mer to direct seven amino acid changes) (1). This contrasts favorably with grafting restriction fragments from one clone to another as in the replacement of a recognition helix of 434 repressor by that of the 434 cro protein (22). Where several sites to be mutated are distant, this can be achieved either by several rounds of mutagenesis or by using several mutagenic primers at the same time. In the latter case, the frequency of a double mutant appears to be no higher than the product of the two single mutants, and it therefore seems an excellent strategy for mutants expected in good yield.

Where it is necessary to change a single amino acid at one site into several possibilities, it may be possible to use a mixed sequence oligonucleotide (19). However, since some point mutations are more readily repaired than others (see above), a few mutants tend to predominate: the use of cells deficient in mismatch repair should avoid this. Given the range of Td values expected for hybridization of the mixed sequence oligonucleotide with each of the mutants, it is best to probe for mutants with a probe for the original sequence and then to identify the mutants as those clones which do not light up (19).

Insertions and Deletions
Insertions may be made with an oligonucleotide complementary to both regions flanking the insertion point, with the extra portion of sequence as an insertion bubble in the primer. Probably the size of the insertion is limited only by the length of the mutagenic oligonucleotide, allowing 5' and 3' flanking regions of some 7-12 nucleotides. For large insertions it may be better to make a unique restriction site and clone in a synthetic duplex. Deletions are made with an oligonucleotide complementary to both regions flanking the deletion point, so that in the annealing reaction the intervening region of template is looped out. Sections as long as 1kb have been deleted specifically using 18-24 mers, but a number of spurious deletions are often seen. These occur where one end of the oligonucleotide has hybridized at the target site and the other has hybridized at another site which shows some fortuitous complementarity to the primer.

CHANGING PROTEIN STRUCTURE AND FUNCTION

Designer proteins, built from first principles for a specific function, are the ultimate goal of protein engineering. Given the difficulty of predicting three-dimensional structure from sequence, it may help to use a protein framework of known structure, such as the framework residues of an antibody molecule, and to insert binding and catalytic groups to fashion an active site. Antibodies offer a particularly attractive route to designer proteins as it may be possible to fashion a crude active site by immunizing an animal with substrate or a transition state analogue as hapten. So far protein engineering has been applied to making small changes in known structures in an attempt to identify the roles of individual amino acid residues. Applications include the tyrosyl tRNA synthetase (2, 7, 8, 13, 23-25), dihydrofolate reductase (20), aspartate transcarbamylase (17), α1-antitrypsin (3, 18), interleukin-2 (21), T4 lysozyme (15), p21 (19), 434 repressor (22), interferon (14), and cytochrome c (16). For example, Phe 87 of cytochrome c is conserved at the surface of the protein and was thought to be directly involved in electron transfer between cytochrome c and cytochrome c peroxidase. Altering Phe 87 to Tyr, Ser, or Gly yields functional protein, although for the Ser 87 and Gly 87 mutants the redox potential is lower. This shows that Phe 87 is not strictly required for electron transport and supports the view that Phe 87 controls the polarity of the heme environment by excluding water.

Defining the roles of individual residues by site-directed mutagenesis can have important spin-offs. For example, it has been shown how the net interaction of substrate and enzyme represents an exchange with water molecules and how the energetics of hydrogen bonds between enzyme and substrate depends on the solvation of electrostatic charge (7). Furthermore, it has been possible to engineer new functions: for example, an improved binding affinity of the enzyme tyrosyl tRNA synthetase for its substrate (23), a more heat-stable T4 phage lysozyme (15), and an oxidation-resistant antitrypsin (3, 18). We now review how protein engineering has been used to change catalytic rate, substrate affinity, and specificity in an enzyme.

Changing Catalytic Rate and Substrate Affinity

It has been proposed that an enzyme can utilize the binding energy with their substrates to increase the rate of chemical catalysis. The structure of an enzyme should thus be complementary to the structure of the transition state of the substrate rather than the unreacted substrate. The improved binding energy, ongoing from substrate to transition state, lowers the activation energy of the reaction and so increases the catalytic rate. It is therefore convenient to divide those residues in the active site of an enzyme into those which bind to the ground state of the substrate (and thereby dramatically increase its local concentration) and those which bind to the transition state. A group which binds purely to the ground state will increase the affinity of the enzyme for the substrate but increase the activation energy barrier (and hence decrease k_{cat}). A group which binds purely to the transition state will decrease the energy barrier and not affect the substrate affinity. However, in principle, a group could bind to both ground state and transition state: thus a group which binds equally to ground state and transition state will increase the affinity of the enzyme but not affect the energy barrier. Thus the first requirement of engineering changes in substrate affinity or catalytic rate is to understand the role of residues in the active site in binding the ground state, transition state, or both. The second requirement is to know how much energy each type of bond (for example, van der Waals, electrostatic, hydrogen bonds) is worth. Protein engineering experiments can illuminate both aspects, as is illustrated by work on the enzyme tyrosyl tRNA synthetase.

The tyrosyl tRNA synthetase from Bacillus stearothermophilus has three substrates: tyrosine, ATP, and $tRNA^{tyr}$. The enzyme catalyzes the transfer of tyrosine to tRNA in a two-stage reaction in which the tyrosine is first activated at its carboxyl groups with ATP and then the tyrosyl adenylate is attacked by the 2' or 3' hydroxyl of the tRNA:

$$E + Tyr + ATP = E \cdot Tyr - AMP + PP_i$$

$$E \cdot Tyr - AMP + tRNA^{tyr} = Tyr - tRNA^{tyr} + AMP + E.$$

The location of tyrosyl adenylate was identified in the crystallographic map by diffusing tyrosine and ATP into the crystals, as the adenylate is then formed in situ. There are several side and main chain atoms which could make hydrogen-bonding contacts to the tyrosyl adenylate. For example, there are three side chain contacts from the enzyme (Cys35, His48, and Thr51) to the ribose moiety of the adenylate, probably three side chain contacts to the amino group of tyrosine (Asp38, Asp78, and Tyr169), and two contacts to the phenolic hydroxyl of the tyrosine side chain (Asp176 and Tyr34). Furthermore, a computer graphic model indicated that although Thr40 and His45 were unlikely to bind to the tyrosyl adenylate, they might bind to the pyrophosphate-leaving group in the transition state.

Binding to transition state/ground state. To evaluate the precise role of these residues in binding to ground state and to transition state, mutants were constructed in which the hydrogen bonding capability of these groups were removed. Thus Cys35, His45, and His48 were converted to Gly, Thr40 to Ala, and Tyr169 and Tyr34 to Phe. The kinetics of wild-type and mutant enzymes were then compared, enabling the energetic contribution of each side chain to both ground state or transition state to be calculated (Table 1). This showed that the binding of tyrosine to Tyr34 or Tyr169 is the same in the ground state and transition state, while the binding of ribose to Cys35 or His48 is much greater in the transition state than in the ground state (Wells and Fersht, unpublished). Likewise, His45 or Thr40 bind only to the transition state as expected from the model (not shown; Leatherbarrow, Fersht, and Winter, unpublished).

Thus the application of protein engineering and enzyme kinetics can show the roles of side chains in binding the ground state of a substrate, the

TABLE 1 - Binding of side chains to substrates and transition states in the activation of tyrosine.

	Binding to substrate kcal/mol	Binding to transition state kcal/mol
Tyr34-Phe	-0.52	-0.57
Tyr169-Phe	-3.11	-3.10
Cys35-Gly	-0.03	-1.25
His48-Gly	-0.44	-1.31

transition state, or both, and is a first step in changing catalytic rate or substrate affinity of an enzyme to order.

Energetics of the contacts. The energies of various hydrogen bonds, salt bridges, or van der Waals contacts to a substrate can be estimated from protein engineering experiments as shown above. For example, hydrogen bonding of enzyme (E-H) and substrate (B-S) in aqueous solution involves competition of water for the hydrogen-bonding sites. Thus the energetics of an individual hydrogen bond, calculated by comparing the kinetics of the wild-type enzyme with a mutant which lacks the hydrogen-bonding group, does not correspond to the direct interaction energy of the contact, rather it is the net energetics of an exchange with water (7):

$$E\text{-}H....OH_2 + HOH....B\text{-}S = E\text{-}H....B\text{-}S + HOH....OH_2.$$

An extensive comparison of these exchange energies for hydrogen-bonding contacts in the tyrosyl tRNA synthetase is given in Table 2 (7). These results show that deletion of a side chain between enzyme and substrate to leave an unpaired, uncharged hydrogen-bond donor or acceptor can weaken binding energy by 0.5-1.5 kcal/mol. But the presence of an unpaired, charged donor or acceptor can weaken binding by a further 3 kcal/mol: thus it is important that electrostatic charge (on enzyme or substrate) is

TABLE 2 - Comparative binding energies of transition states in wild-type and mutant tyrosyl tRNA synthetase with substrates in the activation of tyrosine.

Altered contact in enzyme	Contact in substrate	ΔG(kcal/mol)
Tyr34-Phe	Tyr-OH	-0.52
Tyr169-Phe	Tyr-NH3 +	-3.72
Gln195-Gly	Tyr-CO2	-4.49
Cys35-Gly	ATP 3'-OH	-1.14
His48-Gly	ATP-05	-0.96

solvated either by a water molecule or by a hydrogen-bonding group in the complex. These observations could be utilized in engineering substrate affinities or biological specificity (see below).

Improving enzyme affinity: an example. The construction of several mutants at Thr51 in the tyrosyl tRNA synthetase suggests there are several ways in which we might improve the affinity of the enzyme for its substrate. The side chain γ-hydroxyl makes a long contact to the ring oxygen of the ribose ring of tyrosyl adenylate. This side chain was converted to Gly, Ala, Ser, Cys, or Pro. By comparing the steady-state kinetics of the mutants, it is possible to calculate the energetics of each substituent group (γ-hydroxyl, γ-methyl, and β-methyl) of Thr51 with the transition state of the substrate (Table 3) (13, 23).

TABLE 3 - Contributions of individual groups at position 51 to binding energy.

Comparing enzymes	Group	Contribution to binding in kcal/mol
Thr51/Ser51	Y-methyl	-0.92
Ser51/Ala51	γ-hydroxyl	+ 1.36
Ala51/Gly51	γ-methyl	-0.72

The results show that the contact from the γ-hydroxyl is a liability and therefore the affinity of the enzyme for the transition state is improved (by 0.4 kcal/mol) by mutating Thr51 to Ala. The disadvantage of the bond from the γ-hydroxyl may be due to an unfavorable exchange with water molecules: a strongly bound water molecule must be displaced from the enzyme in return for a poor bond to the substrate. Thus simply removing a "bad" contact can improve the affinity of an enzyme (23). The affinity could be improved further by constructing a good contact. If the bond from the γ-hydroxyl to substrate is too long, then a bond from a γ-sulfhydryl (as in Cys51) should be optimal. Indeed, in the Cys51 mutant there is an increase in affinity of 0.9 kcal/mol over Thr51 (0.5 kcal/mol over Ala51), due to improved local contact.

However, directly altering side chain contacts to the substrate is but one way of altering affinity: this has also been achieved indirectly by distorting the polypeptide backbone and thereby altering the orientation or position of a side chain contact elsewhere. Converting Thr51 to Pro might be expected to distort the active site (23). In the Pro51 mutant there was a dramatic increase in affinity (-1.9 kcal/mol) of the enzyme for transition state. The reasons for this could be threefold. The substrate might dock directly against the pyrolidine ring of Pro51; the distortion introduced by Pro51 into the backbone might forge entirely new contacts with the enzyme; the distortion might improve existing contacts. In the Thr51 enzyme there is probably a hydrogen bond from the main chain amide of residue 51 back to the main chain carbonyl of Gly47 (α-helix) or His48 (3_{10} helix). This is not present in the Pro51 enzyme as the amino group cannot make the bond. Whatever additional effects of the Pro51 mutation, it seemed likely to affect the backbone at His48 and therefore possibly the orientation of the imidazole side chain. It was possible to test whether Pro51 had improved the contacts at His48 by constructing a double mutant, Gly48 Pro51 (Fig. 5).

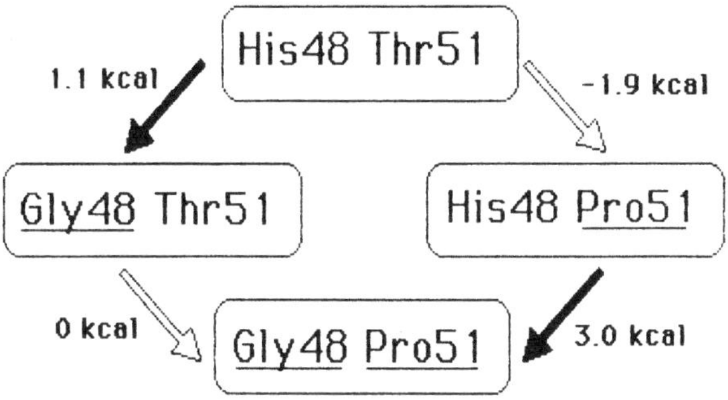

FIG. 5 - Changes in interaction energy of tyrosyl adenylate transition state with mutant enzymes, as derived from steady-state kinetics.

The results suggest that in the wild-type enzyme the imidazole group at position 48 contributes 1.1 kcal/mol to transition-state binding but that in the Pro51 enzyme it contributes 3.0 kcal/mol. Thus Pro51 improves the interaction of His48 with substrate. At the same time these results rule out the possibility that the ribose makes a good docking interaction with the pyrolidine ring, as the affinity of the Gly48 Pro51 enzyme is almost identical to that of the Gly48 Thr51 enzyme (2).

Is the improved interaction of His48 on making Pro51 due simply to an improved hydrogen bond with the substrate, or could there be a new electrostatic or van der Waals interaction of the imidazole ring with substrate? One clue is provided by replacing His48 by Asn, a smaller side chain which might make a hydrogen bond via the amide nitrogen (13). The interaction is relatively unchanged with Thr at position 51 (Fig. 6), implying that Asn48 can indeed make a similar hydrogen bond to the ribose as His48 and also that there is no electrostatic or van der Waals contribution from the imidazole ring of His48. However, the energetics of the double mutant Asn48 Pro51 indicate that although the proline residue improves substrate affinity, the improvement is much less with Asn48 than His48 (Fig. 6). Thus the effect of Pro51 on the His48 interaction with substrate is twofold: first there is an improved hydrogen bond to the ribose and second an enhanced electrostatic or van der Waals contact.

Indirectly altering side chain contacts to the substrate by distorting the polypeptide backbone has a major disadvantage compared to direct substitution of contact residues, as it is more difficult to predict the effect. For example, we suspect that the insertion of Pro51 allows His48 to come closer to the substrate since there appears to be a new van der Waals or electrostatic contact. However, it is difficult to predict this by inspection of the active site.

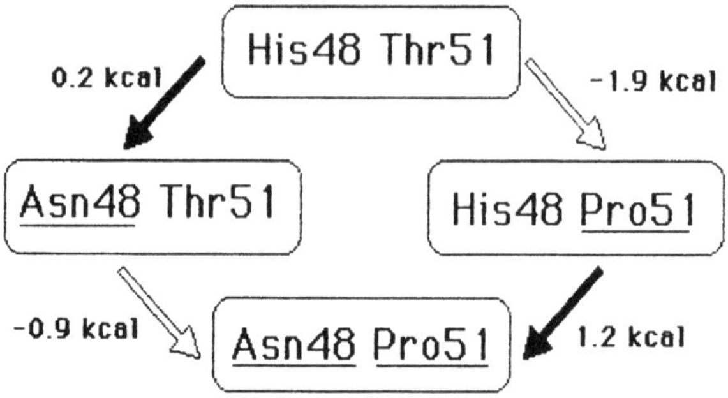

FIG. 6 - Changes in interaction energy of tyrosyl adenylate transition state with mutant enzymes, as derived from steady-state kinetics.

Altering Substrate Specificity

Enzymes must not only recognize and bind to the correct substrate but exclude other related molecules from the active site. Repulsive forces can prevent a large group on the substrate fitting into a small pocket on the enzyme but do not prevent a small group fitting into a larger pocket. Such discrimination may be provided by differing van der Waals contacts, hydrogen bonds, and salt bridges. As discussed earlier, the effect of hydrogen bonds between uncharged centers on the enzyme and substrate can be relatively small, providing 0.5 to 1.5 kcal/mol of binding energy, or a factor of only 2.5 to 15 towards specificity (7). Thus the tyrosyl tRNA synthetase discriminates between tyrosine and phenylalanine (to 1/10,000) on the basis of the phenolic hydroxyl of tyrosine. The crystal structure indicates that a contact is made from Tyr34 (and also from Asp176) to the phenolic hydroxyl of the tyrosine. Altering Tyr34 to Phe (and thus deleting the H-bond) increases the error rate by only 15-fold. However, hydrogen bonding between a charged and uncharged center can affect binding by some 4 kcal/mol, or a factor of 1000 in specificity. In an enzyme-substrate complex, an electrostatic charge should be solvated either by a water molecule or by a hydrogen-bonding group on the other partner (7). Designing an altered specificity for an enzyme could therefore in principle utilize steric repulsion to exclude not only an undesired substrate but also the need to avoid unsolvated charges.

Altering enzyme specificity towards novel substrates may ultimately permit the tailoring of enzymes to order. However, specific interactions between large molecules, such as antibodies with complement or cell receptors, or binding proteins and DNA might also be utilized practically. For example, would it be possible to design DNA-binding proteins specific for any DNA sequence, which could be used to repress the transcription of any gene for which the DNA sequence is known? Insight into how proteins can recognize sites on DNA has emerged from the crystallographic structures of CAP, lambda repressor protein (cl), and lambda cro protein. In each case a helix provides most of the sequence-specific contacts within the major groove of B-DNA. Thus a helix in 434 repressor and a similar helix in 434 cro protein are believed to interact with the major groove of DNA. Grafting the cro helix (434 phage) in place of the repressor helix (434 phage) alters the fine interaction of the altered repressor with the DNA. The new interaction is similar to that of the cro protein (22), indicating that the specificity transfers with the donated helix. Thus the construction of variant helices may permit the design of novel repressors (5, 22, 26).

Altering Folding/Thermostability

In theory stabilizing a native protein to thermal denaturation might require only 1-2 kcal/mol, the energy of an extra hydrogen bond being sufficient. A study of mesophilic and thermophilic bacterial ferredoxins, in which the sequences were fitted to the known crystallographic structure of one particular ferredoxin, indicated that thermophilic ferredoxins have a few extra external salt bridges and hydrogen bonds compared to the mesophiles. Likewise, the introduction of a surface disulfide linking together two distal portions of the sequence in T4 lysozyme (Cys3-Cys97) appeared to give an enhanced thermal stability, although the effect was contingent on blocking another free thiol (Cys54) (15). Thus surface cross-links may help stabilize a structure. However, there is a thermostable variant of the lambda repressor protein which appears to have an improved nonpolar interaction between aromatic side chains in the core of the protein. Therefore it seems that there are several ways of improving thermostability. Which approach should the protein engineer select?

Given the close packing of hydrophobic side chains in the core of the protein, manipulating the surface residues seems preferable. Introducing surface disulfides (although these would provide a strong covalent bond) could create problems as competing disulfides would have to be unscrambled as the protein is folded (14). Furthermore, the geometry of the disulfide link is critical, and bonds with a deviant configuration could equally well introduce strain and destabilize the protein. Surface salt bridges or hydrogen bonds might be easier to engineer in that there are more possibilities, but an experimental study of a known protein by protein engineering techniques is needed to define the best strategy.

Acknowledgement. We wish to thank EMBO and the Royal Society of London for long-term fellowships.

REFERENCES

(1) Carter, P; Bedouelle, H.; and Winter, G. 1985. Improved oligonucleotide site directed mutagenesis. Nucl. Acids Res., 13: 4431-4443

(2) Carter, P.; Winter, G.; Wilkinson, A.J.; and Fersht, A.R. 1984. The use of double mutants to detect structural changes in the active site

of the tyrosyl tRNA synthetase (Bacillus stearothermophilus). Cell **38:** 835-840.

(3) Courtney, M.; Jallat, S.; Tessier, L.-H.; Bonavente, A.; Crystal, R.G.; and Lecocq, J.-P. 1984. Synthesis in E. coli of a1-antitrypsin variants of therapeutic potential for emphysema and thrombosis. Nature **313:** 149-151.

(4) Craik, C.S.; Largman, C.; Fletcher, T.; Roczniak, S.; Barr, P.J.; Fletterick, R.; and Rutter, W.J. 1985. Redesigning trypsin: alteration of substrate specificity. Science **228:** 291-297.

(5) Ebright, R.H.; Cossart, P.; Gicquel-Sanzey, B.; and Beckwith, J. 1984. Mutations that affect the DNA sequence specificity of the catabolite gene activator protein of E. coli. Nature **311:** 232-235.

(6) Enea, V.; Vovis, G.F.; and Zinder, N.D. 1975. Genetic studies with heteroduplex DNA of bacteriophage f1. Asymmetric segregation, base correction and implications for the mechanism of genetic recombination. J. Molec. Biol. **96:** 495-509.

(7) Fersht, A.R.; Shi, J.-P.; Knill-Jones, J.; Lowe, D.M.; Wilkinson, A.J.; Blow, D.M.; Brick, P.; Carter, P.; Waye, M.M.Y.; and Winter, G. 1985. Hydrogen bonding and biological specificity analysed by protein engineering. Nature **314:** 235-238.

(8) Fersht, A.R.; Wilkinson, A.M.; Carter, P.; and Winter, G. 1985. Fine structure-activity analysis of mutations at position 51 of tyrosyl tRNA synthetase. Biochemistry **24:** 5858-5861.

(9) Kramer, B.; Kramer, W.; and Fritz, H.-J. 1984. Different base/base mismatches are corrected with different efficiencies by the methyl-directed DNA mismatch-repair system of E. coli. Cell **38:** 879-887.

(10) Kramer, W.; Drutsa, V.; Jansen, H.-W.; Kramer, B.; Pflugfelder, M.; and Fritz, H.-J. 1984. The gapped duplex DNA approach to oligonucleotide-directed mutation construction. Nucl. Acids Res. **12:** 9441-9456.

(11) Kramer, W.; Schughart, K.; and Fritz, H.-J. 1982. Directed mutagenesis of DNA cloned in filamentous phage: influence of hemimethylated GATC sites on marker recovery from restriction fragments. Nucl. Acids Res. **10:** 6475-6485.

(12) Kunkel, T.A. 1985. Rapid and efficient site specific mutagenesis without phenotypic selection. Proc. Natl. Acad. Sci. USA **82:** 488-492.

(13) Lowe, D.M.; Fersht, A.R.; Wilkinson, A.J.; Carter, P.; and Winter, G. 1985. Probing histidine-substrate interaction in tyrosyl tRNA synthetase using asparagine and glutamine replacements. Biochemistry **24:** 5106-5109.

(14) Mark, D.F.; Lu, S.D.; Creasy, A.A.; Yamamoto, R.; and Lin, L.S. 1984. Site-specific mutagenesis of the human fibroblast interferon gene. Proc. Natl. Acad. Sci. USA **81:** 5662-5666.

(15) Perry, J.L., and Wetzel, R. 1984. Disulphide bond engineered into T4 lysozyme: stabilisation of the protein toward thermal inactivation. Science **226:** 555-557.

(16) Pielak, G.J.; Mauk, A.G.; and Smith, M. 1985. Site-directed mutagenesis of cytochrome c shows that an invariant Phe is not essential for function. Nature **313:** 152-153.

(17) Robey, E.A., and Schachman, H.K. 1984. Site-specific mutagenesis of aspartate transcarbamylase. J. Biol. Chem. **259:** 11180-11183.

(18) Rosenberg, S.; Barr, P.J.; Najarian, R.C.; and Hallewell, R.A. 1984. Synthesis in yeast of a functional oxidation-resistant mutant of human antitrypsin. Nature **312:** 77-80.

(19) Seeburg, P.H.; Colby, W.W.; Capon, D.J.; Goeddel, D.V.; and Levinson, A.D. 1984. Biological properties of c-Ha-ras 1 genes mutated at codon 12. Nature **312:** 71-75.

(20) Villafranca, J.E.; Howell, E.; Voet, D.H.; Strobel, M.S.; Ogdon, R.C.; Abelson, J.N.; and Kraut, J. 1983. Directed mutagenesis of dihydrofolate reductase. Science **222:** 782-788.

(21) Wang, A.; Lu, S.-D.; and Mark, D.F. 1984. Site specific mutagenesis of the human interleukin-2 gene: structure-function analysis of the cysteine residues. Science **224:** 1471-1433.

(22) Wharton, R.P.; Brown, E.L.; and Ptashne, M. 1984. Substituting an α-helix switches the sequence-specific DNA interactions of a repressor. Cell **38:** 361-369.

(23) Wilkinson, A.J.; Fersht, A.R.; Blow, D.M.; Carter, P.; and Winter, G. 1984. A large increase in enzyme-substrate affinity by protein engineering. Nature **307:** 187-188.

(24) Wilkinson, A.J.; Fersht, A.R.; Blow, D.M.; and Winter, G. 1983. Site-directed mutagenesis as a probe of enzyme structure and catalysis: Tyrosyl tRNA synthetase cysteine 35 to glycine 35 mutation. Biochemistry **22:** 3581-3586.

(25) Winter, G.; Fersht, A.R.; Wilkinson, A.J.; Zoller, M.; and Smith, M. 1982. Redesigning enzyme structure by site-directed mutagenesis: tyrosyl tRNA synthetase and ATP binding. Nature **299:** 756-758.

(26) Youderian, P.; Vershon, A.; Bouvier, S.; Sauer, R.T.; and Susskind, M.M. 1983. Changing the DNA-binding specificity of a repressor. Cell **35:** 777-783.

(27) Zoller, M.J., and Smith, M. 1983. Oligonucleotide-directed mutagenesis using M13-derived vectors: an efficient and general procedure for the production of point mutations in any fragment of DNA. Nucl. Acids Res. **10:** 6487-6500.

(28) Zoller, M.J., and Smith, M. 1984. Oligonucleotide-directed mutagenesis: a simple method using two oligonucleotide primers and a single stranded DNA template. DNA **3:** 479-488.

Top row, left to right:
Bo Mattiasson, Hermann Sahm, Jerzy Sobieszczanski, Yair Aharonowitz, Ulf Stahl, John Bu'lock

Middle row, left to right:
Oense Neijssel, Maria-Regina Kula, Gareth Morris, Simon Silver

Bottom row, left to right:
Ken Timmis, Ananda Chakrabarty, David Hopwood

Biotechnology: Potentials and Limitations, ed. S. Silver, pp. 71-81. Dahlem Konferenzen 1986. Berlin, Heidelberg, New York, Tokyo: Springer-Verlag.

Microbiology and Industrial Products

Group Report

M.-R. Kula, Rapporteur
Y. Aharonowitz
J.D. Bu'lock
A.M. Chakrabarty
D.A. Hopwood
B. Mattiasson
J.G. Morris
O.M. Neijssel
H. Sahm
J. Sobieszczański
U. Stahl
K.N. Timmis

INTRODUCTION

The selection of problems for discussion in the group was based on the background papers and the expertise of the people present. Therefore the topics selected should not be considered a priority list with regard to the field in general; especially some aspects in biochemical engineering and industrial fermentation were underrepresented. In view of industrial application it was felt that existing curricula in basic science at universities and organizational barriers in industry have not encouraged the communication between different disciplines now needed to establish complex biotechnological processes of the next generation. An integration of the knowledge available in several fields and a systems approach will eventually save time and money and yield more economical processes than attempts to optimize single stages only. To improve communication and overcome language barriers between disciplines, training grants for postdoctorates, visiting scientists, etc., should be made available to allow work for a period of time in a laboratory primarily concerned with a different discipline. In the future, teaching of major subjects such as genetics, microbiology, biochemistry, and biochemical engineering should also be complemented by some introduction into the other sectors of biotechnology.

UNCOUPLING OF GROWTH FROM PRODUCT FORMATION

Continuous culture techniques provide a means of studying the effects of changes in environmental conditions on the physiology of microorganisms, and together with genetic techniques they can be used to increase product yield and culture performance by manipulating the growth environment.

The formation of an end product (e.g., fermentation product) accomplished by a free energy-conserving process can be sustained stably in a continuous culture since biomass formation is endergonic. The yield of such products may be enhanced by the operation of futile (energy-spilling) cycles elicited by particular growth environments with consequential increased heat production. In this context, the question was raised as to whether microbial oxidation of methanol, for example, could be useful for heat production in the future.

The formation of products that ensue from endergonic reactions (e.g., overproduction of foreign proteins) is unstable under selective growth conditions unless a specific growth advantage operates, e.g., improved substrate utilization. Therefore, such processes are best performed in batch culture in which the growth phase is deliberately separated from that in which the gene is expressed. Multiple regulatory mutants (e.g., for amino acid overproduction) are also counterselected by growth under limiting conditions and are, therefore, exploited in industry in batch processes. Here, and in many other cases, only certain parts of the metabolic network of microorganisms are relevant to product formation. Immobilized cells can probably serve as a good source of excreted products provided their productivity is at least as great as that which can be achieved by alternative technologies such as cell retention or recycling.

To sustain their viability and growth potential, cells expend "maintenance energy." Its magnitude will depend upon the environmental conditions but it is often small compared with that required for biosynthesis of the desired product.

More attention should be paid to understanding biochemical processes in cells under conditions of no net growth ("cryptic growth"). We need to understand these better so as to achieve a high efficiency of product formation over longer periods of time.

In general, there is a lack of knowledge about the physiology of most industrially important microorganisms: biochemical pathways and their regulation, compartmentation and channeling of common intermediates (e.g., amino acids into protein or antibiotic formation), and interaction of transport systems with metabolite production. For example, nothing is known about the export of antibiotics from the producing organism. Such knowledge is especially needed for a rational construction of recombinant organisms with improved performance in the biosynthesis of products that result from coupled sequences of several enzymatic steps. This knowledge would also help to rationalize the optimization of media for current production strains. Therefore, it is strongly recommended that research into the physiology and genetics of organisms other than E. coli be intensified.

BIOTECHNOLOGY IN OPEN SYSTEMS: PROSPECTS AND PROBLEMS

Several existing or envisaged industrial applications of microorganisms require their release into the open environment. Examples are listed below:

Application	Organisms	Goals
Agriculture	Rhizobium Azospirillum Mycorrhizal fungi	- nodulation, nitrogen fixation iron sequestering
	Bacillus	- insecticidal toxin
	Pseudomonas	- prevention of ice nucleation
Pollution control	Pseudomonas Acinetobacter Alcaligenes	- degradation of toxic chemicals, oil spills, and other wastes - removal of heavy metal (mercury, cadmium, lead, arsenic)
Metal leaching	Thiobacillus	- copper and uranium production - recovery of other heavy metals
Desulfurization of coal and oil	Thiobacillus Pseudomonas	- reduction of atmospheric pollution, i.e., acid rain
Oil recovery	Pseudomonas Xanthomonas Corynebacterium	- surfactant, polysaccharide production

Public Acceptance and Regulatory Procedures

The exploitation of endogenous wild-type organisms has in some cases been established for centuries, though the general public has been largely unaware of their usage. Examples of this are found in food fermentations, silage, or wastewater treatment. Application of exogenous wild-type organisms is a universally accepted practice in milk-processing and other food fermentations and is currently being introduced in other areas (e.g., Bacillus thuringensis toxin for selective control of insects or Thiobacillus for metal leaching). For soil detoxification and improvement of wastewater treatment plants, microbial cultures (labeled "natural isolates") are on the market and may contain large numbers of as yet unidentified microbes. Their effectiveness has often not been proven. On the other hand, there exists public controversy as to the possible dangers of releasing recombinant organisms into the environment. Such fears are not expressed with regard to existing organisms carrying the same genes. Regrettably, there is a lack of education concerning the presence and beneficial nature of the vast majority of microbial processes occurring in nature, and public imagination has turned to unrealistic monster scenarios or fears concerned with biological warfare. This has to be overcome by continuous efforts in education and public information.

If processes are desired by the public, as in pollution control, and few or no alternate treatments exist, acceptance of biological processes in general will not be a major problem. However, in order to prove the effectiveness of a process and the application system, field trials are necessary. In the case of recombinant organisms, criteria have to be defined (e.g., survival of organisms, survival of a genotype under selective and nonselective conditions, and the possible impact of enhanced expression of the genotype on the ecosystem and materials) in order to develop a rational basis for comparison of benefits, alternative technology, and risk, so that organisms and processes can eventually be declared safe. With increased experimental evidence from controlled field trials our judgement will be based more on knowledge and less on assumptions. Recent advances in medical microbiology, immunology, ecology, and phytopathology should all be used to minimize the potential risk associated with field trials.

NEW SCREENING SYSTEMS?

Scope

Whereas a few years ago the search for new antibiotics seemed to be approaching the limits of diminishing returns, the picture today is quite different. On one hand, the methods of screening for new antibiotics have been changed with notable positive results (e.g., clavulanic acid avermectins), and on the other, new screening methods are being used to find metabolites with useful properties other than antimicrobial action (antitumor agents, immunoreactive agents, enzyme inhibitors, etc.). What can we say about the scope of such attempts?

1) The limits of the diversity of existing organisms (most especially in the diversity of their secondary metabolites) are, in general, not yet in sight. Screening methods are our guide in exploiting that diversity.

2) It is arguable whether interspecific hybridization contributes to that diversity in nature; certainly laboratory hybridization does so, and screening methods are needed to exploit hybridization techniques fully.

3) No screen based on simple, repeatable, readily observed tests has failed to find new active products. Whether or not they are useful is a secondary question.

4) A primary requirement is for soundly trained, competent general microbiologists to furnish characterized organisms, often from previously unexplored sources.

Limitations and Challenges

Many microbial products have found their eventual application for use to be quite different from the screening objectives that led to their discovery (e.g., cyclosporin). Some directions of screening (e.g., for broad-spectrum antimicrobials) are probably near their limits. Supersensitive screens may otherwise discover presumably nonattainable, inaccessible products which may become challenges for chemists. There is a major problem in finding "key" physiologically relevant activities that can be used as the basis for screening tests, but this is a challenge requiring better understanding of the target physiology.

Screening Systems
Advances from "traditional" antimicrobial screens include applications of resistant microorganisms, supersensitive microorganisms, and exotic microorganisms as indicators. The use of resistant microorganisms parallels the use of enzyme inhibitors in screens for higher levels of enzyme excretion.

Extensions in using microorganisms are found in screening with microbial differentiation systems as test objects. Tests on microbe-host systems, especially microbe-plant interactions, offer potential scope. Higher organisms may also be used as test systems, for example, "whole-animal" screens for nematode infestation, or live insects, nematodes, fish or protozoal systems or plant systems, respectively.

Enzyme-based screens are a radical departure. They depend on selecting a physiologically relevant enzyme system and using this as the test object. The problem is to identify such systems in relation to targets for "useful" microbial products. For example, how would we reduce the immunosuppressor effect of cyclosporin to a readily observed, simple enzyme test? We do not know enough about the molecular basis of physiological processes. Some target systems of particular interest and/or promise would include cell-cell interactions of higher organisms, special functions in the immune system, and hormone receptor-effector interactions.

A radically different alternative is to isolate the microbial product first and investigate its activities in quantity and at leisure. This is not mere speculation, it is technically quite feasible and arguably quite rewarding.

Improving Chances
It was discussed whether experimentally set, stressful situations for microorganisms other than the stationary growth phase would lead to expression of different subsets of genes and therefore to more and different secondary metabolites. Such stress could be exerted by trace element starvation, osmotic pressure, membrane perturbance, low water regime, anoxia, etc.

NEW (GENETIC) STRATEGIES FOR MANIPULATING SECONDARY METABOLITE PRODUCTION

Quantitative ("Yield" Improvement)

The production of a secondary metabolite, such as an antibiotic, by a microorganism is of a multigenic character and may depend on the expression of many genes, including others besides the set of 10-30 structural genes for the enzymes of the biosynthetic pathway itself. Thus, it is more complex, for example, than the production of an amino acid by a bacterium. Nevertheless, knowledge of the main regulatory circuits that control gene expression in antibiotic producers would surely lead to more rational procedures for developing high-yielding strains than the totally empirical mutation and screening regimes used up to now. The main contribution of genetic engineering to this problem in the near future will therefore be to elucidate the regulation of the antibiotic biosynthetic genes. Preliminary work on analyzing the model polyketide antibiotic actinorhodin has indicated that the structural genes for the biosynthetic pathway are clustered together and organized into a series of operons closely linked with at least one regulatory gene, in this case an "activator." A slightly increased copy number of this activator gene in an otherwise wild-type strain leads to dramatic overproduction of the antibiotic. Indications from several other studies are that clustering of the biosynthetic genes for actinomycete antibiotics (but not those of fungi) is the rule rather than the exception (1), arguing well for genetic engineering studies. At present, rational yield improvement is clearly information-limited: gene cloning could rapidly help to obtain the information needed for progress.

Qualitative (New Molecules)

New secondary metabolites could arise as a result of gene transfer between different antibiotic-producing organisms for two different reasons: the activation of the expression of normally "silent" genes and the generation of "hybrid" molecules by the cooperation of biosynthetic genes from two parental strains. The gene transfers could in principle be brought about by in vivo gene exchange systems, but gene cloning is likely to be a much more versatile approach. Both mechanisms are being established through current studies on streptomycetes.

The "silent gene" idea is that a typical actinomycete (and this may apply also to other differentiating microbes) probably has the genetic capability to produce many more antibiotics than are detected under normal laboratory conditions. Some antibiotics probably play a role as hormone-like signals within the colony, helping to coordinate sporulation, spore germination, etc. They are produced in minute concentrations, at particular "windows" of the developmental cycle, and in response to specific physiological triggers. Their production could be deregulated by gene transfers from other strains (as well as by environmental regimes that tend to "confuse" the physiology) so that the products become detectable. The recent activation of a silent gene for phenoxazinone synthase in Streptomyces lividans by specific DNA segments from the actinomycin-producing S. antibioticus is an indication of the validity of this approach (3).

Recent studies (2) in which the actinorhodin gene set of S. coelicolor, or parts of it, were transferred to strains producing medermycin or dihydrogranaticin with the production, in both cases, of novel "hybrid" antibiotics, have established this strategy also as a promising approach to antibiotic discovery. The precise outcome of such recombinations will depend on the relative affinities and specificities of the enzymes of the two pathways, any possible "channeling" of the intermediates of the pathways, etc., and a large empirical element will characterize this approach for some time to come. Nevertheless, it seems certain that interesting new molecules will be produced by this route. It could be that they will represent a some what different "spectrum" of molecules compared with those routinely found to be produced naturally by microorganisms since they could include compounds whose production is actually disadvantageous to the producing organism, a situation not likely to be encountered with natural metabolites.

New chemical structures of antibiotics, or other secondary metabolites, can also be produced by a bioconversion process. In such a process a compound, natural or synthetic, is converted into a structurally related product by one or a small number of enzymes within a cell. These reactions can be carried out with growing cells, resting cells or spores, or alternatively by purified enzymes. They can be carried out by organisms of close or remote phylogeny. Reactions such as methylation, hydroxylation, dehydrogenation, amination and deamination, halogenation, and phosphorylation are just a few feasible activities. Employing the new recombinant DNA techniques, one might envisage the possibility of introducing genes coding for such activities into producer cells in order to construct new modified compounds.

Mutational biosynthesis of antibiotics is a further way to make new unnatural secondary metabolites. In this approach a mutation is induced in a gene coding for one of the precursors of an organism's natural secondary metabolite. The mutation leads to the synthesis of an incomplete, inactive molecule lacking a single chemical constituent. When the missing precursor is added to the medium in which the organism is grown, the natural antibiotic is produced. However, when precursors with slightly different structures (analogues) are added to the medium, new compounds may be synthesized.

Specific antibiotic synthetases can also be used to form novel structures of unnatural antibiotics. Although very few pathways for antibiotic biosynthesis have been studied in detail, some of these studies have started to generate new ideas and compounds. The enzymes involved in the biosynthesis of penicillins and cephalosporins have recently been characterized in great detail and might represent a good model system.

One of the enzymes involved in early penicillin and cephalosporin biosynthesis is a cyclizing enzyme, called cyclase or isopenicillin N synthetase, which oxidizes and leads to the cyclization of the linear natural tripeptide δ-(L-α-aminoadipyl)-L-cysteinyl-D-valine (ACV) into the bioactive molecule isopenicillin N. This enzymatic step constitutes the only β-lactam ring-forming reaction which leads from the amino acid oxidation level to penicillins and cephalosporins. This oxidative reaction is carried out in both prokaryotes (streptomyces) and eukaryotes (Cephalosporium acremonium). The reaction is very similar in both cases and for activity the cyclase requires Fe^{2+}, ascorbate, and oxygen. Although this cyclase reaction is specific for the tripeptide substrate (ACV), analogous molecules have been synthesized which were recognized as substrates. Thus, they went through cyclization and new β-lactam nuclei were formed (6).

Since enzymes isolated from Streptomyces strains were found to be quite stable, it has become possible to employ a crude enzyme preparation to convert a tripeptide to a cephalosporin molecule in one operation. If the synthesis can be carried out efficiently in vitro starting with amino acids, then genetic engineering could be used to improve enzyme production and apply an enzyme reactor to avoid complications arising from the channeling of common intermediates or from transport that presently plague in vivo biosynthesis of multigenic products from simple carbon and nitrogen sources.

INDUSTRIAL APPLICATIONS OF CATABOLIC PATHWAY MANIPULATION

Objectives: Single and multiple-step bioconversion (e.g., degradation of benzoate or substituted benzoates to the corresponding dihydrodiols by strains carrying cloned genes for benzoate dioxygenase) and detoxification of toxic compounds or degradation of toxic or persistent compounds (e.g., 2,4,5-trichlorophenoxyacetic acid - 2,4,5-T) are some of the objectives.

Strategies: Direct isolation of organisms from the environment can be attempted. Where this is not possible, enrichment and/or chemostat selection are further routes to find suitable organisms. Plasmid-mediated horizontal gene transfer is probably of substantial importance for the evolution of new catabolic functions, e.g., evolution of 2,4,5-T degrader. The strategy for in vitro construction of a hybrid metabolic pathway calls for design of feasible pathways, identification of organisms with appropriate segments of pathways, cloning of their genes and regulatory determinants, and a combination of these in a suitable strain (e.g., evolution of an orthopathway for the degradation of both methyl- and chloro-substituted aromatics).

More knowledge is needed, however, about the biochemical bases of recalcitrance and toxicity for soil microbes and the strategies microbes employ to evolve new catabolic pathways, transport systems, and protection systems for such molecules. We also need to combine evolved pathways with others appropriate for "real," as opposed to laboratory, situations.

STRAIN STABILITY

As a general comment it can be said that strains will generally be stable if formation of the desired product confers a selective advantage upon the cell. This is not true for a number of important products which are of two types: those that have little or no influence on cell physiology and those that confer a significant disadvantage upon cell physiology. In the latter case, there is a strong selection for elimination of the gene(s) specifying the product of interest. This selection is manifested most clearly under conditions where production is coupled to growth, particularly prolonged growth in continuous culture. In general, the problem will be minimized if gene expression is uncoupled from growth, e.g., by judicious use of vectors with regulatable promoters, batch fermentation, and expression of the cloned gene at the end of log phase/stationary phase, or by use of immobilized cells.

In a limited number of situations, judicious choice of conditions in a coupled, continuous culture fermentation will cause the product to provide a selective advantage, e.g., directly by using protein as a carbon source in protease fermentations, or indirectly through the use of N-limitation to produce extracellular polysaccharide.

Known Causes of Instability

Assuming that the vector itself is stable, plasmid loss is in some cases due to an active promoter in the cloned segment. From this, transcription events are initiated that proceed into maintenance determinants and perturb their normal functioning. Solution: insert cloned genes into the chromosome; remove the promoter or insert a strong transcription terminator between the cloned gene and the vector maintenance determinants.

Inactivation of a desired gene by deletion, insertion, or point mutation is another cause of instability. The frequency of deletion and insertion events is determined by the presence or absence of insertion elements or repeated sequences. Solution: remove the offending sequences; reduce recombinational events by using recA-type mutants as hosts (disadvantage: recA mutants grow slowly).

Another possible source of gene instability is the apparent induction of a recombination-repair system in Pseudomonas by certain foreign DNA segments. Nothing is known about the mechanism, but future work on this system may enable its inactivation.

It is clear that our current knowledge about causes and mechanisms of instability of cloned genes is fragmentary. In order to solve the problem in a general way, more information is urgently needed about plasmid maintenance functions that play crucial roles in long-term plasmid stability under fermentation conditions, about the cellular mechanisms involved in inactivation of cloned genes, and about the ways in which cloned gene products perturb cellular physiology. In the meantime, it is prudent to incorporate where possible the following principles into cloning strategies: a) use very stable vectors with determinants that provide strong selections, e.g., antibiotic resistance, nutritional requirement complementation, or flocculation in yeast; b) link desired gene(s) close to the selectable determinants and subsequently eliminate as much nonessential DNA as possible; c) eliminate the native promoter of cloned genes and subclone into an expression vector with strong but regulatable promoters that express the desired gene only when an appropriate signal (e.g., temperature shift) is given at the end of the culture growth phase; and d) if plasmid instability is still a problem, insert the desired gene into the chromosome.

PRODUCTION AND RECOVERY

In industry, biotechnological processes are carried out for a purpose: often this is the manufacture of a defined product, be it single cell protein (SCP), ethanol, penicillin, protease, or insulin. The production and recovery stages of such a process have to be considered together. Estimates from industry state that recovery costs contribute 50-80% of the manufacturing cost. Further improvement of the biosynthesis of a given product by genetic engineering techniques and/or cultivation technology will increase the relative proportion of the recovery cost even further. This leads rapidly to a situation where requirements for recovery will influence and even dictate the choice of the biological production system. Efforts to integrate production and downstream processing go in the same direction.

Indirect Measures

The size of the organism and the rheological properties of the culture determine the primary separation step, that is, the removal of cells from spent medium. The latter in turn is a question of the physiological conditions of cultivation and shear regime. Much energy can be saved if the size of the organism employed is at least 5 μm, if the flow behavior is Newtonian, and if the viscosity is low.

Selective excretion of the product simplifies further purification. It allows the use of immobilized cells in cases where product formation can be uncoupled from growth. However, excretion of the proteins during growth under aerobic conditions is potentially dangerous since there is a risk of denaturation of the protein on the gas-liquid interface as well as proteolysis of the product.

Intracellular protease activity should be controlled to allow accumulation of undegraded proteins in the cytoplasm in a soluble state. "Inclusion bodies," the insoluble aggregates formed during high expression of most mammalian proteins in E. coli, are not desirable despite the possibility of rather easy selective purification from soluble intracellular products. Renaturation requires (besides an empirical search for suitable conditions) very low protein concentrations in the range of 1-50 μg/ml, depending on the molecular weight and quarternary structure of the protein involved. Therefore, huge volumes have to be handled in order to produce sizable amounts, and problems in adequate mixing can be anticipated during scale-up to achieve sufficiently fast dilution of the denatured product.

Uniform products simplify purification. Therefore, fully processed proteins are requested. The modification capacity of the host should be considered and improved if necessary. Similar requirements are noted for secondary metabolites. An increase in the titer of an antibiotic is no improvement if the mixture becomes much more complex.

Pyrogens must be avoided for all intravenously applied drugs. This consideration influences the choice of the most desirable host cell.

Different host cells or even the same cell will express a different spectrum of proteins if physiological conditions are changed. If we could identify the two or three proteins most difficult to separate from a desired one, e.g., by 2D-electrophoresis, we could try to avoid the presence of these proteins altogether by a proper choice of cell or growth conditions.

Direct Measures

Genetic engineering techniques can also be applied directly to improve downstream processing, e.g., introduction of flocculating genes for better cell separation, excretion of the product as discussed above, introduction of genes for the controlled expression of lytic enzymes which will contribute to cell disintegration in the manufacture of intracellular products, or tagging of an easily separable portion to the desired product. The first example of tagging employs protein A as an excretion vector for proteins (4). The fusion proteins can then be differentially purified on antibody-bearing affinity columns interacting with protein A. In another approach, several arginine residues were added in a row to the C-terminal end of a protein, which altered its ionic properties in a predictable way (5). In both cases additional enzyme-catalyzed processes are required to cleave the tag from the desired product before final purification.

REFERENCES

(1) Hopwood, D.A.; Malpartida, F.; and Chater, K.F. 1985. In Regulation of Secondary Metabolite Formation, eds. H. Kleinkauf, H. von Döhren, H. Nornauer, and F. Nesemann. Weinheim: VCH Verlagsgesellschaft, in press.

(2) Hopwood, D.A.; Malpartida, F.; Kieser, H.M.; Ikeda, H.; Duncan, J.; Fujii, J.; Rudd, B.A.M.; Floss, H.G.; and Omura, S. 1985. Production of "hybrid" antibiotics by genetic engineering. Nature **314**: 642-644.

(3) Jones, G.H., and Hopwood, D.A. 1984. Activation of phenoxazinone synthase expression in Streptomyces lividans by cloned DNA sequences from Streptomyces antibioticus. J. Biol. Chem. **259**: 14158-14164.

(4) Nilson, B.; Holmgren, E.; Josephson, S.; Gatenbeck, S.; Philipson, L.; and Uhlen, M. 1985. Efficient secretion and purification of human insulin-like growth factor I with a gene fusion vector in Staphylococci. Nucl. Acid Res. **13**: 1151-1161.

(5) Sassenfeld, H.M., and Brewer, S.J. 1984. Polypeptide fusion designed for the purification of recombinant proteins. Bio-Technol. **2**: 76-81.

(6) Wolfe, S.; Demain, A.L.; Jensen, S.E.; and Westlake, D.W.S. 1984. Enzymatic approach to synthesis of unnatural beta-lactams. Science **226**: 1386-1392.

Biotechnology: Potentials and Limitations, ed. S. Silver, pp. 83-97. Dahlem Konferenzen 1986. Berlin, Heidelberg, New York, Tokyo: Springer-Verlag.

Fermentation Products: Physiological and Bioenergetic Considerations

O.M. Neijssel and D.W. Tempest
Laboratorium voor Microbiologie, Universiteit van Amsterdam
1000 HE Amsterdam, Netherlands

Abstract. The overproduction of substances by microbial cultures is discussed in a bioenergetic context. Since many fermentation products are generated by energy-yielding reactions, it is clear that the rate of energy consumption of a cell will influence the rate of product formation. It is shown that the nature of the growth environment influences to a great extent the magnitude of these energy-consuming reactions. On the other hand, the energetic efficiency of the energy-generating reactions can be reduced and this will lead, assuming the same rate of energy consumption, to an increased rate of product formation. Another class of products, in particular proteins, can only be produced by energy-consuming reactions. Therefore, the extra energy consumption caused by the overproduction of proteins is in direct competition with the energy consumption that the cell requires for maintenance and growth. The physiological implications of this metabolic conflict are discussed and it is suggested that this type of overproduction will be inherently unstable, unless the overproduction of a protein increases the fitness of the producer organism.

INTRODUCTION

Microorganisms play a vital role in biotechnology for two main reasons. The first is their capacity to produce useful compounds, ranging from amino acids, solvents, and antibiotics to enzymes and polysaccharides. The second reason is that many organisms are able to degrade a wide variety of organic compounds; this capacity is used in wastewater treatment and in the biodegradation of xenobiotics. With the advent of the recombinant DNA technique it has become possible to force microbes to produce special substances of animal, plant, or viral origin that were previously only available in low quantities. To all this one can add the relatively simple nutritional requirements of most microorganisms and the mild process conditions, and one can perceive easily the enormous potential of microbes in industrial processes.

An important question in the use of microorganisms in industrial processes is how to optimize the performance of these creatures. In principle this can be achieved by genetic techniques and by choosing the appropriate culture conditions. If one wants to avoid a brute force method for optimization, one has to have a detailed knowledge of the behavior of the particular organism, and this is where physiology plays an important role, since this branch of microbiology studies the relation between cell structure and function. We believe that optimizations using mathematical models that treat the microorganism as a "black box" have only a limited success for the very reason that this approach ignores the subtlety of metabolic regulation that occurs inside the cell.

In this contribution we want to formulate some principles for the production of compounds by microorganisms. These principles have been derived from studies with organisms that are themselves not necessarily great producers of important substances. But we feel that one can use this type of information in formulating strategies with which a more fundamental optimization of process conditions will be feasible. In the title of this contribution we use the words "fermentation products." One could interpret these words in a strict sense, i.e., that we will discuss only the production of the classic fermentation products such as ethanol, lactate, acetone, or butanol. But since in industry one usually calls any process in which microbes play a role a "fermentation," we will not limit ourselves to anaerobic processes or the products of anaerobic metabolism but rather consider microbial products in general.

THE ENERGETICS OF GROWTH OF MICROORGANISMS

When we define as "viable" the physiological state in which the cell needs to be in order to grow and multiply, it should be clear that every viable cell needs energy to maintain itself in that state, if only for thermodynamic reasons (entropy). Whereas the actual amount of energy that is needed for this purpose, the so-called maintenance energy requirement, is not known, it is beyond doubt that it is heavily dependent on the environmental conditions in which the cell lives (9, 19, 21). Thus, the concentrations of several ionic species in the cytoplasm of a living cell have to be maintained at some optimal value. When the cell membrane is not totally impermeable to a particular ion, and when the extracellular concentration of this ion is different from that in the cytoplasm, a leakage current will result and the cytoplasmic concentration will change. The cell has to expend energy to counteract this leakage current. When, however, the gradient of an ion across the membrane is small or zero, there will be only a small or no leakage current and the energy effect will be similarly small or zero. Another type of maintenance energy-consuming reaction is the turnover of macromolecules. Although it has not been proven unequivocally, it is very likely that the growth environment influences the magnitude of these reactions also. A detailed analysis of all maintenance energy-requiring reactions is outside the scope of this contribution, but we can conclude by stating that the maintenance energy requirement of a cell always has to be fulfilled with a top priority. If this condition is not met, through a lack of energy, the cell will starve and, ultimately, die.

When more energy is available to the cell than is necessary to cover the requirements for maintenance, growth is possible provided that the necessary elements for the synthesis of new cell material are present. And again, the growth environment plays an important role in prescribing the magnitude of the energy needed for growth. Many authors have postulated that there is a linear relationship between energy supply and growth rate (9, 13, 15, 19, 21), and whereas this may not be invariably true (19, 21), it is sufficient to note that growth will depend in some manner on the rate of energy provision. Thus, the growth requirement ranks second in the energy-requiring reactions of the cell.

Finally, it is possible that the cell produces more energy than can be used for growth and maintenance. Since storage of energy in the form of ATP is not possible, one can observe that under these circumstances storage polymers are synthesized (glycogen, lipid, polyphosphate, etc.) (2). This way a major part of the extra energy is stored. When further synthesis of storage polymers is not possible, or when a cell is unable to synthesize such polymers in the first place, the excess energy is wasted through energy-spilling reactions (9, 19). In this connection, it is interesting to note that, whereas growth is only possible when enough energy is available, the reverse is not true: cells do not need to grow in order to consume energy at a high rate. The presence of energy-spilling reactions will cause a rapid dissipation of any excess of energy, whether the cell is growing or not.

We will show below that there is a direct relationship between the generation of energy in the cell and the synthesis of a great number of products. It will, therefore, depend on the energy status of the cell whether synthesis of these products is stimulated or not. It is our conviction that in living cells the energetic priorities will dominate all other processes and that when the production of a compound is in conflict with these priorities, this will lead to a decline in production either by a decline of cell viability or by the generation of daughter cells that are mutated in such a way that production of the compound has ceased.

PRODUCTS THAT ARE LINKED WITH ENERGY PRODUCTION

In this category fall, among others, the classic fermentation products such as ethanol or lactic acid, but also oxidation products from glucose such as gluconic acid or citric acid. The physiology of the production of these compounds is relatively simple. As an example we take the production of lactic acid from glucose.

glucose + $2NAD^+$ + 2ADP + $2P_i$ → 2 pyruvate + 2ATP + 2NADH + $2H^+$

2NADH + $2H^+$ + 2 pyruvate → 2 lactate.

The key factors in these reactions are the adenine and pyridine nucleotides. The intracellular concentration of these cofactors is in the millimolar region. This means that the conversion of glucose to pyruvate would stop immediately if ADP and NAD^+ were not rapidly regenerated from ATP and

NADH, respectively. In this case recycling of NAD^+ is carried out by the reduction of pyruvate to lactate. The production of lactate therefore serves two functions: production of energy and regeneration of NAD^+. For the recycling of ADP, the considerations discussed in the previous section are relevant: ATP-consuming reactions are responsible for carrying out this process. Thus, even a resting cell suspension will be able to convert glucose into lactate due to the maintenance of energy requirement. But this leads us to the second important point in product formation: the rate at which this process occurs. It was argued above that the cells' energy demand will increase when they are growing and that in growing cultures a positive correlation will exist between growth rate and energy consumption. This was indeed found to be the case (Table 1). It must be noted that at low growth rates the conversion of glucose into lactate is rather poor (0.78 moles of lactate formed/mole of glucose consumed, compared with a theoretical value of 2). Although some of the glucose must have been used for cell synthesis, this cannot be the sole explanation, since at high growth rates the ratio of lactate formed/glucose consumed is substantially higher. However, it has been observed that putatively homolactic bacteria do not form lactate from glucose under a stringent glucose limitation; they form formate, acetate, and ethanol instead (20). This could also have been the case in these experiments.

TABLE 1 - Specific rates of glucose consumption and lactate production by chemostat cultures of Streptococcus faecalis that were, respectively, glucose-limited and tryptophan-limited (data recalculated from (14)).

Limitation	Dilution rate (h^{-1})	Specific rates of consumption and formation (mmol $g^{-1}h^{-1}$)		Ratio lactate / glucose
		glucose	lactate	
Glucose	0.22	4.9	3.8	0.78
	0.31	7.5	6.8	0.91
	0.43	9.6	17.6	1.83
Tryptophan	0.22	12.9	13.4	1.04
	0.31	16.6	20.2	1.22
	0.43	16.1	30.6	1.90

The interesting fact is that it should be possible to optimize the production of this type of metabolite further, and we propose the use of the following two methods: a) decrease of the number of moles of ATP formed per mole of product formed, and b) increase of the cells' energetic demands. With respect to the first point, it can be predicted that if the ATP/lactate ratio could be caused to decrease from 1 to, say, 0.5, the cells would need to produce twice as much lactate to cover the same energetic demand. That this is not pure speculation is obvious from the present interest in Zymomonas mobilis as a potential ethanol producer. When we compare the ethanol-yielding reactions of Z. mobilis with those of Saccharomyces cerevisiae:

Z. mobilis	glucose → 2 ethanol + $2CO_2$ + 1ATP
S. cerevisiae	glucose → 2 ethanol + $2CO_2$ + 2ATP,

it will be clear why Z. mobilis should be, at least in principle, a better ethanol producer. If we assume that the amount of ATP needed to synthesize biomass from glucose is about the same for Z. mobilis as for S. cerevisiae, the first organism will need twice the amount of glucose and produce twice the amount of ethanol from it in order to be able to generate the same amount of biomass as S. cerevisiae. The reason for this difference in ATP production between the two organisms lies in the fact that Z. mobilis uses the Entner-Doudoroff pathway of glucose metabolism, yielding only 1 mole of ATP per mole of glucose degraded, whereas S. cerevisiae uses the "normal" Embden-Meyerhof glycolysis.

To what extent would it be possible to decrease the ATP yield per mole of lactate in streptococci? In some members of the Enterobacteriaceae and Pseudomonadaceae the presence of yet another glucose-degrading pathway has been observed, the methylglyoxal bypass (1):

dihydroxyacetone phosphate → methylglyoxal P_i → D-lactate → pyruvate.

From this scheme it follows that D-lactate production from glucose is possible without the concomitant formation of ATP. The physiological function of this pathway has become apparent from studies with Klebsiella aerogenes. When this organism was grown anaerobically in a glucose-limited chemostat culture it consumed glucose via the Embden-Meyerhof pathway, but when these cells were offered a sudden excess of glucose, they converted a major part of this excess glucose to D-lactate, apparently via the methylglyoxal bypass (17). From this observation one can conclude that, at least in this organism, this pathway is used as a mechanism to avoid the obligatory coupling between glucose breakdown and energy generation which occurs in the Embden-Meyerhof pathway (i.e., it is an energy-spilling reaction). This way monomers for cell synthesis become quickly available without creating an imbalance in the ATP/ADP ratio. However, the use of this pathway must be strictly regulated, since a complete and continuous glucose degradation via the methylglyoxal bypass would lead to energy depletion.

The presence of this bypass reaction in streptococci has, to the best of our knowledge, not been described, although production of methylglyoxal by S. faecalis has been observed (1). Nevertheless, one can argue that a streptococcus possessing a regulated methylglyoxal bypass would be a superior lactate producer. That streptococci must possess some mechanism for spilling energy is obvious again from the data in Table 1, since when S. faecalis was grown tryptophan-limited it consumed glucose at a higher rate compared to glucose-limited cells growing at the same rate.

The ATP requirement of cells can be increased by several means. In the section on growth energetics we described the effect of ion leakage currents through the membrane on the energetic requirements of the cell. Clearly, then, one can use this phenomenon to promote product formation. One of the most promising approaches would be to use potassium-limited cultures. The cytoplasmic concentration of this ion is about 40-100 mM in Gram—negative organisms and above 200 mM in Gram-positives (18). The extracellular concentration of K^+ can be kept very low in K^+-limited cultures (in the µmol/l region). That this enormous K^+ gradient across the membrane will induce a potassium leakage current can be derived from the observation that K^+-limited cultures of K. aerogenes and Escherichia coli expressed a strongly increased respiration rate compared to glucose-limited cultures growing at the same rate ((4), and Neijssel and Tempest, unpublished results). When the K^+ limitation was suddenly relieved by adding K^+ (20 mM end concentration) to a K^+-limited K. aerogenes culture, the respiration rate dropped instantaneously by about 40% (4). Since potassium transport is driven by both the membrane potential and ATP, it is not surprising that it is the respiration rate that is influenced in such a dramatic way by the extracellular K^+ concentration.

In this connection, it is interesting to note that K^+-limited K. aerogenes cells produce large quantities of gluconic and 2-ketogluconic acid from glucose (10). These compounds were synthesized via direct oxidation of glucose by a pyrrolo-quinoline quinone-linked glucose dehydrogenase and of gluconate by a gluconate dehydrogenase (a flavoprotein) (11). Since these dehydrogenases can react directly with the respiratory chain, the production of gluconate and 2-ketogluconate will stimulate electron transport without extensive carbon source metabolism, and this is exactly in tune with the cells' energetic demands. Although we do not recommend K. aerogenes as a producer organism for gluconate, it will be obvious that we predict that Gluconobacter spp. will produce gluconate optimally when they are kept under a K^+ limitation.

A similar reasoning can be applied to phosphate- or magnesium-limited growth conditions. These types of cultures of K. aerogenes behaved qualitatively in much the same way as potassium-limited cultures, although the rates of oxygen consumption and of gluconate and 2-ketogluconate production were somewhat lower (10). This is probably a reflection of the lower cytoplasmic concentrations of phosphate and magnesium when compared with the potassium concentration (18), or of lower leakage currents of these ions.

Finally, one can induce a proton leakage current through the membrane by adding an uncoupler of oxidative phosphorylation (protonophore). From studies with K. aerogenes it has become apparent that this procedure will stimulate product formation (7). When this organism was grown in an ammonia-limited chemostat culture, it produced a number of organic acids and polysaccharide in considerable quantities (see below) but no gluconic or

2-ketogluconic acid. Addition of 1 mM 2,4-dinitrophenol (DNP) to these cultures changed the production pattern drastically: polysaccharides were no longer synthesized, but the quantity of organic acids produced and the respiration rate increased and now gluconic and 2-ketogluconic acid were also produced. This observation was confirmed by showing that de novo synthesis of glucose dehydrogenase occurred after the addition of DNP to the culture. A similar observation was made with sulfate-limited cultures (Neijssel and Tempest, unpublished results). Clearly, these cultures behave now similarly to potassium-limited cultures.

Concluding this section, we can predict that when the synthesis of a product is linked with energy generation in the cell, its production will be stimulated either by decreasing the efficiency of energy generation or by increasing the cells' energy demand. Moreover, an excess of the carbon source will generally be necessary to improve production rates.

PRODUCTION OF COMPOUNDS AT THE EXPENSE OF ENERGY

Among this type of product we can classify proteins, some amino acids and polysaccharides, and many secondary metabolites. We have argued above that there is a certain hierarchy in the energy-consuming reactions of the cell. This means that the energetic demand for the synthesis of the desired product will have its own place in this hierarchy and that this ranking will be dependent upon the role that the product plays in the survival of the microbe. When the product is, for instance, an enzyme, and when this enzyme plays a vital role in the metabolism of the organism, production of this enzyme will be mandatory. When, on the other hand, the product is a human hormone, production of this substance is superfluous, since it is a nonsense protein from the organism's point of view. Overproduction has to be considered in the same context: only when overproduction of a compound has a positive effect on the survival characteristics of the producer organism will this phenomenon be a stable property with a high energetic priority. The effect of genetic manipulation is frequently that the cells are forced to give a higher priority to the production of the desired compound than they would have done without the intervention of the genetic engineer. A more detailed analysis of the growth kinetics of microorganisms will serve to make this clear.

When a wild-type strain of E. coli is growing in a glucose-limited chemostat culture it will synthesize new cell material from the nutrients in the medium. The relationship between the concentration of the growth-limiting nutrient (in this case glucose) and the growth rate expressed by the culture can be described by the formula of Monod:

$$\mu = \mu_{max} \cdot \frac{s}{K_s + s} .$$

In this formula, which is of course an approximation of reality, two important parameters can be discerned: μ_{max} and K_S. The first is the maximal growth rate, which is expressed when all nutrients are present in excess of the growth requirement. It depends on the composition of the medium and the type of strain that is used but is a constant once the medium and organism have been chosen. The other, K_s, is an affinity constant; it indicates at which concentration of the growth-limiting substrate the growth rate of the culture will attain half its maximal value. This formula predicts that when in the culture a variant arises with a higher μ_{max} value, or a lower K_s value, it will grow faster than its parent strain and will gradually dominate the culture. Any variant with a lower μ_{max} or higher K_s value will be unable to compete successfully with the parent strain. When we compare the properties of the wild-type E. coli with a genetically engineered mutant which overproduces a foreign protein to a level of, for instance, 10% of its cellular protein, the following considerations apply.

When the wild-type strain is growing in the chemostat it will synthesize cells that will contain the minimum amounts of those proteins and other cell components that are necessary for survival under these circumstances, in other words, the cells will use the carbon source with maximum economy. The engineered mutant, however, will have to expend a considerable amount of energy and carbon in synthesizing the foreign protein over and above the minimum amount of those cell components necessary for growth. The effect of this overproduction on the maximal growth rate can be estimated by approximation. Under optimal conditions, E. coli will grow on glucose with a yield (Y) of about 90 g/mole glucose. Since $\mu = Y \cdot q$, this implies that when $\mu_{max} = 1\ h^{-1}$ the specific rate of glucose consumption (q) has to be 11 mmole glucose$\cdot g^{-1} \cdot h^{-1}$. Since the average bacterial cell growing in a mineral medium at $\mu = 1 h^{-1}$ will contain about 60% protein (relative to dry weight), the mutant organism will have to synthesize a cell containing 66.6% protein (the protein of the parent cell plus the desired product). This means that the glucose consumption rate has to increase by about 7%. However, it is generally found that at μ_{max} the uptake system for glucose, the phosphotransferase system, is fully saturated (8, 12). From this it follows that, the uptake rate of 11 mmole glucose$\cdot g^{-1} \cdot h^{-1}$ being a maximum value, the amount of glucose available for cell synthesis (apart from the desired product) will be lowered by about 7%. The equation $\mu = Y \cdot q$ then prescribes that the actual growth rate of the mutant will also be 7% lower, i.e., 0.93 h^{-1}.

We have to emphasize that this calculation is based on a number of assumptions. One of these is that production of the foreign protein will not lead to an increase in the glucose uptake rate. It is, however, difficult to envisage how production of the foreign protein will induce a glucose uptake system. We have also ignored the effect of maintenance energy in our calculations, but it is highly likely that this will cause a further drop in the growth rate of the mutant organism, since there will also be turnover of the product, leading to extra energy consumption. Furthermore, one can criticize these calculations because they are based on the behavior of cells in a glucose-limited chemostat and one should use glucose excess conditions. This is justified, but other limitations also cause problems for the production of the foreign protein: ammonia and sulfate limitation cannot be used, since these two elements are necessary for protein synthesis. Phosphate, magnesium, or potassium limitation create energetic problems for the cell, as we argued above. The conclusion, then, must be that this type of production will be inherently unstable in chemostat culture, since any revertant of the mutant strain will have an increased growth rate and will therefore quickly dominate the culture. This means that a batch culture will be the most satisfactory culture technique, first, because there is no nutrient limitation until the culture is in the stationary phase, and second, revertants will have only a small chance to dominate the culture when the inoculum is the pure mutant organism, since the culture period is restricted.

For the production of amino acids and polysaccharides similar reasonings apply. Although amino acids have to be produced by an organism (when they are not provided in the medium), <u>over</u>production will have the same serious energetic consequences as overproduction of proteins. Extracellular polysaccharide production by an organism will be a positive feature in some natural environments, since it is widely known that capsulated organisms are more resistant to antimicrobial agents and more virulent to their host (plant or animal). In a pure culture this selective advantage is lost, and it is therefore not surprising that it frequently has been observed that a capsule-forming organism loses this property after a number of subcultures in the laboratory. If this reasoning is correct and the production of extracellular polysaccharide provides the organism with a permeability barrier, one may speculate that the addition of certain antimicrobial agents to the medium would select for the polysaccharide-producing cells. The concentration of the antibiotic has to be such that the nonproducers are killed and the producer cells are not. Finally, we have to emphasize that the production of some amino acids and some types of polysaccharides does <u>not</u> consume energy but is an exergonic process (5). These types of products therefore fall outside the scope of this section and belong in the previous category of products.

Until now we have painted a rather somber picture of the overproduction of proteins. Yet there are a number of interesting cases where the outlook is not so bleak. Sikyta et al. have selected in a lactose-limited chemostat culture an E. coli strain that produced 27.5 times more β-galactosidase than the parent organism and, in a similar way (serine limitation), an overproducer of serine deaminase (26.5 times more enzyme) was isolated by these authors (16). This phenomenon is at first sight surprising, since we have argued above that overproduction of proteins is not a stable property in chemostat cultures. However, in this case the enzymes that were overproduced were involved in the uptake of the growth-limiting nutrient (lactose and serine, respectively). Thus, whereas the maximum growth rates may have been lowered in these overproducing variants (these values were not reported by the authors), an increase in V_{max} of the uptake system for the growth-limiting nutrient will lower the K_s value of the organisms (12), and this will give the mutant a selective advantage over the parent organism below certain growth rates (Fig. 1).

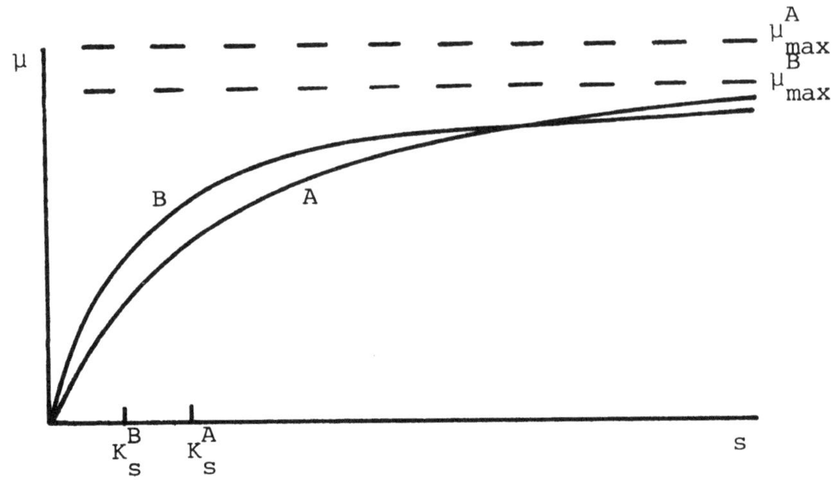

FIG. 1 - Relationship between specific growth rate (μ) and growth-limiting substrate concentration (s) of a parent strain (A) and a hypothetical mutant (B) with a 10% lower μ_{max} value and a 50% lower K_s value.

We have to emphasize that this is a theoretical analysis, since we do not know a number of the relevant data. Yet we believe that the same strategy will lead to increased production of, for instance, proteases (limiting factor: protein as carbon and nitrogen source), lipases (limiting factor: lipid as carbon source), and similarly, amylases and cellulases (see also (22)). In all these cases a chemostat culture with a low dilution rate will provide a growth environment in which overproducing mutants will be selected.

It can be concluded that overproduction of compounds that are synthesized with an input of metabolic energy is not easily achieved. In a number of cases one may be able to stimulate production by applying specific selective pressures on the producer organism. Regrettably, this is not always possible and it is therefore one of the great challenges to the microbial physiologist to try to provide optimal culture conditions to those organisms that have been so ingeniously constructed by the geneticist!

OVERFLOW METABOLISM

Overproduction of metabolites and polymers is frequently the result of bottlenecks in metabolism: the activity of one or more key enzymes has been modified by genetic means in such a way that the flow of carbon is steered in the desired direction. This approach is a classic one in the optimization of the production of antibiotics and amino acids. It is insufficiently realized that one can also create metabolic bottlenecks by physiological means. These latter types of bottleneck have the advantage that, since they are created by the culture conditions, they cannot be circumvented, and that therefore genetic stability does not play a role. A trivial example is the production of ethanol by yeast (absence of oxygen, hence respiration is blocked). But there are more possibilities and some of them have been discussed above: the application of potassium, magnesium, or phosphate limitation in the production of energy-yielding compounds. As an example we now discuss metabolite production in aerobic cultures of K. aerogenes and E. coli.

When these organisms are grown carbon-limited in a fully aerobic chemostat culture, they convert the carbon source (in this case glucose) exclusively into new cell material and CO_2:

K. aerogenes: 12 glucose + $26O_2$ + $9NH_3$ $\rightarrow$ 1 kg cells + $31CO_2$ + $50H_2O$

E. coli: 13 glucose + $28O_2$ + $9NH_3$ $\rightarrow$ 1 kg cells + $34CO_2$ + $55H_2O$.

(Approximate equations are based on experiments carried out at D (= μ) = 0.17 h^{-1}, pH = 6.8, and temperature 35°C.)

When one grows these organisms ammonia-limited and glucose is present in excess of the growth requirement, a number of products are formed. These products are clearly not the result of fermentative reactions (they often are more oxidized than glucose):

K. aerogenes: 36 glucose + $44O_2$ + $9NH_3$ $\rightarrow$ 1 kg cells + $40CO_2$ + $72H_2O$ + 4 acetate + 3.5 pyruvate + 11 2-oxoglutarate + 1.9 kg polysaccharide

E. coli: 28 glucose + $58O_2$ + $9NH_3$ $\rightarrow$ 1 kg cells + $58CO_2$ + $95H_2O$ + 10 acetate + 3 pyruvate + 2.5 2-oxoglutarate + 0.6 kg polysaccharide.

(Approximate equations are based on experiments carried out at D (= μ) = 0.17 h^{-1}, pH = 6.8, and temperature 35°C.)

The synthesis of polysaccharide is a common observation when organisms are grown in a medium with a high C/N ratio (2). More interesting is the synthesis of 2-oxoglutarate, pyruvate, and acetate. By limitation with ammonia, a bottleneck is created at the level of the ammonia incorporating enzyme system, glutamine synthetase, and glutamate synthase. The result is that 2-oxoglutarate is accumulated intracellularly and excreted. Probably because of this accumulation, precursors of 2-oxoglutarate also accumulate in the cell and one of these, pyruvate, and another degradation product, acetate, are excreted.

Another interesting limitation is sulfate limitation:

K. aerogenes: 33 glucose + $46O_2$ + $9NH_3$ $\rightarrow$ 1 kg cells + $42CO_2$ + $117H_2O$ + 10 acetate + 15 pyruvate + 1 2-oxoglutarate + 0.4 kg polysaccharide.

In this case the incorporation of sulfate will be impeded; this is not apparent from the excreted products, probably because the intermediates involved in sulfate assimilation cannot be excreted. Due to the sulfate limitation, the sulfur-containing cofactors will be present in the cell in limiting amounts and these cofactors play a vital role in the pyruvate and 2-oxoglutarate dehydrogenase complexes. This can explain the production of, again, pyruvate and 2-oxoglutarate.

A general conclusion of these studies can be that frequently the uptake and metabolism of glucose is not strictly regulated and that glucose transport is not the rate-limiting step in the growth of an organism. The result is excretion of oxidized products of glucose metabolism by the organism. This type of metabolite production is fundamentally different from the production of compounds evoked by potassium limitation. Under sulfate- or ammonia-limited growth conditions the limiting nutrient is taken up and is subsequently converted into a different chemical form. In this respect one may consider ions such as potassium, magnesium, zinc, iron, etc., as growth catalysts rather than as precursors of structural cell material. Phosphate plays an intermediate role; it is incorporated into nucleic acids and phospholipids but, on the other hand, is a catalyst in much the same way as ATP and ADP.

Other carbon sources can cause the excretion of different oxidation products. Thus, sulfate-limited cultures of K. aerogenes growing on malate produced and excreted oxaloacetate into the culture medium (Neijssel and Tempest, unpublished data). On the other hand, sulfate-limited cultures of Paracoccus denitrificans did not show extensive overflow metabolism, indicating that the type of organism is also an important factor (Neijssel and Tempest, unpublished results). It is highly tempting to consider also citrate production by Aspergillus niger in this context. The production of this acid is provoked by a limitation of iron in the growth medium, and aconitase needs an iron-citrate complex for activity (3). Thus, a low intracellular concentration of iron would create a decrease in the aconitase activity and therefore lead to an intracellular accumulation of citrate. However, it has been reported that a low availability of zinc or manganese also promotes citrate production by this fungus (6). This indicates that either the mechanism involved in citrate production is more complex or else the aconitase reaction is somehow regulated by the availability of zinc or manganese.

Summarizing this section, we can conclude that overproduction of substances can be provoked by specific growth limitations other than the carbon source and that the type and quantity of the product that is formed will depend on the substrate, the nature of the limitation, the type of organism, and the permeability of the cell membrane to intracellularly accumulated metabolites.

CONCLUDING REMARKS

In this short survey we have tried to explain the phenomenon of metabolite overproduction in a physiological context. A key factor in this approach is a better understanding of the role that overproduction of substances plays in the maintenance of viability and growth potential of the organism. Without this insight an inevitable consequence must be that the organism has to be treated as a "black box" and that process optimization must be based on trial and error. This leads us to the formulation of another bottleneck: the lack of physiological knowledge of the behavior of producer organisms. Whereas much literature is available on E. coli and a few other bacteria and fungi, the existing knowledge about the physiology of interesting organisms such as the actinomycetes, streptomycetes, thermophilic organisms, and many fungi is rather limited. Simple questions such as "Why does a Penicillium sp. produce penicillin?" remain to be answered. We are convinced that organisms in their natural environment are (and have been for millions of years) subject to a severe selective pressure and that, therefore, the properties they express in the laboratory or in an industrial process have to be viewed in this perspective. As a consequence of this view, one has to reject the notion that some properties (such as pigment formation, production of secondary metabolites, etc.) are fortuitous. We would like to put forward the idea that every microbial process has its own relevance to the organism and that a lack of physiological understanding sometimes leads us to believe that a process is irrelevant for the survival of the organism.

Further progress in biotechnology will depend on the continuing acquisition of fundamental knowledge of the genetics and physiology of microbes. It is essential that scientists of both these disciplines interact with one another and that they are made aware of problems and of the progress made in the other field. One can therefore understand the timeliness of this workshop!

REFERENCES

(1) Cooper, R.A. 1984. Metabolism of methylglyoxal in microorganisms. Ann. Rev. Microbiol. **38:** 49-68.

(2) Dawes, E.A., and Senior, P.J. 1973. The role and regulation of energy reserve polymers in micro-organisms. In Advances in Microbial Physiology, eds. A.H. Rose and D.W. Tempest, pp. 135-266. London: Academic Press.

(3) Glusker, J.P. 1968. Mechanism of aconitase action deduced from crystallographic studies of its substrates. J. Molec. Biol. **38:** 149-162.

(4) Hueting, S.; de Lange, T.; and Tempest, D.W. 1979. Energy requirement for maintenance of the transmembrane potassium gradient in Klebsiella aerogenes NCTC 418: a continuous culture study. Arch. Microbiol. **123:** 183-188.

(5) Jarman, T.R., and Pace, G.W. 1984. Energy requirements for microbial exopolysaccharide synthesis. Arch. Microbiol. **137:** 231-235.

(6) Kubicek, C.P., and Röhr, M. 1977. Influence of manganese on enzyme synthesis and citric acid accumulation in Aspergillus niger. Eur. J. Appl. Microbiol. **4:** 167-175.

(7) Neijssel, O.M. 1977. The effect of 2,4-dinitrophenol on the growth of Klebsiella aerogenes NCTC 418 in aerobic chemostat cultures. FEMS Microbiol. Lett. **1:** 47-50.

(8) Neijssel, O.M.; Hardy, G.P.M.A.; Lansbergen, J.C.; Tempest, D.W.; and O'Brien, R.W. 1980. Influence of growth environment on the phosphoenolpyruvate: glucose phosphotransferase activities of

Escherichia coli and Klebsiella aerogenes: a comparative study. Arch. Microbiol. **125:** 175-179.

(9) Neijssel, O.M., and Tempest, D.W. 1976. Bioenergetic aspects of aerobic growth of Klebsiella aerogenes NCTC 418 in carbon-limited and carbon-sufficient chemostat culture. Arch. Microbiol. **107:** 215-221.

(10) Neijssel, O.M., and Tempest, D.W. 1979. The physiology of metabolite overproduction. In Microbial Technology: Current State, Future Prospects, eds. A.T. Bull, D.C. Ellwood, and C. Ratledge, 29th Symposium of the Society of General Microbiology, pp. 53-82. Cambridge: Cambridge University Press.

(11) Neijssel, O.M.; Tempest, D.W.; Postma, P.W.; Duine, J.A.; and Frank Jzn, J. 1983. Glucose metabolism by K^+ -limited Klebsiella aerogenes: evidence for the involvement of a quinoprotein glucose dehydrogenase. FEMS Microbiol. Lett. **20:** 35-39.

(12) O'Brien, R.W.; Neijssel, O.M.; and Tempest, D.W. 1980. Glucose phosphoenolpyruvate phosphotransferase activity and glucose uptake rate of Klebsiella aerogenes growing in chemostat culture. J. Gen. Microbiol. **116:** 305-314.

(13) Pirt, S.J. 1965. The maintenances energy of bacteria in growing cultures. Proc. Roy. Soc. Lond. B **163:** 224-231.

(14) Rosenberger, R.F., and Elsden, S.R. 1960. The yields of Streptococcus faecalis grown in continuous culture. J. Gen. Microbiol. **22:** 726-739.

(15) Schulze, K.L., and Lipe, R.S. 1964. Relationship between substrate concentration, growth rate, and respiration rate of Escherichia coli in continuous culture. Arch. Microbiol. **48:** 1-20.

(16) Sikyta, B.; Kyslík, P.; Voleský, B.; Pavlasová, E.; and Stejskalová, E. 1982. Over-production of endoenzymes in Escherichia coli - selection of hyperproducing strains in a chemostat. In Over-production of Microbial Products, eds. V. Krumphanzl, B. Sikyta, and Z. Vanek, pp. 593-599. London: Academic Press.

(17) Teixeiria de Mattos, M.J.; Streekstra, H.; and Tempest, D.W. 1984. Metabolic uncoupling of substrate level phosphorylation in anaerobic glucose-limited chemostat cultures of Klebsiella aerogenes NCTC 418. Arch. Microbiol. **139:** 260-264.

(18) Tempest, D.W.; Dicks, J.W.; and Hunter, J.R. 1966. The interrelationship between potassium, magnesium, and phosphorus in potassium-limited chemostat cultures in Aerobacter aerogenes. J. Gen. Microbiol. **45:** 135-146.

(19) Tempest, D.W., and Neijssel, O.M. 1984. The status of Y_{ATP} and maintenance energy as biologically interpretable phenomena. Ann. Rev. Microbiol. **38:** 459-486.

(20) Thomas, T.D.; Ellwood, D.C.; and Longyear, V.M.C. 1979. Change from homo- to heterolactic fermentation by Streptococcus lactis resulting from glucose limitation in anaerobic chemostat cultures. J. Bacteriol. **138:** 109-117.

(21) van Verseveld, H.W.; Chesbro, W.R.; Braster, M.; and Stouthamer, A.H. 1984. Eubacteria have 3 growth modes keyed to nutrient flow.

Consequences for the concept of maintenance and maximal growth yield. Arch. Microbiol. **137**: 176-184.

(22) Wouters, J.T.M., and Buysman, P.J. 1977. Production of some exocellular enzymes by Bacillus licheniformis 749/C in chemostat cultures. FEMS Microbiol. Lett. **1**: 109-112.

Biotechnology: Potentials and Limitations, ed. S. Silver, pp. 99-112. Dahlem Konferenzen 1986. Berlin, Heidelberg, New York, Tokyo: Springer-Verlag.

Bioactive Microbial Secondary Metabolites

Y. Aharonowitz and G. Cohen
Tel-Aviv University
George S. Wise Faculty of Life Sciences
Dept. of Microbiology
Ramat-Aviv 69978, Israel

Abstract. Microorganisms have become a veritable resource of biologically active compounds. Thousands of such metabolites have been identified thus far and show a diversity of chemical structure, complexity, and physiological function. The types of bioactivity characterized have depended ultimately on the operational approach taken for their detection and isolation. Traditionally, the ability to antagonize the growth of bacteria, fungi, viruses, and other forms of life has been the major requirement. Therefore, it has resulted in the isolation of metabolites possessing mainly antibiotic activity. However, this trend is now being changed. It has become possible to design screening procedures that turn up active molecules having entirely different properties. The availability of powerful tools for increasing genetic variability, appreciation of the physiological characteristics of the producing organisms, as well as advances in biochemical understanding have been of enormous importance in developing new microbial strains for industrial application and basic research and in allowing microbial cells to be subverted for the purpose of making novel metabolites. However, the question of why microorganisms make all these metabolites remains unanswered. Little is known about the natural role of these important compounds. They have, therefore, been termed "secondary metabolites" so as to distinguish them from the primary metabolites of the cell that are known to be essential for growth of the organism.

USE OF THE TERM "SECONDARY METABOLITES"

The term "secondary metabolites" has been used mainly to define in a practical way a variety of unique natural biosynthetic products in order to distinguish them from the well characterized "primary metabolites" - those cellular metabolites known to be essential for the growth, development, and maintenance of an organism. Discussions of secondary metabolites usually include aspects of their evolutionary significance: their role in the life of the producing organism, their role in the ecosystem, and whether they are mere random products of the cell's degradative or turnover processes. There are no conclusive answers yet to such questions. The specific function of most secondary metabolites in nature is still obscure, and although many such functions have been proposed, they remain largely hypothetical (18). When the subject of secondary metabolism is pursued further, criteria such as their appearance in a narrow, systematically defined group of organisms (mainly differentiating and soil microorganisms), the complexity and variability of their chemical structure, the timing of their production during different phases of the growth cycle, and their biosynthetic origin and the relationship between chemical structure and their bioactivity are often mentioned (11). The introduction of new, highly sensitive detection methods, new principles of screening, systematic studies based on ecological approaches, and the design of extreme environmental conditions for microbial growth have all been very fruitful in leading to the discovery of new secondary metabolites. New bioactive compounds have been found as well as new types of bioactivities. Production of secondary metabolites has been found more widespread in nature than was previously believed. Furthermore, known organisms have been shown to produce new kinds of secondary metabolites, and known secondary metabolites have been shown to be produced by a larger variety of organisms.

One of the major obstacles in studying the nature of any given secondary metabolite is that loss of the ability to produce secondary metabolites does not entail loss of ability to grow in pure culture. Therefore, mutations causing non-, over-, or underproduction of a secondary metabolite offer no obvious selective advantage to the producer cell. This is in sharp contrast to mutants blocked in biosynthesis of primary metabolites, which generally lead to auxotrophy. However, once the requirement for a given secondary metabolite is proven, its natural function understood, and a selective advantage for the producer cell has been demonstrated, this substance might then be categorized as a "primary metabolite." Therefore, we would suggest that the term secondary metabolite be used for naturally occurring metabolites, only on a temporary basis.

ANTIBIOTICS, THE LARGEST GROUP OF KNOWN SECONDARY METABOLITES

Antibiotics constitute the largest group among the known bioactive secondary metabolites. This does not necessarily reflect the actual situation in nature. The study of secondary metabolism has been dominated, for historical reasons, by a distinct operational approach. Soon after the initial excitement of the discovery of such microbially produced chemotherapeutic agents as the penicillins and streptomycins, enormous efforts were made both in academic institutions and in industry to isolate new bioactive compounds by systematically screening thousands of different microorganisms from a variety of natural habitats. Waksman defined antibiotics as "chemical substances produced by microorganisms and possessing the ability to kill or to inhibit the growth of bacteria and other microorganisms." The impact of Waksman's definition was that methods or screens employed to identify bioactive compounds were all based on the ability to antagonize the growth of microorganisms. By further restricting the screens to select antibiotics possessing high activity against pathogenic microorganisms and relatively low toxicity for humans and animals, we have found mainly those biologically active metabolites with antibacterial, antifungal, and more recently, antitumorogenic action. The antibiotic property of special interest has been that which is expressed outside the producer cell, against organisms other than those found in the natural ecological niche of the producer.

Thousands of natural antibiotics have been thus obtained (of which only a small fraction has been used medically). It has become apparent that many different bacteria, fungi, algae, and lichens as well as green plants can produce antibiotically active secondary metabolites. Perhaps the most characteristic property of microbial secondary metabolites and of antibiotics in particular is the sheer diversity of their chemical structures, encompassing virtually every class of compound known to the organic chemist. Antibiotics have been found that affect almost every key cellular process, ranging from DNA replication, transcription, and translation to a variety of biosynthetic processes such as fatty acid synthesis and cell wall peptidoglycan or membrane synthesis and function (1). In spite of the fact that many of these cellular processes are fundamentally similar in organisms that vary from the simple prokaryotic bacteria up the evolutionary scale to higher eukaryotes, there are characteristic differences that allow antibiotics to discriminate between the activity in one organism and another. This ability to discriminate lies at the very heart of the notion of selective toxicity and will therefore determine the particular application of an antibiotic as an antibacterial, antifungal, antiparasitic, antihelminthic, antiviral, or antitumor agent.

While extensive studies on the mode of action have revealed the mechanisms mentioned above, it does not necessarily follow that the same or similar functions are exhibited by the antibiotics in their producer organisms. However, some experimental data indicate a possible function of this kind. Most studies have involved peptide antibiotics produced by bacilli, particularly Bacillus brevis strains (10). Thus, gramicidin S has been shown to regulate spore outgrowth and to inhibit RNA synthesis during the outgrowth phase. Tyrocidine, a membrane-active peptide antibiotic known also for its ability to bind to DNA and affect transcription, has been shown to induce sporulation of B. brevis cells starved for nitrogen. Natural functions have also been suggested for antibiotics produced by streptomycetes and myxobacteria, but the genetic and biochemical analysis of these systems is still just beginning.

Another interesting aspect of the relationship between cells and the antibiotics they produce is the inhibitory activity against their producers that many antibiotics display. Such inhibition can be observed when antibiotics are added to culture broths at high external concentrations, or during certain phases of the growth cycle, e.g., during spore germination or outgrowth and during early vegetative growth. Moreover, certain antibiotics inhibit growth when added at low external concentration before onset of antibiotic production, although the strain is able to accumulate the antibiotic in large quantities. Clearly microorganisms are competent to deal with such a situation, but what mechanisms do they use to acquire tolerance? Demain (5) phrased this question as "how do antibiotic-producing strains avoid suicide?" It has been found that antibiotic-producing organisms have well-defined biochemical mechanisms of resistance to the antibiotics they produce (4). In many cases resistance is due to the elaboration of either a chromosomal or a plasmid-encoded enzyme that modifies the antibiotic molecule in order to inactivate it. Such an enzyme is the aminocyclitol phosphotransferase found in streptomycetes-producing aminocyclitol antibiotics. Resistance to molecules such as erythromycin and thiostrepton arises by another mechanism, alteration of the target ribosome by ribosomal methylases. Another suggested mechanism for resistance is alteration of membrane permeability, which may limit the access of the antibiotic to its target enzyme. There are similarities between antibiotic resistance mechanisms in clinical isolates of bacteria and those found in antibiotic-producing organisms, suggesting the possibility that resistance to antibiotics originated in the producer organisms. Plasmids and transposons may mediate the spread of antibiotic resistance genes both within and between microbial species. These facts are of great practical importance. The continuous use of antibiotics to control bacterial infection creates a powerful selective pressure for survival of strains that acquire resistance. Therefore, an understanding of natural resistance mechanisms should promote a rational approach to the development and isolation of new and improved drugs. As suggested by Davis (4), recognition of resistance mechanisms in producing organisms might be used to predict the mechanisms that will develop during subsequent clinical use of the antibiotic. The elimination of antibiotic resistance problems is also necessary to ensure the continuing clinical utility of successful antibiotics as antibacterial agents.

The Search for New Antibiotics

Broadening the range of antimicrobial activity against resistant organisms is not the only objective in developing a new drug. Increased potency, a wider spectrum of activity, lower toxicity, improved stability, and better pharmacokinetic characteristics are all important factors. This is best exemplified with β-lactam antibiotics. Since the β-lactams are by far the most important and effective group of antibiotics, we will devote most of the remaining discussion of antibacterial agents to this group without intending to diminish the importance of such other groups as the aminocyclitols, tetracyclines, macrolides, ansamycins, polyethers, and peptide antibiotics.

In most cases resistance to penicillins and cephalosporins is due to β-lactamases, enzymes that hydrolyze the cyclic amide bond in the β-lactam ring. How can one counteract the presence of β-lactamase in resistant target strains? The most straightforward solution would be to employ another type of antibiotic, such as the macrolide erythromycin. A second approach might be to modify chemically the β-lactam antibiotic, or to find a new chemically related antibiotic resistant to the β-lactamase activity yet retaining antibiotic potency. Alternatively, one could look for compounds that specifically inhibit or inactivate β-lactamases and use these compounds in synergy with the β-lactamase-sensitive but otherwise powerful antibiotic. In this latter case such compounds could be identified in screens by their ability to render resistant strains sensitive again to β-lactam antibiotics. Indeed, an inactivator of β-lactamase, clavulanic acid, was discovered in fermentation broths of Streptomyces clavuligerus in this way. While this compound is in fact a β-lactam antibiotic in its own right, its main potential is as a β-lactamase inhibitor. Recent studies on the mechanism of β-lactamase inhibition have shown that clavulanic acid acts in three ways: it is a substrate for the β-lactamase, it acts as a transient inhibitor of the enzyme, and it leads to irreversible enzyme inactivation. This property of clavulanic acid is presently being exploited to protect β-lactamase-sensitive penicillins from inactivation by β-lactamase-containing resistant organisms. The drug, augmentin, now manufactured by Beecham Pharmaceuticals U.K., is a combination of the β-lactam antibiotic amoxicillin and clavulanic acid.

Perhaps the most promising new class of β-lactam antibiotics are the newly discovered carbapenem antibiotics, of which thienamicin, from Streptomyces cattleya, was the first to be discovered. These compounds are extremely potent against a broad spectrum of bacteria and are also highly effective β-lactamase inhibitors. Because of this combination of properties, they would seem to be ideal chemotherapeutic agents for attacking the serious problem of β-lactam resistance pathogens. However, the natural thienamycins are too unstable to be used directly and must be chemically modified for successful clinical use.

One of the most successful approaches to satisfy the requirements for new antibiotics has been the modification, by chemical synthesis, of natural antibiotics. "Semisynthesis" of new antibiotics has become common practice, and the number of such drugs in use now exceeds the number of microbial metabolites used directly in chemotherapy (1, 3). The clearest illustration of the value of this strategy again comes from the field of β-lactams. The biological activity of all these antibiotics is closely associated with the β-lactam moiety. In the case of penicillin the deacylated molecule, 6-aminopenicillanic acid (6-APA), represents the minimum structural requirement for bioactivity. Incorporation of certain acyl side chains increases the potency manyfold and also confers new characteristics. In the most widely used fermentation process, the natural penicillin-producing organism, generally Penicillium chrysogenum, is grown in the presence of phenylacetic acid to give benzylpenicillin (penicillin G). The antibiotic can be efficiently extracted from the broth and exposed to microorganisms elaborating acylase enzymes that selectively remove the acyl (benzyl) group from penicillin G, thereby yielding 6-APA. The availability of this nucleus now permits the chemist to design a variety of new side chains and attach them chemically or enzymatically to form new and improved penicillins which currently dominate the β-lactam drug market.

The semisynthetic approach has been especially successful for generating new antibiotics from poorly effective natural metabolites structurally related to known and useful antibiotics, as was the case with cephalosporin C. This natural antibiotic from Acremonium chrysogenum possesses a β-lactam moiety like that of penicillin. It exhibits a rather weak activity against bacteria and as such is unsuitable for clinical use. The amide side chain of cephalosporin C can be cleaved away leaving a nucleus (7-ACA) similar to that of 6-APA, and as with the penicillin derivative, specific chemical acylations can be carried out to give new drugs possessing much better pharmacological properties. Semisynthesis has been widely used for making new derivatives of other antibiotics such as streptomycins, kanamycins, gentamycins, tetracyclines, rifamycins, and antitumor antibiotics. These few examples of the semisynthetic approach reflect the potential for further research into modification of the numerous and very effective natural antibiotics that have never been introduced into human therapy.

Antibiotics as Antineoplastic Agents

Systemic screening of microbial metabolites as a source of antitumor agents began around 1953 and continues to be one of the most exciting and demanding areas of research on bioactive metabolites. Among the early compounds meriting serious appraisal were the bleomycins, glycopeptide antibiotics, isolated by Umezawa and his co-workers from culture broths of Streptomyces verticillus (15). The activity of these metabolites stems from their ability to bind DNA and to induce strand scission. Bleomycins are a group of compounds that differ from one another mainly in the nature of an amine substituent. By cultivating the producer organism in media containing different amines and by semisynthetic methods, it has been possible to obtain new analogs, some of which have proven clinically useful in treatment of malignancy.

A second important group of antineoplastic drugs are the anthracyclines, complex compounds built from an aminoglycoside moiety and an anthraquinone chromophore (16). The anthracyclines exhibit a broad spectrum of biological activity against bacteria, fungi, viruses, and animal cells. Like the bleomycins they also bind to DNA, probably by intercalating between the strands of the helix and thereby interfering with DNA replication and RNA synthesis. The selective inhibition of tumors by both of these groups of DNA-binding drugs may then be a consequence of the highly active replication and transcription processes in tumor cells. Adriamycin and daunorubicin are clinically important anthracyclines produced by streptomycetes. The closely related carminomycin is the product of an Actinomadura strain. The anthracycline group of antibiotics is one of the most promising in current use for treatment of acute and chronic cancers.

The major drawback with all these drugs is the incidence of serious side effects, especially cardiotoxicity. Unfortunately, because we know so little about the molecular basis for the differences between normal and tumor cells, a search for antitumor drugs becomes essentially a matter of trial and error, the empirical testing of new compounds. In this respect, the proven value of the anthracyclines and bleomycins in antitumor chemotherapy justifies the search for new compounds, especially when we recall the immense variety of compounds belonging to the microbial world. Some idea of the present status of this field is indicated by the current antineoplastic development programs carried out by the NCI and industry. Their basic aim is to generate an extensive range of new microbial metabolites with novel chemical features, which can subsequently be evaluated through comprehensive screening systems for potential antineoplastic use. These screens are designed to be rapid, sensitive, inexpensive, predictive, and effective for the analysis of a broad range of compounds. The first part of such projects generally employs a broad spectrum of new and established microorganisms which are fermented under unusual conditions so as to increase the variety of metabolites they will produce. The fermentation broths are then tested with a battery of in vitro pre-screen assays which may include cytotoxicity in mammalian tumor tissue cultures, effects on cell differentiation, cell surface changes, immunogen stimulation, bacteriophage induction, and employment of "engineered" bacterial strains by which direct interaction of metabolites with DNA can be detected. The SOS chromotest (13), which assays the induction of an SOS function in E. coli to measure genotoxicity, is an elegant example for this latter procedure. In addition, specific chemical and physical assays that use purified DNA molecules are included to give an insight into the mechanism of action of the putative therapeutic agents. Active organisms are re-fermented, and if active again, further in vivo tested are performed with experimental animals bearing tumors. Providing that a metabolite passes all these tests, it is further tested in a hierarchy of assays reflecting the major classes of human tumors. Additional techniques, such as chemical modifications and bioconversions, can now be used to modify subtly these products to yield more desirable properties such as lower toxicity and better pharmacological properties.

BIOACTIVE SECONDARY METABOLITES OTHER THAN ANTIBIOTICS

The isolation of microbial inhibitors of β-lactamases has been mentioned already as an example of how clever screening of fermentation broths can yield non-antibiotic pharmacologically important metabolites. Since many pathophysiological disorders have their origin in the abnormal activity of simple regulatory chemicals such as hormones, neurotransmitters, and immunomodulators, one approach to treating these diseases might be through the controlled use of specific molecules (e.g., enzyme inhibitors) that affect the biosynthesis of these compounds or, alternatively, by modifying the target or receptors on which they act. Setting up screening strategies designed to find materials interfering favorably with pathological conditions offers a chance to reveal a whole new set of bioactive compounds which do not have to be primarily antibiotic in nature. This original approach was pioneered by Umezawa.

In 1965, Umezawa and colleagues at the Institute of Microbial Chemistry and the Department of Antibiotics, National Institute of Health, Tokyo, initiated new screens for bioactive microbial products (14). Their approach was guided by two different lines of observation. The first was the discovery that animal and plant tissue enzymes can be inhibited by substances occurring naturally in the same tissue. Such inhibitors against different proteolytic activities were formed like trypsin, chymotrypsin, urokinase, and rennin, as well as against other enzymes such as amylase, phospholipase, ribonuclease, and tyrosine hydroxylase. The inhibitors generally had molecular weights higher than 5000 and chemical structures of an oligopeptide nature. The second observation came from studies on antibiotic modes of action. At this time it was known that certain antibiotic activities could be attributed to inhibition of specific enzymatic activities in the target organism. The antibiotics that were also enzyme inhibitors were of low molecular weight and inhibited enzyme targets involved in basic growth and multiplication processes. Realizing that enzymes and their specific inhibitors coexist in higher organisms, that antibiotics can act as enzyme inhibitors, and that microorganisms produce a variety of metabolites, Umezawa's group postulated that microbial metabolites possessing no antibiotic activity might possess enzyme inhibitory activity which could be detected if a proper enzyme assay were designed. This has actually become a different operational approach in screening for bioactive microbial products. Compared to the screening for new antibiotics which at that time had become relatively unproductive, it resulted in immediate success. Proteolytic enzyme inhibitors of microbial origin, such as leupeptin which inhibits trypsin, plasmin, and papain activity, chymostatin which inhibits chymotrypsin, and pepstatin which inhibits pepsin were among the first to be discovered. All were found to be of low molecular weight and to be produced by a variety of actinomycetes. Soon afterwards, low molecular weight inhibitors of two other enzymes, tyrosine hydroxylase and dopamine β-hydroxylase, that mediate the synthesis of the neurotransmitter norepinephrine were obtained. Since 1965 many such microbial metabolites with extremely interesting bioactivities (6) have been discovered by Umezawa's group and other groups in industry and academia. Similar reasoning suggests that inhibitors of enzymes such as reverse transcriptase, implicated in viral oncogenesis, and cAMP phosphodiesterase, found at high concentration in brain malignancy, might be potentially useful antitumor drugs. Other areas where this approach might be fruitful are the control of hormone release, fertility, anti-inflammatory activity, and cardiovascular effects.

Of particular interest is the possible use of enzyme inhibitors and other microbial metabolites for modulating the immune response. It is well established that microbial products can modify the induction or expression of cellular and/or humoral immunity. If microbial products were able to affect significantly either the number, distribution, or activity of cells engaged in immune reactions, and as a consequence to modulate their response to tumor target cells, such a compound might well be a potential candidate as an antitumor drug. Bestatin, a low molecular weight metabolite isolated by Umezawa and his co-workers from Streptomyces olivoreticuli, appears encouraging in this respect. Bestatin inhibits several enzymes, including aminopeptidases and esterases, located on the cell surface of macrophages, lymphocytes, and a number of other cells. Even more striking is its activity against slow-growing tumors, especially when combined with other antitumor drugs. In the search for new antitumor antibiotics we may, then, have only just begun to explore the full richness of our microbial resources.

A FERMENTATION PROCESS DEVELOPMENT

The initial discovery of a microbial metabolite with interesting bioactive properties is just the first step in a long and complex process of fermentation development that ultimately leads to manufacture of the drug in commercial amounts. The overriding objective of such a program is to make a drug that is safe, potent, and cheap. Obviously this requires the carefully planned coordination of many different disciplines. Here we will discuss a few of the significant biological factors influencing secondary metabolite production and omit the chemical engineering and economic considerations that go into product development.

Antibiotic or other secondary metabolite synthesis is first and foremost a highly controlled process in which gene expression is regulated by several different mechanisms acting in concert and affected by the anabolic, catabolic, and energy metabolism of the cell. The principal targets for these regulatory mechanisms are a) specific structural genes that govern the antibiotic synthetases, enzymes that directly mediate synthesis of the antibiotic molecule; these determine qualitative parameters such as chemical structure and the variety of the antibiotic molecules produced by a single organism; b) specific antibiotic regulatory genes which will determine quantitative parameters such as timing, rate, and specific production capacity; c) the primary metabolic pathways which provide the building blocks or direct precursors for the antibiotic synthetases; and d) the expression of antibiotic-modifying or -degrading enzymes.

Genetic manipulation provides the key methodology for strain improvement in industrial microorganisms. Basically, genetic tools provide a rational way to increase secondary metabolite yields through the isolation of mutant strains with elevated levels of biosynthetic enzymes, cofactors, and precursors or, alternatively, by eliminating unwanted characteristics from a particular strain. Amplification of antibiotic genes and alterations in the mechanisms which regulate their expression are but two ways in which this may be achieved. However, little is known at present about the location of antibiotic genes and less about the nature and regulation of transcription and translation in antibiotic-producing organisms. Furthermore, although genetic techniques for manipulating many of the producing strains have now been described, they are still not yet sufficiently well developed to allow for subtle modification of controlling elements. For this reason classical mutation and mutant selection continue to be the most fruitful methods used in the fermentation industry. Since the mutation process is essentially random and the desired mutation causing overproduction provides no obvious selective advantage for the strain, it becomes necessary to screen large populations of microbial colonies, each colony being taken through a sequence of steps to determine its ability to make antibiotics. Despite the lack of any real understanding of the function of antibiotics in the organisms producing them, it has been possible to associate empirically certain morphological and biochemical variants with an ability to overproduce antibiotics. Improved bioassay systems as well as sophisticated analytical procedures have contributed to the chance of successful screens. In this respect the development of computerized automated processes for dispensing and monitoring very large numbers of single cells and microbial colonies ought to reduce dramatically the work involved in searching for new traits.

Biochemical, physiological, and genetic studies of secondary metabolite synthesis reveal that secondary metabolism is critically dependent on the level of primary metabolites. From a biosynthetic point of view, secondary metabolites are derived from the intermediates of primary metabolism which provide the organism with energy and the precursors for macromolecular synthesis. Thus, carbohydrates, amino acids, purines and pyrimidine bases, fatty acids, and "active" acyl groups are among the basic substrates acted upon by specific antibiotic synthetases. This link between primary and secondary metabolism strongly suggests the possibility of obtaining high producer strains through the selection of mutants deregulated in the biosynthesis of primary metabolites directly involved as precursors of secondary metabolites.

A most significant advance in the genetic approach to strain improvement and creation of new metabolites has been the recent development of new and powerful techniques for gene transfer and cloning, culminating in the successful cloning of a number of antibiotic genes. This subject has been reviewed recently by Hopwood et al. (8) and by Martin and Gil (12); we will refer to just three examples. Feitelson and Hopwood (7) employed molecular cloning techniques to isolate Streptomyces chromosomal DNA segments complementing a mutant defective in a step of the prodigiosin biosynthetic pathway. Gil and Hopwood used two different strategies to clone a gene (pab) coding for p-aminobenzoic acid synthase, an enzyme involved in biosynthesis of the polyene antibiotic candicidin. In the latter experiments both shotgun cloning of donor strain DNA and direct selection of clones carrying the pab gene of the donor strain, followed by complementation of a PABA-requiring strain were employed. Baily et al. (2) have reported recently the cloning of a Streptomyces clavuligerus genetic locus involved in clavulanic acid biosynthesis. They cloned wild-type DNA into nonproducer mutant strains and screened transformants for restoration of the ability to synthesize clavulanic acid.

The most straightforward general approach to selecting clones harboring antibiotic biosynthetic genes is by seeking to restore production in mutants impaired in the antibiotic biosynthetic pathway. The immediate difficulty arising from such an approach is the lack of direct selection for antibiotic production in the cloned population. Therefore, positive clone recognition again requires the laborious screening of a large number of transformants. Moreover, since regulatory mutations and pleiotrophic effects might convert mutants to nonproducers, restoring production among such mutants might not always indicate the introduction of clones for a specific antibiotic gene. Finally, the use of well-defined standard host strains for the cloning of antibiotic genes is generally impractical when antibiotic production is anticipated; antibiotic synthesis is a multi-enzyme-catalyzed process and will require the simultaneous cloning and expression of a complete set of biosynthetic genes (8). Even when this is achieved, such organisms may not tolerate the introduction of genes for toxic metabolite biosynthesis.

A different approach for selecting clones harboring antibiotic genes might rely on the construction of DNA probes based on amino acid sequences found in specific antibiotic synthetases or, alternatively, the employment of immunological techniques for probing such proteins resulting from cloned antibiotic gene expression in the host strains. However, this approach requires a deep understanding of the biochemistry of antibiotic synthesis.

ANTIBIOTIC BIOSYNTHESIS - A BIOCHEMICAL APPROACH

The biosynthetic pathways for most secondary metabolites are not yet known in detail, although for some of them, e.g., the peptide antibiotics gramicidin, tyrocidine, actinomycin, the tetracyclines, certain aminocyclitols, and some β-lactam antibiotics (penicillins and cephalosporins), the biosynthetic pathways are well established. A clear illustration of the importance of knowing the biosynthetic pathway and its significance not only for the genetics of antibiotic biosynthesis but also for the development of new drugs again comes from the field of β-lactam antibiotics.

The main features of the biosynthesis of penicillins and cephalosporins have now been worked out (9). The initial steps are formation of the dipeptide L-α-aminoadipyl-L-cysteine (AC) followed by its condensation with L-valine to give with epimerization of the L-valine the tripeptide LLD-ACV. The ACV tripeptide can be chemically synthesized and used as a substrate for the next reaction - cyclization to isopenicillin N. This latter reaction is catalyzed by a single enzyme which has been isolated and purified from both fungi and Streptomyces strains. The cyclization activity is an oxidative reaction involving the removal of four protons and the closure of two ring systems and requires the SH form of the tripeptide, ascorbate, and ferrous ions as cofactors as well as molecular oxygen. The biosynthesis of penicillins and cephalosporins diverges at isopenicillin N; cephalosporins are formed in fungi such as Acremonium chrysogenum and Streptomyces species, whereas penicillins with nonpolar side chains are formed in fungi such as Penicillium chrysogenum. Penicillin N is formed from isopenicillin N by epimerization of the L-α-aminoadipyl side chain. The epimerase from fungi is very labile. Once penicillin N is formed, a ring expansion can take place and transforms the molecule to desacetoxycephalosporin C in an oxidative reaction involving the loss of two more protons. Expandase activity also requires ferrous ions and ascorbate as cofactors and, in addition, α- ketoglutarate. By employing test organisms which respond differently towards penicillins and cephalosporins in bioassays, as well as high pressure liquid chromatography, the products and reactants of these reactions can be detected and measured. Since the epimerase activity in Streptomyces extracts is quite stable (cf. fungi), direct chemical transformation of the precursor peptide into a cephalosporin type of molecule can be achieved with a crude enzyme preparation from this prokaryotic organism.

This detailed study has involved mainly Demain's laboratory at MIT, Baldwin's and Abraham's laboratories at Oxford, and those of Wolfe, Jensen, and Westlake in Canada. For the first time, there is sufficient understanding of a process to advance our studies on antibiotics at four major levels: a) identification of mutants lacking an antibiotic biosynthetic enzyme and therefore useful as a host for gene cloning by scoring for complementation of the missing character(s). (b) It enables the study of biochemical and physiological effects of intracellular antibiotic deprivation. (c) It enables one to study enzyme structure in detail and consequently to form genetic probes. d) It provides an enzymatic approach to synthesize new, unnatural β-lactam compounds. This latter approach is based on the observation that when chemically modified tripeptides analogous to the natural ACV were supplied as substrates to the cyclase and expandase enzymes of cephalosporin biosynthesis, novel β-lactam antibiotic structures were seen (17).

CONCLUDING REMARKS

The characteristic properties of naturally occurring secondary metabolites, both in the variety of complex chemical structures and in their bioactive properties, reflect the almost limitless synthetic capacity of microorganisms to create complex metabolites from simple substrates. Secondary metabolites are likely produced in high frequency by microorganisms but not always in sufficient amounts to facilitate their detection and isolation. Apparently, the types of secondary metabolites discovered are indicative of the types of assay and test methods used in screening programs rather than representative of the real biosynthetic potential of the organism. The only factors limiting the discovery of useful compounds among the secondary metabolites are our own commitment, ingenuity, and ability to devise simple in vitro screening procedures for desirable activities (6).

Although it is evident that the biosynthetic machinery of producing organisms involved in secondary metabolism consists of some tens or even hundreds of specific structural and regulatory proteins, the function in nature and the selective advantage that secondary metabolites confer on the producer organism are still obscure. The necessity to screen large populations when strain development programs are carried out, due to the absence of rational selection procedures, is perhaps the immediate difficulty that results from our lack of knowledge concerning secondary metabolism. Categorization of these "nonfunctional" metabolites as secondary in nature can probably be justified only as a temporary measure; once their role in nature is understood, they may well be termed primary metabolites. Metabolites, for which the genetic information involved in their biosynthesis derives entirely by natural means or by recombinant DNA technology from a foreign organism, are perhaps the only true secondary metabolites.

Under natural conditions the metabolic activities of isolated strains are generally well coordinated by regulatory mechanisms. In contrast, the most favorable conditions for industrial use of a producer organism will be those in which almost all of the metabolic functions are actively engaged in synthesis of the product. Achievement of this goal requires concerted input from several disciplines such as physiology, biochemistry, genetics, and biochemical engineering. Progress in this direction is increasing rapidly. Perhaps the most important recent development is the successful cloning of antibiotic genes. Employing gene cloning, as well as characterization of the specific enzymes involved in secondary metabolism, will enable us to study some of the central problems (including qualitative and quantitative parameters) of secondary metabolite biosynthesis. How genes involved in secondary metabolism are organized, what determines the onset of secondary metabolite synthesis during the developmental cycle of the organism, how the normal controls of expression can be modulated or overcome, and what might be the effect of introducing genes for secondary metabolism from one organism to another are only some of the questions which become targets for research. The intriguing questions also still remain to be answered: what is the physiological role of secondary metabolites? What are the mechanisms of their action in producer cells?

Acknowledgements. I wish to thank L.C. Vining and M.A. Pickard for reading the manuscript and for their valuable critical comments and V. Kervin for the expert typing of the manuscript.

REFERENCES

(1) Aharonowitz, Y., and Cohen, G. 1981. The microbial production of pharmaceuticals. Sci. Am. **245:** 141-152.

(2) Bailey, C.; Butler, M.J.; Normansell, I.D.; Rowlands, R.T.; and Winstanley, D.J. 1984. Cloning a Streptomyces clavuligerus genetic locus involved in clavulanic acid biosynthesis. Bio/Technology **2:** 808-811.

(3) Berdy, J. 1980. Recent advances in and prospects of antibiotic research. Proc. Biochem. **Oct./Nov.:** 28-35.

(4) Davis, J.E. 1982. Plasmids in antibiotic-producing organisms: their role in biosynthesis and resistance. In Genetics of Industrial Microor-

ganisms, eds. Y. Ikeda and T. Beppu, pp. 103-111. Tokyo: Kodansha Ltd.

(5) Demain, A.L. 1974. How do antibiotic-producing microorganisms avoid suicide? Ann. NY Acad. Sci. **235:** 601-612.

(6) Demain, A.L. 1983. New applications of microbial products. Science **219:** 703-714.

(7) Feitelson, J.S., and Hopwood, D.A. 1983. Cloning of a Streptomyces gene for an o-methyltransferase involved in antibiotic biosynthesis. Molec. Gen. Genet. **190:** 394-398.

(8) Hopwood, D.A.; Bibb, A.; Bruton, M.J.; Chater, K.F.; Feitelson, J.S.; and Gil, J.A. 1983. Cloning Streptomyces genes for antibiotic production. Trends Biotechnol. **1:** 42-48.

(9) Jensen, S.E. 1986. Biosynthesis of cephalosporins. Crit. Rev. Biotechnol., in press.

(10) Kleinkauf, H., and Von Dohren, H. 1983. Peptides. In Biochemistry and Genetic Regulation of Commercially Important Antibiotics, ed. L.C. Vining, pp. 95-145. Reading, MA: Addison-Wesley Publishing Company, Inc.

(11) Lancini, G., and Parenti, F. 1982. Antibiotics: An Integrated View. New York: Springer-Verlag.

(12) Martin, J.F., and Gil, J.A. 1984. Cloning and expression of antibiotic production genes. Bio/Technology **2:** 63-72.

(13) Quillardet, P.; Huisman, O.; D'ari, R.; and Hofnung, M. 1982. SOS chromotest, a direct assay of induction of an SOS function in Escherichia coli K-12 to measure genotoxicity. Proc. Natl. Acad. Sci. USA. **79:** 5971-5975.

(14) Umezawa, H. 1972. Enzyme Inhibitors of Microbial Origin. Tokyo: University of Tokyo Press.

(15) Umezawa, H. 1981. Problems and trends in the development of new antibiotics and other useful microbial products. In Advances in Biotechnology, Fermentation Products, eds. C. Vezina and K. Singh, Vol. III, pp. xv-xxiv. Toronto, Canada: Pergamon Press Ltd.

(16) White, R.J. 1983. Anthracyclines. In Biochemistry and Genetic Regulation of Commercially Important Antibiotics, ed. L.C. Vining, pp. 277-291. Reading, MA: Addison-Wesley Publishing Company, Inc.

(17) Wolfe, S.; Demain, A.L.; Jensen, S.E.; and Westlake, D.W.S. 1984. Enzymatic approaches to syntheses of unnatural beta-lactams. Science **226:** 1386-1392.

(18) Zahner, H.; Drautz, H.; and Weber, H. 1982. Novel approaches to metabolite screening. In Bioactive Microbial Products: Search and Discovery, eds. J.D. Bu'lock, L.J. Nisbet, and D.J. Winstanley, pp. 51-70. London: Academic Press.

Biotechnology: Potentials and Limitations, ed. S. Silver, pp. 113-125. Dahlem Konferenzen 1986. Berlin, Heidelberg, New York, Tokyo: Springer-Verlag.

Technological Processes for Biotechnological Utilization of Microorganisms

B. Mattiasson
Dept. of Biotechnology
Chemical Center, University of Lund
221 00 Lund, Sweden

Abstract. When developing new processes for production of various biochemicals the creation of the organism, often by gene technology, is a very important step which must be accompanied by development of cultivation as well as downstream steps. There is a lot to gain by treating the whole process, from creation of the organism to isolation of the product, as one process that has to be optimized in the overall mode.

The present paper deals with recent developments in process technology for utilizing microorganisms and with some efforts to integrate bioconversion and upgrading.

INTRODUCTION

The large proportion of the public interest in biotechnology has been focused on genetic engineering and the potential of this new technology. The scientific society has also given genetic engineering much attention and, in some cases, argued that gene technology and biotechnology are synonymous. When genetically modified cells actually appeared, however, it was soon realized that this was only the first step. Culturing, harvesting, and purification problems all still remain to be solved and optimized. Since the new cells, in some cases, are not stable and in other cases have a reduced ability to compete with wild-type cells or to grow under changing conditions, these process technology factors are very important.

The development of these more engineering-based aspects of biotechnology was initially given second priority, but today it is realized that if we want to truly capitalize on the advances of gene technology, more attention has to be given to the design and performance of the biotechnological processes.

In that context, improvements in process technology have not been of the highest priority. Gene technologists have in the past mainly been occupied with the production of high-priced products intended for the medical market. Utilization of modified organisms for production of speciality and bulk chemicals will be the next area where gene technology will be utilized. In these latter cases, the biotechnological processes will have to compete with conventional chemical processes. Biotechnology has, generally speaking, been characterized by conversions at low temperature, low pressure, in water solutions, and at high dilutions. The catalyst density has normally been low and thus extended periods of reaction time have been needed. Furthermore, as the performance of the catalyst is strongly influenced by environmental factors, there is a need to maintain these factors.

From an engineering point of view, biotechnological processes can be considered rather primitive. There is much room for process improvement in biotechnology. Improved productivities in biotechnological processes may be achieved by increasing catalyst densities in the reactor, better control over reaction parameters, and removal of inhibitory substances from the reactor. In the long-range perspective, physiological control of the organism may be included.

Finally, it should be remembered that the isolation and purification of biotechnological product is a major part of the reaction scheme. The upgrading of the diluted product is very expensive. In several processes it has been calculated to be 40-60% of the total processing cost.

This paper will discuss the current state of the art in process intensification as well as pointing out some of the bottlenecks and questions that will have to be addressed if biotechnology is to advance on the process technology side.

A biotechnological process can be illustrated as:

a) organism → cultivation → harvesting → cell disrupture → purification

b) organism → cultivation → bioconversion → isolation of product.

Genetic engineering has dealt with the organism itself, the initial step in the process. In the last few years, however, reports have become available on the use of gene manipulation to improve downstream processing. Two examples are the introduction of genes for flocculation (5) and for the addition of affinity groups on the peptides produced (32). Affinity groups are used in the purification process to facilitate product isolation. Before the product is finished, the group is split off and removed.

Improvements of biotechnological processes may take place on several levels, the major ones being listed below.

1) Process intensification by increased catalyst density

2) Extractive fermentations

3) Integration of bioconversion and downstream processing

4) Process control with the use of

   a) conventional sensors
   b) biosensors
5) Physiological control
6) New environments for biocatalysts:
   a) organic solvents
   b) gas phase

PROCESS INTENSIFICATION BY INCREASE OF CATALYST DENSITY

An obvious approach to improve the speed of a reaction is to increase the density of the participating reactants. In catalytic reactions an increased catalyst concentration will speed up the reaction. In batch processes an elevated initial substrate concentration results in higher productivity. For continuous processes, a faster supply of substrate gives similar results. In the latter process configuration, the risks of washout of cells sets the limit. There are methods which make it possible to operate at increased cell densities as well as at high dilution rates (see Table 1).

TABLE 1 - Methods to increase cell density in a bioreactor.

immobilization
adhesive fermentation
cell recycling
membrane technology
centrifugal separators
flocculating cells

Immobilized Cells

By entrapping cells in or adsorbing them to solid supports, a physical shape which can be handled conveniently can be obtained in, e.g., packed beds or fluidized bed reactors.

Much work has been done on the development of immobilization methods which has been applied to the immobilization of bacteria, fungi, plant cells, and mammalian cells (1, 9, 20). Very little has yet been presented on genetically engineered cells, but it seems natural to expect no differences in behavior from that of wild-type cells.

Industrial applications of immobilized cells are very rare, although several processes have been tested on the pilot plant scale.

The physiological condition of the entrapped cells may differ markedly. The status of immobilized cells can be: homogenized cross-linked cells, dead cells, permeabilized cells, viable cells, resting cells, and spores.

It has been reported that immobilized viable cells are more active or generate other products than the corresponding free cells (Table 2).

TABLE 2 - Changes induced by immobilization or similar treatment.

Organism	Change Reported	Reference
Saccharomyces cerevisiae	improved ethanol productivity	7, 26, 31
Nostoc commune	downshift in nitrogenase activity	28
Bacillus amylo-liquefaciens	improved production of α-amylase	29
Claviceps purpurea	increase in alkaloid production	12, 13
Penicillium chrysogenum	increased production of penicillin G or V	12
Streptomyces aureofaciens	increased production of chlorotetracychine	12
Clostridium acetobutylicum	changed metabolism	25
Morinda citrifolia	increase in antocyanins	2

The basic mechanisms behind these observations are not fully elucidated, although some working hypotheses have been presented. The effects of the support material as well as the environment created by the biocatalytic reactions in the support have been studied in great detail for immobilized enzymes, but very little has been done concerning immobilized cells. The immobilization leads to a transfer of the formerly free cells to an environment with changed water potential. This reduction in water potential is thought to cause some of the effects observed (22, 28). Changed concentrations of essential nutrients and/or inhibitors may also be very important.

A variation on immobilization developed in nature long ago is adhesive growth. Cells adhering to and growing on solid surfaces offer certain advantages. The immobilization is cheap - just supply a suitable support and adhesion starts spontaneously. From microbial ecology it is known that at low substrate concentrations the substance is unevenly distributed with the best supply found at the surface of the support. This has been argued as one of the reasons why cells which are normally free-living upon starvation become sticky and adherent (11). In studies on adhesive fermentations, it has been shown that the cells stayed in the reactor also at very high dilution rates. However, marked morphological changes occurred in that the adhesive cells became substantially larger than normal free-living cells (27). This phenomenon has not yet been explained.

RECYCLING OF CELLS IN CONTINUOUS PROCESSES

The general process scheme is as shown in Fig. 1.

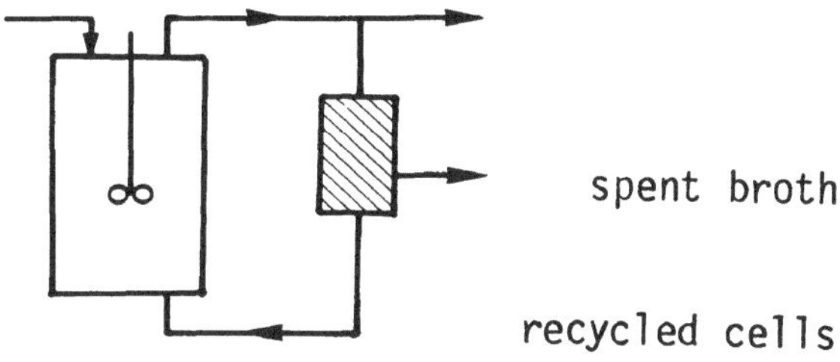

FIG. 1 - Schematic presentation of a cell-recycling process.

The separation unit may be any mechanism that under sterile conditions can separate spent medium from a dense cell suspension which is fed back to the fermenter. In the past, continuous centrifugal separators have been employed to accomplish this. Today membrane technology also offers an alternative to achieve cell recycling. To improve the separation efficiency, spontaneously flocculating strains that can sediment without centrifugation, just by means of a static settler, have been used.

Most work so far has been carried out on bacteria and yeast since the pumping involved may partly damage more fragile structures. For microorganisms this approach seems to offer certain advantages in that very high cell densities can be attained, yet the cells are still free and thus operate under very low diffusion restrictions. A low bleed of cells must be taken out since there is a tendency for inhibitory products to be enriched in the retentate (16, 23). The ratio between the bleed and retentate volumes must be varied from application to application.

For more labile cells, membrane technology may be very useful, but here the cells are kept in one place by the membrane and the substrate/product stream is pumped. Several examples with cells kept behind a membrane have been presented, e.g., on the shell side of a hollow fiber ultrafiltration unit while substrate is fed via the fiber lumen (10, 24). For fast-growing cells this solution may be troublesome because as they grow in extremely dense masses in the cavities of anisotropic membranes, pressure is generated that eventually breaks the membrane (10, 24). Slower-growing cells such as plant or mammalian cells are better suited for this kind of cultivation. This technology may also be useful for the production of high-molecular-weight products.

EXTRACTIVE BIOCONVERSIONS

The efficiency of a bioconversion can be bettered either by increasing the concentration of the starting reagents or by reducing those of the products in the reaction mixture. In extractive fermentations this is done in a mode which makes continuous operations possible. The extractive step can be performed by mixing the extractant into the bioreactor and then removing it in a separation step (see Table 3).

TABLE 3 - Extractants used in extractive bioconversions.

organic solvents
aqueous two-phase systems
solid adsorbents
vacuum fermentation
supercritical CO_2

Organic Solvents

In bioprocesses where hydrophobic substrates or products are involved, the low solubility of the reactant in the aqueous phase where the biocatalyst is found will result in very slow reactions. By adding an organic solvent, either miscible or immiscible with water, the solubility is increased and thereby the concentrations of the reactants as well. An efficient mass transfer is then obtained in dispersions. When using water-immiscible organic solvents, the organic phase can easily be separated from the aqueous phase and the product recovered.

This approach has been used mainly for operation with bioconversions involving only one or a few enzymatic steps and not for viable cells, because of the solvent toxicity which rapidly destroys membrane potentials by dissolving membrane components.

Careful treatment with organic solvents may cause permeabilization and thus make substrates for intracellular enzymes more accessible. Since bioorganic chemistry is a rapidly growing discipline within biotechnology, one can foresee a dramatic expansion of this kind of treatment in order to create permeabilized cells that will function as packages of enzymes stabilized in their native environment.

Aqueous Two-phase Systems

Polymer-polymer aqueous two-phase systems consist of two phases of essentially water (80-95% water in each phase). The interfacial tension is extremely low compared to that between organic solvent and water, making it easy to form and maintain an emulsion of the aqueous two-phase system.

Biological material tends to partition in aqueous two-phase systems and it is quite realistic to create systems with cells or cell particles exclusively partitioned to one of the phases. If a bioprocess is carried out in such a system, it is possible to keep the cells in one phase and extract the product to the other phase (19). This is a very gentle extraction system. Partition constants may not be as high as those in organic solvent water systems but, on the other hand, these systems are biocompatible. Due to the low energy input needed to keep the emulsion, one can expect such systems also to be good for use with fragile cells.

Solid Adsorbents

If solid adsorbent particles are added to a fermentation broth some substances will bind to the sorbent. By selecting adsorbents of suitable properties, a certain product may be adsorbed selectively from the fermentation broth, thereby reducing product inhibition and thus favoring a continued production (see Table 4).

Several examples of successful applications of such systems have been reported (Table 4). In some cases, however, experiments have shown that the cell-adsorbent contact may be detrimental to the process, and in such cases the adsorbent has to be protected by, for example, entrapment in hydrogels or placement behind a membrane.

TABLE 4 - Examples of processes where solid adsorbents have been used.

Microorganism	Product	Sorbent	Productivity	Reference
Clostridium acetobutylicum	butanol/acetone	Bonopore[1]	2.2[2]	14
Saccharomyces cerevisiae	ethanol	activated carbon	25 g/1h	15
Streptomyces griseus	cyclohexamide	Amberlite XAD-4	2.25[2]	33
Monascus sp.	red pigment	Amberlite XAD-7	1.5[2,3]	6
Pseudomonas aeruginosa	salicyclic acid	Amberlite IRA-400	5.5[2]	31
Corynebacterium renale	salicyclic acid	Amberlite IRA-400	2[2]	30

[1] A nitrated divinylbenzene-styrene copolymer
[2] Increased productivity compared to conventional batch fermentation
[3] Extracellular production

All the above examples deal with rather unspecific adsorbents. The technology is, of course, applicable to the use of affinity adsorbents. This makes this system suitable for production of macromolecules such as protein and antibodies. In these cases product removal is employed not to avoid product inhibition per se, since it has not yet been fully established whether inhibition exists in these conditions, but rather to remove the product from harmful conditions (high shear forces, proteases, oxygen bubbling, etc.) found in the fermenter.

Vacuum Fermentation

This technique has been used for the production of ethanol (4). By reducing the pressure in the fermenter it has been possible to boil off the product at reasonable temperatures and thus to run fermentation and distillation simultaneously. This technique might have a wider application in bioorganic synthesis.

INTEGRATION OF BIOCONVERSION AND DOWNSTREAM PROCESSING

When developing bioprocesses, several categories of scientists are usually employed. The microbiologist takes care of the organism, and a fermentation engineer runs the fermentation before people specially trained in downstream processing then take part. This has led to a situation where each specialist has optimized his/her process step without consideration of the other steps. By optimizing over the whole process, which often means that several of the small steps must be run under suboptimal conditions, clear improvements of the process performance may be attained. An awareness of these facts is beginning to grow and one can foresee increased work in the future along these lines.

When selecting a strain for production it may be advantageous to use a low-producing strain that gives a simplified downstream processing instead of a higher-producing strain with products involving difficult downstream processing. One example may be the use of antifoam agents in the fermentation step. If membrane separation is to be utilized in the downstream processing, then one can foresee problems. Mechanical arrangements to deal with foaming may be preferable.

When developing new processes a tempting alternative is to try to integrate some of the unit operations discussed earlier, such as cell recycling and specific adsorbents (14, 15), membrane separation and extraction (18), or immobilization with extraction (Wagner, personal communication), etc. In such a way it may be possible to operate under high cell densities with a continuous removal of products. Perhaps the volumetric productivities may not always be as high as the top values reported by other methods, but the initial separation step will create a product that is already partially purified.

PROCESS CONTROL

A prerequisite for successful process control is the availability of suitable sensors to give the relevant information and also a solid understanding of the process studied. This understanding will enable one to set up a mathematical model for the process which can be used as a model in evaluating information from the sensors in controlling the process.

Biotechnological processes differ from conventional chemical processes in, among other things, their complexity. At present we have a very poor understanding of most biotechnological processes, which is why accurate models do not exist. This means that crude approximations are used in our models and operating systems are constructed from the initial characterization of the systems in terms of time constants for sensors, efficiencies in control loops, etc.

The sensor area has been attracting more interest in recent years. In the scientific literature a broad spectrum of sensors are described. However, very few stand up to the realities outside the research lab and standard buffered solutions. This latter observation also holds for conventional sensors such as pH and pO_2 electrodes.

A key question that has been addressed by several scientists without resulting in a clear-cut answer is whether one should insert a sensor in the bioreactor or instead withdraw a stream that can be analyzed outside the bioreactor using either biosensors or other analytical machines. There are arguments for both strategies. In Table 5 are listed the pros and cons for the two alternatives. It should be stressed that the number of entries for a certain strategy does not render it superior to the other.

TABLE 5 - Reasons for in-fermenter or off-fermenter analysis.

	In-fermenter	Off-fermenter
Pros	- measures in situ	- easy to make many different measurements
	- real time	- easy to use many different analytical principles
		- easy to use biochemical macromolecules in the assay
Cons	- need to make several transfer holes in the reactor	- delayed measurement - changes may occur during the transport
	- restricted number of techniques suitable for "in-fermenter" operation	- difficulties in taking a representative sample
	- sterility problems	- growth of cells onto the sampling device may cause severe problems
	- need for autoclaveable equipment	
	- growth of cells onto the sensor cause severe disturbances	

It is theoretically possible to monitor substrates as well as products continuously or by frequent discrete analysis in simple bioconversion reactions and in processes leading to the formation of low-molecular-weight products. This has been discussed recently in connection with high-molecular-weight products (21). Another relevant value to monitor is cell density. Several attempts to do this have been presented, but as yet there is no general method available (3, 17).

If data concerning the concentration of substrate and product as well as the cell density and, preferably, the metabolic status of the cells were available, it would still in most cases be difficult to make a good computer program for controlling the process. The basic knowledge about the biotechnological multi-step process run at nonequilibrium conditions is too poor. Only rough simplifications can be used (see Table 6).

TABLE 6 - Requirements for the successful development of process control of bioprocesses.

-	development of reliable, robust biosensors
-	development of methods to take representative samples from a fermenter or strategies to measure at representative locations in the fermenter
-	improved understanding of the biochemical machinery exploited in the process
-	improved interaction between personnel responsible for computer, biochemistry, and microbiology

Most work with biosensors has so far been directed towards clinical analytical applications and thus has not been designed to meet the demands of the "in-fermenter" approach, whereas most systems can quite easily be adapted to "off-fermenter" analysis. A sampling device has to be used if off-fermenter analyses are used. Several attempts to develop such a sensor have been presented (17). A technology transfer from the clinical analytical field to "in-fermenter" analysis does not seem likely. This is because of two reasons: a) the necessity of sterilizing all equipment used in the fermenter hampers such a transfer, and b) many assays of macromolecules and low concentrations of small molecules involve competitive or sequential binding reactions that must take place under strict control. The presently available sensors are useful for monitoring most soluble substances in a fermentation broth. In combination with ion-selective electrodes and off-gas analyses there is the potential of monitoring most soluble substances of relevance for the fermentation.

In gene technology, Escherichia coli has been the main host for the production of new proteins. The bacterial walls keep the material within the cell and thus protect it from the harmful conditions in the fermenter. A method of monitoring the enrichment of a certain substance in the cell in a non-invasive way would be very interesting. However, no such technology has been developed so far. One possibility might be to introduce a strong reporter group on the peptide, a group that is removed before the peptide is fully isolated. Non-invasive analysis is a prerequisite for such analyses.

The biosensor area must develop substantially before fermentation control based on biosignals can be carried out more generally. This means as well that the transfer from batch processes to continuous processes will be delayed.

PHYSIOLOGICAL CONTROL

Earlier in this chapter the effects of various treatments on the physiology of cells was discussed. These effects can of course be exploited, in the sense that the cell physiology might be induced towards a certain state in order to obtain higher productivities. Microbial physiology is an overlooked area that may have much to offer in the area of process intensification.

NEW ENVIRONMENTS FOR BIOCATALYSTS

Gene technology may make enzymes available that in former days were extremely cumbersome to prepare. As the number of enzymes and microorganisms used industrially is very low, an interesting expansion should take place when it becomes possible to get cells loaded with enzymes that were formerly rare. The use of cells/enzymes for bioconversions will increase in importance and totally new applications may be expected. When cells are used, permeabilization will be a frequently used technique. Bioconversions in organic solvent with very low water content might open new possibilities of exposing biocatalysts to totally new substrates, and new reactions may be created.

CONCLUSION

There is a need for process engineering work in biotechnology. All improvement of already existing processes may, of course, be applicable to those based on genetically modified organisms. Since these latter cells often are more fragile, better control of processes must be established. Furthermore, the introduction of new enzymatic activities and changed physiological characters may increase the demands on totally new process technologies. In many cases process intensification per se may give effects in the same magnitude as that of genetic modification alone. An integration of new process technology together with the use of new improved cell lines and overall process optimization may drastically change the effectiveness of biotechnological processes. In order to reach these goals and to exploit fully the advances in gene technology, process engineering must get more attention in the future.

REFERENCES

(1) Brodelius, P. 1983. Immobilized plant cells. In Immobilized Cells and Organelles, ed. B. Mattiasson, pp. 27-56. Boca Raton, FL: CRC-Press.

(2) Brodelius, P.; Deus, B.; Mosbach, K.; and Zenk, M.H. 1979. Immobilized plant cells for the production and transformation of natural products. FEBS Lett. **103:** 93-97.

(3) Blake-Coleman, B.C.; Calder, M.R.; Carr, R.J.G.; Moody, S.C.; and Clarke, D.J. 1984. Direct monitoring of reactor biomass in fermentation control. Trends Anal. Chem. **9:** 229-235.

(4) Cysewski, G.R., and Wilke, C.R. 1977. Rapid ethanol fermentation using vacuum and cell recycle. Biotechnol. Bioeng. **24:** 1125-1143.

(5) de Figueroa, L.I.; de Richard, M.F.; and de van Broock, M.R. 1984. Transfer of the flocculation property to the baker's yeast Saccharo-

myces cerevisiae by conventional genetic manipulation. Biotechnol. Lett. **6(3):** 171-176.

(6) Evans, P.J., and Wang, H.Y. 1984. Pigment production from immobilized Monascus sp. utilizing polymeric resin adsorption. Appl. Envir. Microbiol. **47:** 1323-1326.

(7) Hahn-Hägerdal, B.; Larsson, M.; and Mattiasson, B. 1982. Shift in metabolism towards ethanol production in Saccharomyces cerevisiae using alterations of the physical-chemical environment. Biotechnol. Bioeng. Symp. **12:** 199-202.

(8) Harris, M.C., and Kell, B.K. 1985. The estimation of microbial biomass. Biosensors **1:** 17-84.

(9) Hopkinson, J. 1983. Hollow fiber cell culture: applications in industry. **In** Immobilized Cells and Organelles, ed. B. Mattiasson, pp. 89-99. Boca Raton, FL: CRC-Press.

(10) Inloes, D.S.; Taylor, D.P.; Cohen, S.N.; Michaels, A.S.; and Robertsson, C.R. 1983. Ethanol production by Saccharomyces cerevisia immobilized in hollow fiber bioreactors. Appl. Envir. Microbiol. **46:** 264-278.

(11) Keffort, B.; Kjelleberg, S.; and Marshall, K.C. 1982. Bacterial scavenging: utilization of fatty acids localized at a solid-liquid interface. Arch. Microbiol. **133:** 250-260.

(12) Kopp, B.; El-Sayed, A.-H.; Mahmoud, W.; and Rehm, H.J. 1984. Production of ergot alkaloids, penicillins and chlortetramyeline by immobilized microorganisms. Preprint, 3rd European Congress on Biotechnology, September 1984, pp. 281-286. Munich, Weinheim: Verlag Chemie.

(13) Kopp, B., and Rehm, H.J. 1983. Alkaloid production by immobilized mycelia of Claviceps purpurea. Eur. J. Appl. Microbiol. Biotechnol. **18:** 257-263.

(14) Larsson, M.; Holst, O.; and Mattiasson, B. 1984. Butanol fermentation using a selective adsorbent for product recovery. Preprint, 3rd European Congress on Biotechnology, September 1984, pp. 313-316. Munich, Weinheim: Verlag Chemie.

(15) Lee, S.S., and Wang, H.Y. 1982. Repeated fed-batch rapid fermentations using yeast cells and activated carbon extraction system. Biotechnol. Bioeng. Symp. **12:** 221-231.

(16) Maiorella, B.; Blanch, H.W.; and Wilke, C.R. 1983. By-product inhibition effects on ethanolic fermentation by Saccharomyces cerevisiae. Biotechnol. Bioeng. **25:** 103-121.

(17) Mandenius, C.-F.; Danielsson, B.; and Mattiasson, B. 1984. Evaluation of a dialysis probe for continuous sampling in fermentor and in complex media. Anal. Chim. Acta **163:** 135-141.

(18) Matsumura, M., and Märkl, H. 1984. Elimination of ethanol inhibition by pertraction. Preprint, 3rd European Congress on Biotechnology, September 1984, Vol. II, pp. 415-423. Munich, Weinheim: Verlag Chemie.

(19) Mattiasson, B. 1983. Application of aqueous two phase systems in biotechnology. Trends Biotechnol. **1(1):** 16-20.

(20) Mattiasson, B., ed. 1983. Immobilized Cells and Organelles, vols. I and II. Boca Raton, FL: CRC-Press.

(21) Mattiasson, B. 1984. Immunochemical assays for process control: potentials and limitations. Trends Anal. Chem **3:** 245-250.

(22) Mattiasson, B., and Hahn-Hägerdal, B. 1982. Microenvironmental effects on metabolic behaviour of immobilized cells. A hypothesis. Eur. J. Appl. Microbiol. Biotechnol. **16:** 52-55.

(23) Mattiasson, B., and Larsson, M. 1985. Extractive bioconversions with emphasis on solvent production. **In** Biotechnology and Genetic Engineering Reviews, ed. G.G. Russel, vol. 3. Newcastle-upon-Tyne, England: Intercept Ltd, in press.

(24) Mattiasson, B.; Ramstorp, M.; Nilsson, I.; and Hahn-Hägerdal, B. 1981. Comparison of the performance of hollow-fiber microbe reactor with a reactor containing alginate entrapped cells. Denitrification of water using Pseudomonas denitrificans. Biotechnol. Lett. **3:** 561-566.

(25) Mattiasson, B.; Suominen, M.; Andersson, E.; Häggström, L.; Albertsson, P.-Å.; and Hahn-Hägerdal, B. 1982. Solvent production by Clostridium acetobutylium in aqueous two-phase systems. **In** Enzyme Engineering 6, eds. J. Chibata, S. Fukui, and L.B. Wingard, Jr., pp. 153- 155. New York: Plenum Press.

(26) Navarro, J.M., and Durand, G. 1977. Modification of yeast metabolism by immobilization into porous glass. Eur. J. Appl. Microbiol. Biotechnol. **4:** 243-254.

(27) Nilsson, I. 1984. Influence of entrapment and surface adsorption on the conversion capacity of Pseudomonas spp. Ph.D. Thesis, University of Lund.

(28) Potts, M.; Bowman, M.A.; and Morrison, N.S. 1984. Control of matric water potential (Ψ_m) in immobilized cultures of cyanobacteria. FEMS Microbiol. Lett. **24:** 193-196.

(29) Shinmyo, A.; Kimura, H.; and Okada, H. 1982. Physiology of $- amylase production by immobilized Bacillus amyloliquifacieus. Eur. J. Appl. Microbiol. Biotechnol. **14:** 7-12.

(30) Tangan, S.K., and Ghose, T.K. 1981. Environmental manipulations in salicylic acid fermentation. Proc. Biochem. **16:** 24-27.

(31) Tone, H.; Kitai, A.; and Ozaki, A. 1968. A new method for removal of inhibitory fermentation products. Biotechnol. Bioeng. **10:** 689-692.

(32) Uhlén, M.; Nilsson, B.; Guss, B.; Lindberg, M.; Gatenbeck, S.; and Philipson, L. 1983. Gene fusion vectors based on the gene for straphylococcal protein A. Gene **23:** 369-378.

(33) Wang, H.Y. 1981. Integrating biochemical separation and purification steps in fermentation processes. Ann. NY Acad. Sci. **413:** 313-321.

Top row, left to right:
Albrecht Sippel, Gerhard Siewert, Heinrich Betz, Heinz Schaller

Middle row, left to right:
Walter Fiers, Julian Davies, John Collins, Enzo Paoletti

Bottom row, left to right:
Eberhard Pfaff, David Rowlands, Erwin Wagner

Biotechnology: Potentials and Limitations, ed. S. Silver, pp. 127-138. Dahlem Konferenzen 1986. Berlin, Heidelberg, New York, Tokyo: Springer-Verlag.

Animals and Medicines

Group Report

J. Collins, Rapporteur
H. Betz
J.E. Davies
W. Fiers
E. Paoletti
E. Pfaff
D.J. Rowlands
H. Schaller
G. Siewert
A.E. Sippel
E.F. Wagner

INTRODUCTION

In the discussion of the impact of gene technology, synthetic peptides, studies on protein structure, and new immunological tools (such as monoclonal antibodies) in the area of animals and medicines, we identified an underlying theme which can be described as follows. At present new tools are rapidly being applied to biological systems which have been extensively defined at the biochemical level (e.g., blood homeostasis, many peptide hormones). The future trends can, however, already be discerned in which the isolation of new factors tells us more about the original biological system (e.g., discovery of new peptide hormones on ACTH-precursors; biological modifiers in the blood pressure regulation cascade; interferon gene family members; dissection of antigenic sites of pathogens). The further rapid advancement of this field (which is to be expected) will result from intensive interdisciplinary research in which the main emphasis will rest at the level of basic research on biological systems (complex systems) and protein structure.

For a more detailed prognosis, the topics of "better" proteins, vaccines, diagnostics, gene transfer into higher eukaryotes, and social impact have been addressed.

"BETTER" PROTEINS

Under this heading we considered two main subtopics: a) the search for useful new proteins or peptides, and b) the production of better quality products from recombinant organisms.

New Proteins
Classical approaches to the discovery of new antibiotics, or enzymes for their synthesis, were evaluated as underdeveloped, and continued improvements were considered essential. Discussion crystallized around peptides for clinical use.

Tables 1 and 2 in addition to the background paper by Fiers (this volume) supplies a brief overview of the rapid expansion of this field in the last five years and a definition of the major areas of interest for human application (vaccines and diagnostics are treated separately). Including DNA markers for genetic counseling (gene mapping), some 300 human "genes" and fragments have presently been cloned and new clones are now being reported at the rate of several every month.

The following qualitatively new aspects of this research are emphasized:
- through gene cloning some compounds were obtainable for the first time in form pure enough for their complete characterization (Interferon

TABLE 1 - Examples of pharmaceutical peptides presently being developed via recombinant DNA approaches.

Serum proteins	
	Serum albumin
	Clotting Factors VIII, IX, XII
	Tissue plasminogen activator (TPA)
	Urokinase
	Elastase
	Protease inhibitors: α1 antitrypsin
Growth and differentiation regulators	
Positive:	Interleukin 1 and 2
	Colony stimulating factors
	Insulin
	Insulin-like growth hormone
	Growth hormone
	Growth hormone releasing factor
	Platelet-derived growth factor
	Epidermal growth factor
	Transforming growth factor
	Somatomedins
	Skeletal growth factor
	Parathyroid hormone
Negative:	Interferons (α, β, γ)
	Tumor necrosis factors
Other hormones	
	Calcitonin
	Rennin, Bradykinin, Angiotensinogen, Kinninogen
	Atrial Natriuretic Peptide, Cardiodilatin
	Substance P
	Opiomelanotropic Hormones
	Endorphins

- subtypes, tissue plasminogen activator (TPA), tumor necrosis factor (TNF), components of the AIDS virus).
- for the first time large enough amounts of pure substance were made available for preclinical and clinical studies (TNF, Interferon γ; this led to the detection of a synergistic tumor cell-directed killing activity of these two factors).
- with the rapid discovery of new biological response modifiers, the formulation of the drug and preclinical studies of the biological activity or mechanism of action become rate-limiting steps for clinical application and will form the basis of increased research effort in this area.
- with the availability of natural biological response modifiers (BRM) as pharmaceutical drugs one does not expect toxic side effects as is often the case with "foreign" molecules such as synthetic drugs.
- since the study of the BRM itself clarifies its activity and site of action, indeed, providing a "missing" link in our concept of the functioning of a complex system, it can be expected that this knowledge will provide the basis for more rational treatment protocols - the opportunity to work safely with components of highly infectious agents (FMDV, Lassa, Hepatitis B, AIDS).

TABLE 2 - Diseases for which recombinant vaccines are being developed.

Animal	Human
FMDV (Foot and Mouth)	EBV (Epstein Barr)
Scour	Polio
Feline Leukemia	Herpes
Pseudorabies	Influenza
Infectious bronchitis virus (chicken)	AIDS
Rabies	Hepatitis A + B
Eastern equine encephalitis	Measles
Mareck's disease	Pertussis
Mycoplasma	Malaria
Piglet gastroencephalitis	Rabies
Infectious bovine Rhinotrachitus	Streptococcal Endocarditis
Transmissable gastroenteritis	Filariasis
	Schistosomiasis
	Gonorrhea
	Leprosy
	Cholera
	Caries
	Cytomegalie

Where to Look for New Proteins: Complex Systems

Fruitful areas for further developments were considered in the category "complex" systems. Thus, under the heading cell growth and differentiation, the development of cancer can be considered a transient aberration. Study of oncogenes has revealed that growth factor analogues (the viral oncogene, v-sis; α-tumor growth factor, α-TGF) and growth factor-receptor analogues (e.g., the proto-oncogene erb B) lead to self-stimulation (autocrine) of growth and thus identify specific targets for treatment of particular types of cancer.

In metastasis the dissociation of cancer cells and their (active or passive?) translocation to other sites implies an altered cell-cell interaction which may be associated with altered cell surface antigens or possibly proteases, hyaluronidases, etc., again possible targets for tumor therapy.

Tissue regeneration, combining the problems of growth and differentiation, can be approached. Is the regeneration capacity of rabbit cornea a function simply of growth factor receptors in this tissue compared to their absence in the human cornea? The observation that goldfish and newborn rabbits can regenerate their central nervous system cells gives hope that results from animal studies can be extrapolated to application in man. The analysis of neuropeptides which occur in vanishingly small amounts will be significantly stimulated by gene technology. Will the study of Altzeimer's disease and the detection of autoantibodies directed against a muscle peptide apparently required for neuronal growth lead to identification of nerve cell regeneration factors? To stimulate bone calcification and regeneration after breakage, skeletal growth hormone and parathyroid hormone will play an important role. Rapid advancement in our understanding of blood homeostasis and the immune system, particularly the role of cell-cell interaction (clonal expansion, antigen presentation, platelet aggregation, HLA antigens, T cell receptors) in addition to the multitude of soluble factors involved (lymphokines, interferons), is required not only for the definition of targets for new bioresponse modifiers but for the development of more effective vaccines and treatment of cardiovascular disease and autoimmune disease.

The development of potential pharmaceuticals is inseparable from intensive basic research in the complex biological systems just defined, in which recombinant DNA technology will continue to play a key role.

Second Generation Proteins

The availability of the cloned gene provides a further possibility: the rapid creation by site-directed mutagenesis of derivatives, the so-called second generation products with "improved" properties.

The required "improvements" were defined as follows:

- decreased toxicity (e.g., reduced pyrogenicity of interferons)
- higher stability (e.g., Salmon calcitonin is more stable than human calcitonin in humans; glycosylated Interferon γ has a threefold longer

half-life than the non-glycosylated form in human serum; oxygen resistant α-1 antitrypsin inhibitor)
- better pharmokinetic properties
- higher activity or changed specificity (e.g., protease inhibitor directed to specific sites in the coagulation cascade, or blood pressure peptide network)
- on purely economic grounds, new patentable derivatives
- dissection of activities (e.g., separation of receptor binding and growth stimulatory activities of growth factors to produce competitive growth inhibitors).

In the consideration of the prerequisites for successful development of second or third generation products based on protein engineering, it was agreed that progress in computer-aided prediction of protein folding was necessary. This aspect was covered by Smith et al. (this volume).

Isolation of Biologically Active Proteins from Cells Containing Recombinant DNA

Protein renaturation. The high-level production of the majority of eukaryotic or viral proteins from DNA cloned into bacteria leads to the production of insoluble "denatured" aggregates (inclusion bodies). During subsequent isolation and renaturation, methods are required with which the extent of correct protein folding can be assayed. These include classical methods such as analysis of s-s bridged tryptic peptides and newly introduced procedures such as monitoring denatured contaminants with monoclonal antibodies directed to internal epitopes only exposed in the denatured molecules.

It should be pointed out that the level of purity of recombinant DNA products (also required by the regulatory bodies) exceeds that of classical natural preparations where some denatured product is also to be expected on the grounds of the purification procedures. Therefore, on this basis, no special objections can be raised to the quality of the recombinant products. Monitoring methods still require improvement. In addition, it is suggested that research into the enzyme-induced protein folding by enzymes such as Protein disulfide isomerase (1) be intensified. This should both accelerate renaturation and increase the yield of correctly folded product. The development of secretion vectors (B. subtilis, Staphylococcus, and yeast) or the production in higher eukaryotic cells will partially alleviate this problem.

Post-translational Modification

Eukaryotic peptides are often post-translationally glycosylated, carboxylated, phosphorylated, sulfated, carboxy-amidated, etc. These modifications are often essential for the function of the eukaryotic peptide (e.g., Factor VIII carboxylation). None of these modifications can be carried out in bacterial cells although Cellulomonas and Trichoderma cellulases are reported to be glycosylated. Initial hopes that yeast cells may carry out sufficiently similar glycosylation have not been upheld (3).

The optimal condition would be to use engineered cells of the same type as those initially producing the product. Although this possibility may become reality through the use of tissue-specific tumor inducer vectors, at present only a limited number of cell systems are in use: constitutive production - Chinese hamster ovary (CHO) cells, mouse fibroblasts, or monkey kidney cells (Vero); or batch production - insect cells or COS7 cells (SV40 vectors). Using amplifiable vectors with viral enhancers, high levels of production have been obtained, and CHO cells at present represent the optimal system. As far as glycosylation is concerned, products from such heterologous cells (mouse, monkey, CHO) appear to be quantitatively glycosylated to the same extent as the natural (human) product. Further study of the exact structure of the sugar side chains are necessary in addition to studies on the many other types of post-translational modifications in heterologous cells.

VACCINES

In an evaluation of the requirements for vaccines a risk/benefit equation plays an important role. In the case of a widespread debilitating disease, a vaccine with side effects or reversion to pathogenicity which may be initially highly acceptable may lead to early interruption of the program as the disease comes under control, leading to the development of a highly susceptible population with catastropic results. The main background to this field is provided in the paper by Rowlands (this volume).

With the exception of enteric disease, where live bacterial vaccines seem advantageous, the following discussion is confined to viral diseases.

The main gap in our knowledge appears to be in the understanding of the basis of virulence/pathogenicity and of the mechanism of the immune response. Since the immune system is also concerned with monitoring self/non-self reactions internally, this is inseparably connected to studies of tissue-specific antigens, autoimmune disease, and cell-cell interactions.

In terms of economics and effectiveness in eliciting the immune response, live vaccines are highly preferable. In the improvement of classical attenuated live vaccines, a directed reduced virulence aimed at cell attachment-site mutations and/or reduced replication rate would be desirable.

Recombinant Live Vaccines

The potential for the use of vaccinia virus vectors for the production of recombinant live vaccines seems overwhelming, in particular where multivalent vaccines with "cassettes" of genes for, say, geographically delineated diseases, bovine diseases, or venereal diseases are engineered into the vector. The reinitiation of new immunization programs later does not seem to be substantially compromised by earlier inoculation with the same vector. The recombinant virus can be used at low doses and can accept (theoretically) up to several dozen genes. Side effects shown by patients with immune deficiency or eczema can be eliminated by screening and incidence of vaccinia encephalitis is less than one per million. Clinical trials have not yet started, but animal tests seem highly encouraging. Development of new viral vectors seems unnecessary (Paoletti et al., this volume).

Inert Vaccines
Although the ability to elicit a strong broad immune response with inert vaccines (from synthetic peptides or recombinant microorganisms) is low, this may be compensated for by the ability to make large amounts, elicit a rapid response, have a totally noninfectious agent of high purity, and more rapidly adapt new products to antigenic variants developing in the population. In addition, the products are highly stable and easily stored and shipped, and in the case of small peptides may possibly be applied by nasal spray. In this latter case it would be advantageous to couple the antigen to a ligand capable of adsorbing to mucosa (influenza antigen?) so as to elevate the IgA response. A major disadvantage is the requirement for adjuvants and, because of the short-lived response, booster applications. This might be compensated for later by the development of less toxic adjuvants (Muramyl dipeptide coupled to antigen (2)) or more direct agents aimed at specific targets in the immune response system. As mentioned so often already, this will require more basic research on the immune system.

Subunit vaccines from recombinant bacteria have not been encouraging. Although most viral and bacterial as well as malaria proteins can be produced in an active antigenic form in E. coli, Hepatitis B antigen can only be effectively produced either in yeast (where it is, however, apparently not glycosylated) or in cell culture (in Vero cells).

Passive Immunization
This still appears to be of great interest for small populations at risk as, for example, with immune compromised patients (e.g., against Cytomegalovirus) or rare new pathogens (e.g., Lassa fever). One suggestion for future products came from the necessity to protect hospitalized patients in general against Gram-negative pathogens.

A new application of monoclonal antibodies was suggested as a type of passive immunization to compete for virus receptors on cells. Tolerance to tissue transplantation may be improved by antibodies to immune modulators such as Interleukin-2, or possibly by the use of receptor-binding but inactive peptide fragments of Interleukin-2 or other immune modulators.

DIAGNOSTICS
Amongst short-term potential applications of gene technology in the area of biotechnology, diagnostics figure high on the priority list because of the minimal governmental regulatory limitations.

The development of diagnostic tools is intimately coupled to the development of vaccines and has direct applicability in the study of epidemiology, pathogenicity functions, and investigation of the immune response. In particular, the dissection of antigenic determinants for subunit vaccines can be analyzed with this method and, by the use of monoclonal antibodies, diagnostic tools can be developed for disease monitoring simultaneously to treatment with passive immunization vaccine. The coupling of toxins, etc., to particular antibodies for directed targeting of chemotherapy is a potential development.

The main aims are to broaden diagnostic capabilities and develop faster, simpler, and safer (more reliable) methods. Early tumor diagnosis, coupled to testing tumor cells for their sensitivity to chemotherapeutic agents, is high on the priority list.

Antibody Tools

Under this heading we considered the use of antibodies a) against normal cell and serum constituents, and b) against disease-specific antigens.

Since certain specific allelic variants of the major histocompatibility locus (HLA) appear to be associated with susceptibility to certain diseases (e.g., diabetes, rheumatoid arthritis), this type of "tissue-typing" analysis may increase in importance outside of its present application in transplantation surgery. In the analysis of autoimmune disease anti-idiotypic antibodies may act not only as diagnostic tools, but directly as competitive inhibitors of the auto-antibodies. For degenerative diseases, more rapid diagnosis may be possible (Altzeimer's disease's choline acetyltransferase deficiency).

The use of specific antibodies together with electrodes could lead to almost instantaneous measurement of metabolites, hormones, disease antigens, etc., in blood or urine samples, which would have an enormous impact on simplifying and accelerating routine clinical testing. Earlier cancer diagnosis should be possible with diagnostic kits against ectopically produced hormones (in particular, against autocrine analogues discussed previously).

Nucleic Acid Probes

In the development of this methodology, radionuclide-marked oligonucleotides (DNA and RNA) are still preferred to biotin-labelled probes because of their higher sensitivity, although in the long term it would be better to avoid the use of radioactivity in routine clinical testing.

Main applications will be for monitoring infections, genetic counseling, cancer diagnosis, and detection of fouling or disease agents at low concentrations in food and drugs. The methodology for detection of infectious disease agents with nucleic acid probes needs further refining so that it can become a routine, reliable clinical test (e.g., for the detection of viral infection in potatoes, "dot blot" methodology has already been applied).

Other Methods

Although the necessary competence in this field was lacking in our discussions, the apparent advantages of enzyme electrodes in terms of rapid analysis and the possibility of feedback coupling for the dosing of medicinals (e.g., insulin) make this an attractive method. The development of minute integrated circuits with coupled enzyme/antibody disposable electrodes seems to have reached a point where this method may be able to replace classical antibody-or enzyme-based assay methods.

Social Consequences of Improved Screening Methods

There is some concern that the development of diagnostics and their implementation in broad screening programs before a cure for the disease or treatment for an inheritable condition has been developed may cause more problems than it solves (for example, Tay-Sachs, good results; but sickle cell trait, potentially bad results). Current difficulties in the area of screening serum for HTLV III (AIDS virus) and notifying blood donors without certainty that disease will develop highlights this problem. However, since probes of this nature can often lead to a rapid analysis of the causative agent or biochemical defect, delay before obtaining treatment (if one exists) may be reduced.

One major problem with the development of a multitude of DNA—chromosome mapping markers will increasingly be the compilation and correlation of data on clinical application, linkage, etc. Attempts to coordinate this internationally are under way.

Animal Diseases

It was of prime concern that diagnostic tools be developed to monitor epidemics particularly with respect to serotype variation. At certain centers the sophisticated technique of in situ sequencing (Church and Gilbert) could be set up so that new variants could be rapidly sequenced and programs to produce new subunit antigens (or viral vaccines) immediately initiated.

Economic pressures make it likely that the sexing of chickens and prime breeding stock will be accomplished with new types of diagnostic probes. For public health reasons Toxoplasmosis should be screened in pets.

GENE TRANSFER

Under this heading the methods of gene transfer (Wagner et al., this volume) into a) fertilized eggs (by microinjection), b) embryonal stem cells (by retrovirus vectors), and c) bone marrow stem cells (by retrovirus vectors) were considered. In brief, the methods have been developed to answer basic biological problems about the differential gene expression during development, tissue-specific gene regulation, the function of oncogenes in cancer development (particularly suppression of cancerous growth by cell-cell interaction in developing embryos), and the analysis of complex systems as defined previously.

Gene Transfer to Fertilized Eggs: Transgenic Animals

With this method, developed in the mouse, the overall success rates (foreign gene expressed in adult mouse) are at the most a few percent relative to the initial number of fertilized eggs injected. With sheep and pig, for example, the methodology is still in its infancy due to intrinsic problems of identifying the pronucleus in the egg.

Goals for application of this methodology in domestic animals and birds include the stimulation of growth rates of cows and pigs and milk yields in cows by implanting growth hormone genes. Built-in resistance to pathogens (Marek's disease in chickens) can also be envisaged. However, the methodology has not yet been attempted for chickens, where the egg at laying already contains a 50,000 cell embryo.

The analysis of the mechanism for suppression of viral replication and expression during embryonal development should provide interesting insights into cell-mediated immunity. Animal models of human genetic diseases (e.g., Lesch Nyan's Syndrome in primates and β-thallasemia in mouse) are being studied with this "gene complementation" therapy at the germ line level.

Retrovirus Vectors for Gene Transfer into Stem Cell Systems

A great potential is predicted for this methodology in conjunction with embryonic stem cells and bone marrow stem cells by the use of oncogenes under tissue-specific promoters/enhancers. In this fashion, differentiation-specific immortalized cell lines would be produced which could be further manipulated with cloned genes for the production of tissue-specific products, presumably with the tissue-specific post-translation modifications required for optimal activity.

This method has the advantage of high efficiency, easy technical handling, and clean integration into the genome. It not only provides the basis for studies of the development of complex systems in animal models but is a stepping stone to application of somatic cell therapy in humans. At the present time, three experiments are awaiting approval in the U.S. for treatment of Lesch Nyan Syndrome, and the immune defects caused by purine nucleoside phosphorylase deficiency, and adenosine deaminase deficiency, via gene transfer into bone marrow stem cells. Although the incidence of these diseases is minimal (200 cases/year in the U.S. and 9 and 50 cases worldwide, respectively), later application to globin gene defects in β-thallasemia can be expected should the initial experiments be successful. Additional animal experiments are required to give a firmer basis for this technology before it is applied to humans. Due to the analogy with tissue transplantation therapy, no major ethical obstacle is seen here.

Directed Site-specific DNA Recombination

Although homologous recombination in bacteria and yeast is the preferred route to exchange resident and incoming genes, the usual situation in higher eukaryotes is to integrate the incoming DNA randomly in a nonreciprocal fashion in the genome. Rare site-specific integration events have been observed (10^{-5}) and theoretical protocols have been proposed for selecting for this event amongst expanded populations of stem cells which via reintroduction into early embryos might be introduced into the germ line. This method appears at present too uncertain and laborious to be applied, but it shows that in principle gene repair in the human germ line might be possible. Application of current methods with the high risk of disrupting other essential genes via random integration are ethically unacceptable. It should, however, once again be emphasized that the main application of these studies is in the analysis of cell development, gene regulation, and differentiation in animal systems.

CONCLUDING REMARKS

The use of manipulated genes and cells offers enormous potential in human medicine and veterinary science. Recombinant DNA technology will allow the preparation of products of high purity and in high yields.

Diagnostic and detection methods that are simple and inexpensive will be available through the development of new methods for the detection of peptide and nucleic acid sequences. The scientific or the general community must decide on the proper uses of these methods; diagnostic methods form the basis of effective and reasoned genetic counseling, but obligatory genetic screening for employment, life insurance acceptance, etc., would have social implications which are questionable.

The development of human and animal proteins is going to create new forms of therapy and diagnosis. The discovery of new physiologically active proteins and the rational modification and improvement of existing biologically active molecules will provide new pharmaceuticals for the treatment of cancer and diseases of the immune system, nervous system, cardiovascular system, etc. The need for expensive preclinical research and studies of drug formulation and delivery systems will force economic constraints on the use of "better" proteins, but efforts to reduce cost and expand availability to all members of the population are desirable.

Vaccines remain one of the best approaches to the control of infectious disease. From a global point of view, the world and particularly third world countries are still suffering from major endemic diseases, in both the human and food stock animal populations, with bacterial, viral, fungal, amoebal, and trypanosomal causative agents. That a successful eradication program can be accomplished has been demonstrated for smallpox. Gene and cell manipulations, together with a better understanding of the mechanisms of virulence/pathogenicity and immunity should lead to the development of cheaper and more effective vaccines for major pathogens. New presentations of antigens and the use of better adjuvants should enlarge the protected population; recombinant-live vaccines are particularly promising in their wide applicability and possibility of combining multiple antigens. World needs should be considered in vaccine research and distribution; the pharmaceutical industry must work with international organizations to meet these needs without economic constraints.

Many techniques to introduce, replace, or modify genetic material in germ cells and somatic cells have been established. The use of manipulated somatic cells in the therapy of human hereditary diseases is imminent. Beneficial modifications of the germ line of domestic animals are feasible. The enormous potential of this methodology raises ethical considerations, but prudent applications to serious diseases and applications to human medicine and veterinary husbandry will be of broad benefit.

RECOMMENDED READING

General

Bolon, A.P., ed. 1984. Recombinant DNA Products: Insulin, Interferon and Growth Hormone. Boca Raton, FL: CRC Press.

Caskey, C.T., and White, R.L. 1983. Recombinant DNA Applications to Human Disease. Banbury Report No. 14. New York: Cold Spring Harbor Laboratory.

Church, G.M., and Gilbert, W. 1984. Genomic sequencing. Proc. Natl. Acad. Sci. USA **81**: 1991-1995.

Lwoff, A., and Ullmann, A. 1979. Origins of Molecular Biology. New York: Academic Press.

Paul, J.K., ed. 1981. Genetic Engineering Applications for Industry (in particular the chapter entitled Medical aspects of molecular genetics). Park Ridge, NJ: Noyes Data Corporation.

Williamson, R. 1982. Genetic Engineering, vol. 3. London: Academic Press.

Supplementary

Celis, J.E., and Bravo, R. 1982. Gene Expression in Normal and Transformed Cells. New York: Plenum Press.

Chanock, R.M., and Lerner, R.A. 1984. Modern Approaches to Vaccines. Molecular and Chemical Basis of Virus Virulence and Immunogenicity. Cold Spring Harbor, NY: Cold Spring Harbor Press.

De Maeyer, E., and Schellekens, H. 1983. The Biology of the Interferon System. Amsterdam: Elsevier Science Publishers.

Lurguin, P.F., and Kleinhofs, A. 1982. Genetic Engineering in Eukaryotes. New York: Plenum Press.

Schultz, J., and Brada, Z. 1977. Genetic Manipulation as It Affects the Cancer Problem. New York: Academic Press.

REFERENCES

(1) Hillson, D.A.; Lambert, N.; and Freedman, R.B. 1984. Formation and isomerisation of disulphide bonds in proteins: protein disulphide isomerase. Meth. Enzym. **107**: 281-294.

(2) Leclerc, C.; Löwy, I.; and Chedid, L. 1978. Influence of MDP and of some analogous glycopeptides on the in vitro mouse spleen cell viability and immune response to sheep erythrocytes. Cell. Immunol. **38**: 286-293.

(3) Valenzuela, P.; Medina, A.; Rütter, W.J.; Ammerer, G.; and Hall, B.D. 1982. Synthesis and assembly of hepatitis B surface antigen particles in yeast. Nature **298**: 347-351.

Biotechnology: Potentials and Limitations, ed. S. Silver, pp. 139-154. Dahlem Konferenzen 1986. Berlin, Heidelberg, New York, Tokyo: Springer-Verlag.

Vaccines - the Synthetic Antigen Approach

D.J. Rowlands
Wellcome Biotechnology Ltd.
Pirbright, Woking
Surrey GU24 ONQ, England

Abstract. Technological and conceptual advances in the last few years have opened up a variety of new avenues towards improvement of existing vaccines and providing vaccines against diseases for which there is no protection currently available. These new approaches are discussed with particular emphasis on those involving synthetic peptide antigens.

INTRODUCTION

Protection against virus infection by vaccination has been extremely successful with both animals and humans in the prevention and control of a wide range of diseases since its introduction as a deliberate, controlled measure by Jenner in 1798. Since that time many of the most widespread and damaging diseases of humans and domestic livestock have been reduced to very low levels, and the crowning achievement was the elimination of smallpox from the world in 1977. It is a fitting tribute to the pioneering work of Jenner that smallpox should have been the first disease to be eradicated since it was to protect against this disease that he introduced the deliberate infection or vaccination of humans with the antigenically related cowpox or vaccinia virus. This replaced the earlier practice of deliberate infection with smallpox (variolation) which, not surprisingly, caused a high incidence of overt disease. The term vaccination has now become a more general term referring to any deliberate exposure of an animal or human to antigen(s) in order to provoke an immune response.

Cowpox is a naturally occurring virus which has a much lower virulence for humans than smallpox virus but which will replicate sufficiently well to induce a prolonged immunity against smallpox. Although the precise origin of the vaccinia strain used by Jenner is not clear, it can be thought of as a naturally occurring attenuated vaccine strain. Protection of chickens against Marek's disease by infection with turkey herpes virus is another example of the use of a naturally occurring virus to prevent disease caused by a more virulent pathogen. This situation is rare and the great majority of virus vaccines have been derived so as to expose the species to be protected to the relevant immunogens without the virulent effects associated with the disease agent. In all cases so far this has been achieved in one of two ways: a) by artificially attenuating the virus in the laboratory to derive a strain of greatly reduced virulence (e.g., Sabin polio, Theiler's yellow fever, LEP and HEP rabies, and Plowright's rinderpest vaccines), or b) by inactivating the virus so as to destroy its ability to replicate and cause disease while maintaining the important antigenic features intact (e.g., Salk polio and foot and mouth disease vaccines). A variation of the second approach is the production of subunit vaccines in which the important immunogenic proteins of a virus are separated and injected without the remainder of the virus particles. An example of this approach is the production of some human influenza vaccines.

Vaccination, together with the recognition and adoption of basic hygiene measures, has contributed enormously to the improvement of human and animal welfare, at least in the developed world. Most of the great infectious scourges and plagues of the past have been dramatically controlled. However, constant vigilance is required to maintain this situation and there is a lot to be done to improve current vaccines and to develop new ones to control those diseases for which none is available.

Potent vaccines are only one of the requirements for effective disease control and must be supported by accurate diagnostic methods to allow epidemiological studies so that the origin, incidence, and spread of disease can be monitored. This information can then be used to plan effective vaccination programs and, in the case of livestock, to design animal movement control measures to reduce the expansion of epidemics. Vaccination programs should provide sufficient herd immunity to restrict the spread of disease, and ideally a vaccine should not only protect individuals from disease but also prevent any peripheral replication of the disease agent and consequent danger of passing an infection to nonimmune contacts.

Although many vaccines have proved highly effective in the control and prevention of disease, all could be improved to a greater or lesser extent, depending on the particular example, with great potential benefit. The decision whether or not to vaccinate revolves around a risk/benefit analysis in which the costs and related risks are weighed against the risk of contracting disease. This analysis favors nonvaccination when a disease has been largely controlled and can lead to the expansion of a nonimmune population and so an ideal situation for the introduction of a new epidemic. From these considerations it can be seen that inexpensive, high quality vaccines with minimal associated risks of undesirable side effects and complications are vital components of long-term disease control and possible eradication.

Consideration of problems associated with current vaccines gives an indication of areas in which improvements are desirable.

Diseases for Which No Vaccine Is Available

The clearest examples of problems with vaccines are those for which no method has been described for cultivating the disease agent outside the natural host. The best example is the human liver disease caused by hepatitis B virus which affects millions of people throughout the world. The disease is most prevalent in underdeveloped parts of the world and is an enormous human health problem both from acute disease and associated sequelae such as hepatocarcinoma. The only vaccine currently available is produced from the serum of human carriers. The limited supply and very high cost of this vaccine means that it is only used for a few high risk people (e.g., medical and research workers) and is not available for improving the worldwide disease situation. Enormous effort is going towards the development of an effective hepatitis B vaccine with the new technologies now available. Hepatitis A virus is another major human disease agent which cannot be cultivated in amounts sufficient to produce a vaccine.

Attenuated Virus Vaccines

Most virus vaccines in use today are live attenuated strains of the disease agent. In general, such vaccines afford prolonged immunity and stimulate the entire repertoire of immune responses in the host including both humoral antibody- and cell-mediated immunity. Local IgA immunity at mucosal surfaces is produced which helps to prevent superficial replication of virus and hence spread of infection to other individuals.

The derivation of attenuated vaccine strains has hitherto been empirical, usually involving multiple passage of the virus in a foreign host or tissue culture until it is observed that disease is no longer caused by infection of the target host species. In no case are the changes in the virus genome which account for the loss of virulence understood, with the possible exception of reo virus 3. However, this is a purely laboratory model system of no practical significance. Recent sequencing studies have shown that the Sabin attenuated vaccine strains of poliovirus differ from their wild-type progenitors by a number of point mutations. In the case of poliovirus 1 there are 56 nucleotide changes and 21 amino acid changes, but with poliovirus 3 there are only 10 nucleotide changes and 3 amino acid changes. Which of these changes is crucial for the altered virulence phenotype is not yet known, but in vitro recombination involving infectious cDNA clones of the virus should soon solve this problem.

The crucial point is that the wild-type and attenuated viruses only differ by a relatively small number of point mutations and with the very high rate of mutation seen in RNA viruses (about 1 in 10^4), it is not surprising that virulent revertant viruses can sometimes be isolated from vaccines and have been shown to cause paralytic disease. Of course, even though the vaccinate may be protected, the shedding of revertant virulent virus can put nonimmune contacts at risk.

The derivation of attenuated vaccine strains is always a balance between reducing virulence to an acceptably low level while retaining adequate growth characteristics to provide the vaccinate with sufficient antigenic stimulation to mount a good immune response. The use of temperature-sensitive mutants to restrict virus replication to cooler regions of the body such as the nasal mucosa represents a more directed approach to the development of avirulent vaccine strains, but again the high mutation rate means that phenotypic reversion to virulence is likely to occur.

Another practical problem sometimes associated with live virus vaccines is the maintenance of the infectivity of the vaccine, particularly in the field situation. Some viruses are extremely stable during storage. Vaccinia, for example, will survive in the dried state for many years. Most viruses are much less stable and it is often found that the attenuated strains are more fragile than the wild type. Measles and poliovirus vaccines are two examples where storage problems can occur.

Another potential danger with live vaccines is that of contamination with other agents which may be present in the system used to cultivate the vaccine. This has even been known to be a problem with inactivated virus vaccines prepared from material grown in tissue culture cells. Batches of formaldehyde-killed Salk polio vaccine contained SV40 virus which had not been inactivated by the formaldehyde.

Inactivated Virus Vaccines

The use of killed virus vaccines raises a different set of problems, not the least of which is the need to produce a greater antigenic mass than with live vaccines, since the only material to which the immune system can respond is that present in the vaccine dose. Since the virus used is usually of full virulence, sophisticated production facilities may be required to produce and contain the large volumes of virus harvest required. A good example is foot and mouth disease virus (FMDV) for which the annual world requirement is about 2×10^9 doses. This corresponds to the production of about 10^7 liters of this highly contagious virus.

Application of inactivated vaccines is generally more complicated than live vaccines since they have to be injected, whereas many live vaccines can be applied orally or intranasally. In addition, immunity induced by killed virus vaccines is not usually as long-lived as that provided by live vaccines, and frequent booster injections may be required to maintain protection.

Injection of virus antigens does not give rise to a local IgA response at mucosal surfaces. Consequently, although an individual may be immune from disease, it is still possible that it may support replication of virus at limited sites and thus shed virus to the environment. For example, it has been shown that virus is shed following exposure to FMDV from the pharynx of cattle which have been immunized with killed FMDV vaccine but not from animals which have recovered from a previous infection.

Another potential problem with inactivated virus vaccines, which can also apply to attenuated live virus vaccines, is the selection of antigenic variants by culture in different hosts. It has been shown for both FMDV and for influenza that culture of the virus in BHK cells in the first case and eggs in the second can result in virus which is significantly different from the starting material. In some cases the variant selected by the culturing conditions may be sufficiently different from the original virus to affect seriously its efficacy as a vaccine.

The most crucial problem with killed vaccines is that of ensuring that all infectivity is destroyed. A tragic example of failure in this respect accounted for a large number of cases of paralytic poliomyelitis associated with the early Salk polio vaccinations. Incomplete inactivation has also been strongly implicated as the origin of at least one outbreak of FMD. Both of these examples involved vaccines inactivated with formaldehyde, which has been shown to follow nonlinear inactivation kinetics. More recent inactivating agents such as aziridines acetylethyleneimine (AEI) and binary-ethyleneimine (BEI) have a better record than formaldehyde, but vigilant monitoring of the vaccine is necessary to guard against disaster.

The ideal method of virus inactivation must not only ensure that infectivity is destroyed but also that antigenic properties are not altered. An example where this was not the case was with measles virus vaccines inactivated with formaldehyde. This altered the immunogenic character of the F protein and the vaccine afforded incomplete immunity. This became evident when it was observed that children receiving this vaccine developed a range of unusual symptoms which were associated with altered pathogenicity of measles infection due to partial immunity.

Antigenic Variation

Many disease agents evade host immune defence mechanisms by antigenic variation. The pool of human rhinoviruses, for example, contains more than 100 different serotypes. Rapid antigenic change is a characteristic of several RNA viruses such as influenza which evolves by both genetic reassortment and mutation, the antigenic shift and drift phenomena, and FMDV which occurs as seven distinct serotypes, each of which continuously evolves new variants. Another striking example of defence by variation is the trypanosome parasite which sequentially expresses a repertoire of a hundred or so antigenically distinct coat proteins.

There is, then, room for improvement in most if not all vaccines in current use and a great need for new vaccines for those diseases for which none is currently available. Dramatic advances in techniques in molecular biology and in immunology over the past few years have enabled the study of the structure and functioning of the immunogenic components of pathogens and of the various components of the immune system in great detail, and our recently acquired knowledge should soon see practical application.

NEW APPROACHES

The newer technologies are being applied to the problem of creating or improving vaccines by a variety of routes.

Cloning and Expressing Immunogenic Proteins

Sophisticated genetic engineering techniques have resulted in the production of large quantities of some foreign proteins in bacteria, principally Escherichia coli. However, the application of this technology to the production of immunogenic proteins has had very limited success. The majority of virus immunogens are membrane glycoproteins, and bacteria do not contain the appropriate mechanism for correct glycosylation and transportation of these proteins. The products are frequently degraded or are toxic to bacterial cells. A notable exception is the VP1 protein of FMDV (which is not a membrane protein) which has been expressed at high levels as a chimeric protein in E. coli and has been demonstrated to protect cattle against challenge with FMDV.

Expression in eukaryotes results in a better quality (i.e., glycosylated) product and hepatitis B surface antigen produced in yeast appears to have good potential as a practical vaccine.

This route for the production of novel vaccines will not be discussed further as the topic has been covered in detail elsewhere (Kay et al. and Fiers, both this volume).

Live Virus Vectors

One of the most exciting recent developments was the demonstration by Paoletti, Moss, and their colleagues that foreign proteins can be inserted into the genome of vaccinia and expressed during virus replication. This method appears to have great potential and is discussed in detail elsewhere (Paoletti et al., this volume).

Directed Attenuation

As mentioned earlier, there are many advantages associated with the use of live attenuated vaccines. However, the derivation of avirulent strains has been empirical and the molecular basis for attenuation is not understood. In some cases the frequency with which vaccine strains revert to full virulence is disquieting (e.g., poliovirus type 3). It is therefore necessary to perform elaborate and costly tests on each batch of vaccine before it can be used. It would be of great advantage to have more stably attenuated vaccine viruses in which the factor(s) responsible for the avirulent phenotype are precisely known and are amenable to direct determination. Theoretically, the best way to achieve this goal would be to make specific deletions in the genome which would maintain the viability and antigenic structure but reduce the pathogenicity of the virus. Alternatively, specific recombinants may be constructed which weld the immunogenic regions of the pathogenic agent into the genetic background of a benign virus to produce hybrid vaccine strains. Both methods would have the great advantage that virulence would not be gained by reversion of point mutations.

Although this approach is in its infancy, two separate examples deserve comment. First, it has been shown that destruction of the function of the vaccinia TK gene results in virus which is of considerably reduced virulence in laboratory tests involving mice and tissue culture (4). Although this is a long way from demonstrating that such virus would have a reduced incidence of complications in humans, it is nevertheless an encouraging observation. The second example involves one of the best studied viruses of eukaryotes, namely, poliovirus. Complete genome sequences of the wild and attenuated strains of all three serotypes of this virus are available and show that the attenuated strains differ from their parents by a number of point mutations. The Sabin strain of polio type 3 differs from its neurovirulent progenitor by only 10 nucleotide substitutions, only 3 of which cause amino acid substitutions (23). The corresponding differences in the case of polio 1 are 56 and 21, respectively (20), and it is probably of significance that polio type 3 vaccines are the most problematical in terms of reversion to virulence. The fact that infectious poliovirus can be recovered from full-length cloned copies of the genome means that specific recombinants of wild and vaccine strains can be made, and this method is being used to define those changes which are responsible for the neurovirulence phenotype. An interesting fact to emerge from this work is that a single nucleotide substitution in the 5' non-coding region of the genome always correlates with the neurovirulence score of the virus. Reversion to the wild-type nucleotide at this position is consistently seen within hours of infection of the human gut with vaccine virus, and this single change increases the neurovirulence score of the virus, although not to wild-type levels (5). From this sort of study it appears possible that less mutable alteration of this region of the poliovirus genome may produce more stably attenuated vaccine strains.

Synthetic Peptides

The observation that short synthetic peptides can induce the production of antibodies that recognize the intact protein from which the peptide sequence was derived has stimulated a lot of interest in the possible use of such materials as vaccines (15). The remainder of this paper will deal with this aspect of vaccine development.

It has been possible chemically to synthesize peptide chains of defined amino acid sequence for many years, and as long ago as 1963 it was shown that short peptides could induce the production of antibodies capable of recognizing the intact protein from which the peptide was derived. Until recently, however, the number of proteins for which the complete amino acid sequence was known was extremely limited. That situation has changed dramatically with recent DNA sequencing, and a huge library of protein sequences, including many of the important immunogenic proteins of pathogenic agents, is now available. As a consequence, there is a great interest in the use of synthetic peptides to study the antigenic structure of proteins, the interaction of the various branches of the immune system, the expression of proteins for which only a predicted sequence from nucleic acid studies might be available, and the potential of peptides as vaccines.

Antibodies induced by a native protein can recognize either sequential determinants comprising linear portions of the peptide chain or conformational determinants made up from two or more regions separated on the linear sequences but brought into juxtaposition on the three-dimensional structure of the native, folded protein. The great majority of synthetic peptide antigens which have been studied represent short linear portions of the primary protein sequence and so cannot mimic complex conformational determinants. These synthetic antigens therefore have the potential to closely resemble natural linear antigenic determinants or to present novel structures, not normally seen as immunodominant by the immune system in the context of the intact protein but capable of inducing antibodies which will react with the native protein.

Choice of peptide sequence

Of the antigenic determinants present on any virus, usually only a restricted number elicit the production of neutralizing antibodies and are of significance in immunization. It is therefore of great importance that the most suitable amino acid tracts are chosen for synthesis, and a number of methods have been used to predict the location of important immunogenic sites. These include:

Antigenic activity of protein fragments. In this approach an immunogenic protein is degraded by specific enzymatic and/or chemical cleavage and the fragments tested separately for immunogenic activity. This can be assayed by immunizing with the fragments or by measuring their reaction with neutralizing monoclonal antibodies. Limited N terminal sequencing of the fragments serves to locate their position on the intact protein, provided that the complete primary sequence is known.

This method was successfully employed by Strohmaier et al. (24) to locate two tracts (amino acids 146-154 and 200-213) on the 213 amino acid sequence of FMDV VP1 which are within immunogenically important regions.

Regions of sequence hypervariability. Many viruses display a high degree of antigenic variation and, to a first approximation, it is reasonable to assume that regions of high sequence variation may equate with regions of immunogenic importance. Comparative analyses of this kind have helped to define a number of important epitopes on the HA molecule of influenza.

Sequence studies of the VP1 protein of different serotypes, subtypes, and strains of FMDV have revealed three major regions of variation located at positions 41-60, 134-158, and 195-213 (16). The 134-158 region is characterized by a high frequency of codon insertions/deletions in addition to amino acid substitution. Further evidence for the major importance of the 134-158 region was obtained from analysis of closely related but serologically distinguishable natural variant viruses which were found to differ at only two positions, 148 and 153 within the entire structural protein coding region. Peptides representing tracts 141-160 and 200-213 induce high levels of virus-reactive antibody, although anti 141-160 serum has far higher specific neutralizing activity (2). However, peptide 41-60 does not elicit antibodies which recognize intact virus particles (see below).

Sequence changes selected by monoclonal antibodies. This method relies on the same rationale as the previous one and differs only in that antigenic variants for sequence comparison are deliberately derived by selection of mutations under neutralizing monoclonal antibody pressure. Point mutations are selected in this way and so would be expected to locate precisely important epitopes. Application of this method to influenza virus has reinforced the location of epitopes deduced from analysis of natural variants (3).

The picture with poliovirus is confusing in that substantially different results have been obtained with different serotypes. With poliovirus type 3, the great majority of mutations selected by a panel of neutralizing monoclonal antibodies map in a small discrete region of VP1 (17). With poliovirus type 1, however, monoclonal antibodies which are known from cross-linking experiments to recognize sequences within VP1 frequently select neutralization-resistant variants which have mutations in structural proteins other than VP1 (7). A similar observation has been made with FMDV where neutralizing monoclonal antibodies which recognize epitopes in the C terminal half of the VP1 protein select viruses with mutations distal to these regions (N. Parry and D.J. Rowlands, unpublished observation). The simplest explanation for these examples in which non-epitopic mutations are selected is that the monoclonal antibodies recognize complex conformational determinants, as has been shown to be the case for the FMDV monoclonal antibodies. Distal mutations may alter the tertiary structure of the virus capsids in such a way as to change the spatial relationships of the component parts of the epitopes and render them unrecognizable by the antibodies. From this argument it may be tentatively concluded that the poliovirus type 3 monoclonal antibodies recognize a predominantly linear determinant. These examples serve to emphasize the caution with which sequence comparison data must be interpreted in the prediction of antigenic sites.

Prediction of antigenic sites from tertiary structure considerations. A complete solution of the tertiary structure is only available for a limited number of proteins and only one, the hemagglutinin of influenza virus, has significance for vaccination. However, empirically derived methods are available for the prediction of important features of protein such as hydrophilicity and regional peptide chain structure such as α-helix, β-sheet and β-turns. Hydrophilic regions of proteins tend to be orientated towards the surface and are therefore more likely to be antigenic determinants than hydrophobic, buried regions. In fact, the majority of synthetic peptides which have been shown to induce antibodies which recognize native proteins represent hydrophilic regions of the intact proteins. Also, α-helical regions and β-turns may provide surface features which increase the antigenicity of such regions. These methods have been used to predict antigenic sites on a number of proteins including FMDV VP1 and hepatitis B surface antigen. In the case of FMDV, an α-helical region adjacent to β-turn was chosen. This region coincides with the major active site predicted by other methods including hydrophilicity plots and was shown to be active as a peptide immunogen (22). In the section on Regions of sequence hyperavailability, it was mentioned that three regions of high sequence variation were found on FMDV VP1 but that peptides to one of these (41-60) did not induce virus-reactive antibody. Hydrophilicity plots of the protein show that both of the regions which are immunologically active as peptides are hydrophilic, whereas the 41-60 tract corresponds to a hydrophobic trough suggesting that it may be buried in the protein. Tertiary structure prediction methods have also been successfully used to locate the active sites on a number of immunogenic proteins, e.g., hepatitis B surface antigen.

Prediction of antigenic sites from overlapping peptide sets. The most comprehensive method for determining the appropriate peptides to mimic the antigenicity and immunogenicity of complete protein is to synthesize the protein in its entirety as a series of overlapping peptides and examine these independently. A complete examination of the immunogenicity of such a set of peptides would be an extremely laborious and costly task; however, a rapid method of synthesizing small quantities of large numbers of peptides for analysis of their antigenic activity with antisera has been devised. The complete sequence of FMDV VP1 has been synthesized as an overlapping series of hexapeptides, each separated from its neighbor by one amino acid. All of the 208 peptides were tested for their ability to react with antiserum raised to the intact virus particle. Positive reactions were only obtained with two overlapping regions at amino acids 146-151 and 147-152 which, it will be recalled, are within the regions shown to have antigenic activity by other methods (12). Since it has been shown by other methods that at least one other region (160-180) has antigenic activity (F. Brown, unpublished observation), it appears that this screening procedure provides only a minimum estimate of antigenically important regions, perhaps due to the short length of the peptide probes.

An ingenious variation on the theme of screening antibodies with peptide probes has been developed by Nunberg et al. (21). They digested cloned cDNA coding for the glycoprotein of feline leukemia virus with DNAse to produce a large number of short random fragments. These were then subcloned into a lamda phage expression vector and the resulting products, composed of carrier protein terminated by glycoprotein peptides, screened for reaction with neutralizing monoclonal antibodies. As the subcloned fragments were derived by random cleavage, an average of one in six should be joined to the vector in the correct reading frame and orientation. Constructions giving a positive reaction with antibody were then sequenced to locate the peptides within the feline leukemia virus genome.

It can be concluded that there is no reliable method for the accurate prediction of peptide sequences which may be efficacious as synthetic vaccines. Recent indications that the most mobile regions within native proteins are the best targets to aim for are very interesting, but there is no simple way at present to locate such sites on proteins.

Peptide conformation

Peptides of 20-30 amino acids or less are too small to hold stable secondary conformations in aqueous solutions and so may present the immune system with a continuous array of antigenic shapes. The extent to which coupling of peptides to carrier molecules may restrain flexibility to produce a more limited variety of potential determinants is unknown, as is the possible restriction to mobility imposed by interaction with cell proteins and membranes in the body. The data available concerning the spectrum of antibodies raised against synthetic peptides compared to those induced by the same sequences as part of the intact protein are confusing. In one study monoclonal antibodies were raised against a synthetic peptide representing part of influenza hemagglutinin and checked for their reaction with the native protein. A surprisingly high proportion (75%) of these antibodies recognized both peptide and intact protein (19). On the other hand, when polyclonal antisera to a FMDV peptide capable of inducing immunity were absorbed with intact virus, no loss of ELISA titer against the peptide was found (N. Parry, unpublished observation). This would indicate that the proportion of peptide antibodies which recognize the same sequence in intact virus was 10% or less of the total population. Also, when these peptide antisera were assayed by ELISA against peptide or intact virus the titer was ≃ 20-fold different in favor of the peptide antigen. These observations are supported by the results of experiments involving replacement sets of peptides. It was found that when antisera to intact FMDV or to a synthetic peptide were assayed for reactivity with a set of peptides based on the original sequence but with each amino acid in turn substituted for the other nineteen, the antivirus serum recognized a much more restricted set of peptide variants than the peptide antisera (13). It can be concluded from these examples that restriction of the peptide structure so that a greater proportion mimicked the secondary structure characteristics of the native protein would greatly improve the synthetic antigen.

Significant improvement has been reported in the immunogenicity and antigenicity of a synthetic peptide representing amino acid residues 122-137 of hepatitis B surface antigen by cyclizing via the thiol groups of two cysteines present in the sequence (8). These cysteines are linked by a disulfide bond in the native protein, so it is not surprising that a peptide including this bond may more closely resemble the native determinant. In another example, that of peptide 89-104 of poliovirus VP1, a significant virus-neutralizing titer was only seen when cysteines not present in the protein sequence were added to each end of the peptide (10).

Clearly there is a lot of scope for assessing the effects on limiting the conformational alternatives open to synthetic peptides by introducing cross-linkages and by anchoring both ends of the tract to a carrier. However, at present there is no way of making rational predictions of the advantageous or deleterious effects such manipulation may have on the value of peptide vaccines.

Induction of antibodies to immunorecessive sites

An impressive property of synthetic peptide antigens is that they can elicit antibodies which react with regions of a protein which are nonimmunogenic in the intact protein. This has been demonstrated for influenza hemagglutinin (14). Antibodies raised to a series of peptides representing 75% of the protein sequence were all found to react with the native protein. These peptides represented both the stem and globular head domains. Antibodies against native protein are restricted to epitopes on the globular domain (3). One of the peptides from the stem region induced reasonable levels of neutralizing antibody which, in contrast to whole hemagglutinin antibodies, displayed broad cross-reactivity with different strains of virus (18). This cross-reactivity reflects the sequence conservation found in the region of the HA represented by the peptide.

These observations may have far-reaching importance for the future of synthetic vaccines, particularly for viruses characterized by rapid antigenic evolution. It would seem reasonable that some domains on virus surfaces must remain relatively invariant to preserve vital functions such as cell attachment. If such regions are normally immunosilent, it may be possible to induce broadly reactive antibodies by presenting them as peptides out of the context of the intact protein or virion structure.

Priming with peptides

Another potential application of peptide vaccines has been revealed by the work of Emini et al. (9). It was found that although peptides representing tracts from the VP1 protein of poliovirus would not induce neutralizing antibodies themselves, they primed animals for an enormous anamnastic response to a subsequent subimmunizing dose of virus. This priming was only seen after several injections of large amounts of peptide which induced a high antipeptide titer. It will be interesting to see whether such priming with peptide will protect against challenge with virus even though the virus neutralization titer is low or nonexistent at the time of challenge.

Somewhat different results have been obtained with FMDV peptides (11). Good neutralization titers and protection can be induced by the VP1 141-160 peptide from this virus, and so much lower doses than those used in the polio experiments are required to induce a priming response. It was found that low doses of peptide would prime the immune system for anamnastic neutralizing antibody response following a subsequent injection of peptide. A boosted response was also seen when peptide was injected into animals primed with virus. However, in contrast to the poliovirus results, peptide did not prime for a boosted response with virus. It can be concluded that there is still a long way to go before we have a clear understanding of the nature of natural antigens and their relationship to peptide analogues.

Requirement for carrier

In most cases synthetic peptide antigens have been coupled to protein carrier before formulation into vaccine. Although a variety of proteins have been used for this purpose, keyhole limpet hemocyanin (KLH) has been most commonly employed as carrier because of its large size and "foreignness" to mammalian species. Typically equal weights of peptide and protein have been complexed, which represents about 30 copies of peptide of about 20 amino acids per protein molecule. The methods of coupling most commonly used are with glutaraldehyde or via the thiol group of a cysteine residue added to one end of the sequence.

In our experience with FMDV peptides, the role of carrier in enhancing immune responses to peptides is simply to increase the size of the immunogenic particle and/or to increase epitope density. This conclusion has been drawn from a variety of observations. First, in experiments in which animals were exposed to a small priming dose of peptide coupled to KLH which did not give a measurable primary response, a good anamnastic boost was observed after a second injection of peptide, whether this was coupled to KLH, tetanus toxoid, or uncoupled (11). The response was not carrier-restricted, unlike classical haptens. In addition, good primary responses have been achieved with polymerized peptide and with peptide incorporated into liposomes.

Totally synthetic branched chain polymers have also been successfully used as carrier molecules for synthetic peptide antigens. It can be concluded that there is no real requirement to use complex protein carriers for peptide vaccines.

Adjuvants and vaccination regimens

Most synthetic peptide antigens have been formulated into vaccines with Freund's complete adjuvant. This preparation of Mycobacterial cells in a water-in-oil emulsion is a powerful stimulant of immune responses but cannot be used in practical vaccine formulations because of undesirable reactions such as abscesses. The potentiating effects of Freund's complete adjuvant can be simulated with small synthetic molecules such as muramyl dipeptide, MDP (1). Incorporation of this compound in the totally synthetic antigen, carrier, and adjuvant complexes has been reported markedly to enhance immune responses. Butyl derivatives of this compound have been shown to retain the adjuvant properties while losing the pyrogenicity associated with MDP (6). We have found with FMDV peptide that good responses can be obtained with Freund's incomplete (a water-in-oil emulsion) or with aluminum hydroxide adjuvants and with no adjuvant when the peptide is incorporated in the liposomes.

Although protection of guinea pigs against challenge with FMDV has been obtained after a single inoculation of 100 μg of synthetic peptide, much higher responses have been obtained with two injections of smaller doses separated by a few weeks. In view of this observation, it is important to examine the possibility of incorporation of synthetic peptide antigens in the delayed release formulations to overcome the disadvantageous requirement for multiple injections to obtain an optimum response.

PROSPECTS

Synthetic peptide vaccines have been shown to produce protective immunity against some diseases. Furthermore, since the nature of antibodies elicited by short, linear sequences differ qualitatively from the antibodies produced in response to intact proteins, the theoretical possibility arises that in some cases the synthetic anti-peptide antibodies may have superior characteristics. To a limited extent this has been demonstrated, in that more cross-reactive antibodies have been produced by some synthetic antigens than are found in sera to the intact protein. A major problem which remains is the choice of sequence for synthesis. The ideal region may or may not be that which is recognized as a dominant antigen on the intact protein. Also, the optimum length of peptide can at present only be deduced empirically.

There are many ways in which the conformational spectrum which can be adopted by a small peptide antigen may be restricted in order to maximize the response to the "correct" shape. These include the introduction of disulfide bridges, linkage of both ends to a substrate, and replacement of amino acids of the natural sequence with, for example, proline. Again, these approaches must be largely empirical at present.

Another aspect of this technology which is beginning to receive attention is the T cell response that peptides induce. A stimulation of T helper cell activity is necessary to optimize antibody production by B cells. In addition, with those diseases in which a cellular immunity is of major importance in recovery, a good stimulation of the cytotoxic arm of the immune system is desirable.

Clearly, there are problems associated with the synthetic peptide approach to vaccines. However, they have a number of features which make this pursuit worthwhile. They would have advantages in terms of stability, purity, and definition of product, and flexibility in that they can be conjugated to a wide range of carrier systems and adjuvants as appropriate. Multivalent peptide vaccines may be incorporated into a single preparation, and response to antigenic variation should be easily achieved by changing sequences appropriately.

REFERENCES

(1) Audibert, F.; Jolivet, M.; Chedid, L.A.; Arnon, R.; and Sela, M. 1982. Successful immunization with a totally synthetic diphtheria vaccine. Proc. Natl. Acad. Sci. USA **79:** 5042-5046.

(2) Bittle, J.L.; Houghten, R.A.; Alexander, H.; Shinnick, T.M.; Sutcliffe, J.G.; Lerner R.A.; Rowlands, D.J.; and Brown, F. 1982. Protection against FMD by immunization with a chemically synthesized peptide predicted from the viral nucleotide sequence. Nature **298:** 30-33.

(3) Both, G.W.; Sleigh, M.J.; Cox, N.J.; and Kendal, A.P. 1983. Antigenic drift in influenza virus H3 haemagglutinin from 1968 to 1980: Multiple evolutionary pathways and sequential amino acid changes at key antigenic sites. J. Virol. **48:** 52-80.

(4) Buller, R.M.L.; Smith, G.L.; Cremer, K.; Notkins, A.L.; and Moss, B. 1985. Infectious vaccinia virus TK minus hybrid recombinants which express foreign genes are less virulent than wild type virus in BALB/cBYs mice. In Vaccines '85, eds. R.M. Chanock, R.A. Lerner, and F. Brown, pp. 163-167. New York: Cold Spring Harbor Laboratory.

(5) Cann, A.J.; Stanway, G.; Hughes, P.J.; Minor, P.D.; Evans, D.M.A.; Schild, G.C.; and Almond, J.W. 1984. Reversion to neurovirulence of the live attenuated Sabin type 3 oral poliovirus vaccine. Nucl. Acids Res. **12:** 7787-7792.

(6) Chedid, L.A.; Parant, M.A.; Audibert, F.M.; Riveau, G.J.; Parant, F.J.; Lederer, E.; Choay, J.P.; and Lefrancier, P.L. 1982. Biological activity of a new synthetic muramyl peptide adjuvant devoid of pyrogenicity. Infect. Immun. **35:** 417-424.

(7) Diamond, D.C.; Jameson, B.A.; Emini, E.A.; and Wimmer, E. 1985. Antibody resistant variants of poliovirus type 1. In Vaccines '85, eds. R.M. Chanock, R.A. Lerner, and F. Brown, pp. 163-167. New York: Cold Spring Harbor Laboratory.

(8) Dreesman, G.R.; Sparrow, J.T.; Kennedy, R.C.; and Melnick, J.L. 1984. Immunogenic and antigenic activities of a cyclic synthetic HBsAg peptide. In Modern Approaches to Vaccines, eds. R.M. Chanock and R.A. Lerner, pp. 115-119. New York: Cold Spring Harbor Laboratory.

(9) Emini, E.A.; Jameson, B.A.; and Wimmer, E. 1983. Priming for and induction of anti poliovirus neutralizing antibodies by synthetic peptides. Nature **304:** 699-703.

(10) Ferguson, M.; Evans, D.M.A.; Magrath, D.I.; Minor, P.D.; Almond, J.W.; and Schild, G.C. 1985. Induction by synthetic peptides of broadly reactive, type specific neutralizing antibody to poliovirus type 3. Virology **143:** 505-515.

(11) Francis, M.J.; Fry, C.M.; Rowlands, D.J.; Brown, F.; Bittle, J.L.; Houghten, R.A.; and Lerner, R.A. 1985. Priming with peptides of foot and mouth disease virus. In Vaccines '85, eds. R.M. Chanock, R.A. Lerner, and F. Brown, pp. 203-210. New York: Cold Spring Harbor Laboratory.

(12) Geyson, H.M.; Meloen, R.H.; and Barteling, S.J. 1984. Use of peptide synthesis to probe viral antigens for epitopes to a resolution of a single amino acid. Proc. Natl. Acad. Sci. USA **81:** 3998-4002.

(13) Geyson, H.M.; Barteling, S.J.; and Meloen, R.H. 1985. Small peptides induce antibodies with a sequence and structure requirement for binding antigen comparable to antibodies raised against the native protein. Proc. Natl. Acad. Sci. USA **82:** 178-182.

(14) Green, N.; Alexander, H.; Olson, A.; Alexander, S.; Shinnick, T.M.; Sutcliffe, J.G.; and Lerner, R.A. 1982. Immunogenic structure of the influenza virus hemagglutinin. Cell **28**: 477-487.

(15) Lerner, R.A. 1982. Tapping the immunological repertoire to produce antibodies of predetermined specificity. Nature **299**: 592-596.

(16) Makoff, A.J.; Paynter, C.A.; Rowlands, D.J.; and Boothroyd, J.C. 1982. Comparison of the amino acid sequence of the major immunogen from three serotypes of foot and mouth disease virus. Nucl. Acids Res. **10**: 8285-8295.

(17) Minor, P.D.; Schild, G.C.; Bootman, J.; Evans, D.M.A.; Ferguson, M.; Reeve, P.; Spitz, M.; Stanway, G.; Cann, A.J.; Hauptmann, R.; Clarke, L.D.; Mountford, R.C.; and Almond, J.N. 1983. Location and primary structure of a major antigenic site for poliovirus neutralization. Nature **301**: 674-679.

(18) Muller, G.M.; Shapiro, M.; and Arnon, R. 1982. Anti influenza response achieved by immunization with a synthetic conjugate. Proc. Natl. Acad. Sci. USA **79**: 569-573.

(19) Niman, H.L.; Houghten, R.A.; Walker, L.E.; Reisfeld, R.A.; Wilson, I.A.; Hogle, J.M.; and Lerner, R.A. 1983. Generation of protein-reactive antibodies by short peptides is an event of high frequency: Implications for the structural basis of immune recognition. Proc. Natl. Acad. Sci. USA **80**: 4949-4953.

(20) Nomato, A.; Omata, T.; Toyoda, H.; Kuge, S.; Hanie, H.; Kataoka, Y.; Genba, Y.; Nakano, Y.; and Imura, N. 1982. Complete nucleotide sequence of the attenuated poliovirus Sabin 1 strain genome. Proc. Natl. Acad. Sci. USA **79**: 5793-5797.

(21) Nunberg, J.H.; Gilbert, J.H.; Rodgers, G.; Snead, R.M.; Nitecki, D.; and Winston, S. 1985. Localization of a determinant of virus neutralization on feline leukaemia virus envelope protein GP70. In Vaccines '85, eds. R.M. Chanock, R.A. Lerner, and F. Brown, pp. 221-226. New York: Cold Spring Harbor Laboratory.

(22) Pfaff, E.; Mussgay, M.; Bohm, H.O.; Schulz, G.E.; and Schaller, H. 1982. Antibodies against a preselected peptide recognize and neutralize FMDV. EMBO J. **1**: 869-874.

(23) Stanway, G.; Hughes, P.J.; Mountford, R.; Reeve, R.; Minor, P.D.; Schild, G.C.; and Almond, J.W. 1984. Comparison of the complete nucleotide sequences of the genomes of the neurovirulent poliovirus P3/Leon/37 and its attenuated Sabin vaccine derivative P3/Leon 12a,b. Proc. Natl. Acad. Sci. USA **81**: 1539-1540.

(24) Strohmaier, K.; Franze, R.; and Adam, K.-H. 1982. Location and characterization of the antigenic portion of the FMDV immunizing protein. J. Gen. Virol. **59**: 295-306.

Biotechnology: Potentials and Limitations, ed. S. Silver, pp. 155-164. Dahlem Konferenzen 1986. Berlin, Heidelberg, New York, Tokyo: Springer-Verlag.

A Modern Approach to Live Vaccines: Recombinant Poxviruses

E. Paoletti, M.E. Perkus, A. Piccini, B.R. Lipinskas, and S.R. Mercer
Laboratory of Immunology
Wadsworth Center for Laboratories and Research
New York State Dept. of Health
Albany, New York 12201, USA

Abstract. A technique for preparing live recombinant vaccines is described. The technique is a blend of old and new technologies. Vaccinia virus, used for almost two hundred years in the immunoprophylaxis of smallpox, has been engineered by recombinant DNA technologies to express foreign genetic information derived from heterologous pathogens. This recombinant live vaccine virus has been shown to elicit important immunological responses to these foreign antigens on inoculation of the recombinant virus into animals. Significantly, a number of studies have shown that vaccination of laboratory animals with these recombinant viruses results in protecting these animals against disease on subsequent challenge with the heterologous infectious agent. Vaccinia virus recombinants expressing the influenza virus hemagglutinin, the herpes simplex virus glycoprotein D, the hepatitis B virus surface antigen, the rabies virus glycoprotein, and a malarial parasite antigen are described and the biological properties of these recombinant viruses as live immunogens are detailed. A brief description of the problems and future prospects is included.

INTRODUCTION

In 1980 the World Health Assembly adopted a resolution stating that smallpox, as an infectious disease, had been eradicated from the world. This unique event in human history, the deliberate global eradication of an important infectious human disease, was made possible by two givens: a) the causative agent of smallpox, variola, was infectious only for humans; no other animal reservoir from which the virus could recrudesce existed; and b) the availability of the smallpox vaccine virus, vaccinia. Vaccinia, from which the words vaccine and vaccination derive, was introduced in 1796 by Edward Jenner, who observed that it was sufficiently related to the causative agent of smallpox that inoculation with this virus would prevent the smallpox disease. Indeed, it was the unique properties of the vaccinia virus vaccine that allowed the success of the global smallpox eradication program. The vaccine was inexpensive to produce and it could be stabilized as a freeze-dried preparation that could be shipped to all parts of the world in the absence of refrigeration. Administration of the vaccine did not require highly trained medical personnel nor sophisticated, expensive medical equipment. The vaccine was administered as a simple dermal abrasion by use of a bifurcated needle.

Additional vaccines directed against other infectious diseases were slow in coming and, indeed, not until the late 1950s was another live virus vaccine introduced for general human use: Sabin's live vaccine for poliomyelitis.

The proposal has been made and has initially been supported by laboratory experimentation that the smallpox vaccine virus is yet useful as a genetically engineered vaccine directed against a host of heterologous infectious diseases. This protocol involves the translocation into the genome of infectious vaccinia virus, genetic information from heterologous disease agents. This foreign genetic information would be expressed under vaccinia virus regulation into foreign antigenic material which would in turn elicit an immunological response in the inoculated host, rendering the host immune to the heterologous infectious agent. A variety of foreign genes have now been expressed by vaccinia virus recombinants. Appropriate immunological responses to these foreign antigens and protection of vaccinated laboratory animals against disease by subsequent challenge with the pathogens have been demonstrated in a number of cases showing the feasibility of this approach.

In this paper we will try to give the pertinent background information on the molecular biology of vaccinia virus, the basic genetic engineering protocols required for expression of foreign genes in vaccinia virus vectors, examples of recombinant vaccinia viruses with emphasis on their use as live vaccines, and finally, a brief discussion of future prospects and potential pitfalls pertinent to this kind of approach.

BIOLOGICAL PROPERTIES OF THE SMALLPOX VACCINE VIRUS, VACCINIA

A number of strains of vaccinia virus exist but the riddle of the origin of vaccinia virus has not been solved (1). Vaccinia virus is a rather unique virus. It does not circulate in nature but rather is maintained under laboratory cultivation. The absence of any reservoir allows the distribution of this virus outside the laboratory to be a matter of choice. The virus is dermatotrophic, giving a localized infection resulting in the well-known smallpox vaccination scar. As a vaccine the virus is exceptionally safe, although a few but serious side reactions can occur. Two of the adverse side effects which can occur in immunosuppressed individuals or in individuals with skin conditions such as eczema can be greatly reduced or virtually eliminated simply by good medical practice. Contact transmission to unvaccinated individuals is also a possible side effect and is particularly serious when transmitted to individuals cited above or as an ocular intrusion. However the relatively low infectivity of the virus speaks against large-scale contact transmission complications. Perhaps the most serious complication of smallpox vaccination is post-vaccinal encephalitis wherein fatal incidences can be documented. The availability of a number of vaccine strains which have been attenuated to reduce further the incidence of adverse reactions and the potential ability to engineer genetically the vaccina virus to even more attenuated levels provide reasonable approaches for addressing the issue of safety. Furthermore, one must now factor into the equation the cumulative benefit to be derived in successfully vaccinating an individual against a number of heterologous pathogens versus the risk of a single vaccination intrusion as would be obtained by the use of a recombinant vaccina virus expressing multiple foreign antigens. Vaccinia virus has a very broad host range providing potential usage as a vaccine vector not only for humans but also for veterinary applications.

At the molecular level, vaccinia virus is clearly a unique species of the poxvirus family. The genetic information of the virus is encoded within approximately 180Kb of DNA. On infection of a cell, the viral genetic information is expressed in a very tightly regulated fashion. Early proteins are synthesized from RNA transcribed by a virion-associated DNA-dependent RNA polymerase. These early protein functions allow for the replication of the viral DNA which then triggers another stage of macromolecular expression, the synthesis of a series of late viral proteins. The temporal regulation of expression of viral functions culminate in the maturation and packaging of the virus into infectious particles. Most of these infectious progeny viruses remain intracellular until the cell lysis, but a small percent of the infectious virus is released from the cell prior to lysis. Significantly, the small amount of virus released prior to cellular lysis has a recognizably different membrane composition than the intracellular virus.

Unlike many other animal viruses, the naked DNA of vaccinia is not by itself infectious when introduced into cells. This is probably due to several unique characteristics of the virus: a) the virion contains its own DNA-dependent RNA polymerase and other pertinent enzymatic functions that allow its genetic expression, b) there apparently exists a rather stringent barrier due to unique regulatory signals found in the virus and those found in the cell, therefore, reciprocal recognition and utilization of these regulatory signals is diminished or prevented. The cellular enzymes apparently cannot express genetic information presented to the cell in the form of naked vaccinia DNA and therefore initiate infection, and c) there are apparently considerable differences in codon utilization between vaccinia and the cell.

These properties dictated the development of protocols unique to poxviruses for their genetic manipulation such as eukaryotic cloning and expression vectors (5, 10, 16).

A GENERAL PROTOCOL FOR THE CONSTRUCTION OF POXVIRUS VECTORS

The recovery of inactive genetic elements into infectious progeny vaccinia by marker rescue techniques (8, 17, 23) suggested that extensions of these protocols would allow the insertion and expression of foreign genetic elements by recombinant vaccinia viruses. The large size and fragility of the intact vaccinia genome precludes its facile manipulation in the test tube. Therefore, a cloned subgenomic fragment of vaccina is selected. A foreign genetic element, ideally containing only the open reading frame of interest, is inserted at a locus of the cloned vaccinia fragment which is known to contain genetic information not essential for viral replication. Furthermore, at the extremities of the foreign coding sequence appropriate regulatory signals which ascribe transcriptional initiation and termination functions are engineered so as to allow functional expression of the foreign DNA insert. This chimeric construct is introduced into eukaryotic cells by transfection procedures. The cells are additionally infected with vaccinia virus. When a replicating DNA molecule from the input virus finds itself in close proximity to the chimeric transfected donor DNA, then in vivo recombination occurs between the vaccinia DNA sequences flanking the foreign DNA and the homologous DNA sequences on the vaccinia genome. This recombinational event allows for the insertion of the foreign DNA into a novel recombinant DNA molecule. This novel DNA molecule enters the pool of replicating DNA molecules in the infected cell and can itself participate in the subsequent maturational steps and be packaged into a novel, live recombinant vaccinia virus.

These novel recombinant vaccinia viruses can be identified by a variety of techniques and pure populations prepared readily. The recombinant vaccinia viruses can be identified on the basis of expression of the foreign gene: for example, a) if the foreign gene has a biochemically selectable function such as the expression or lack of expression of thymidine kinase activity, the recombinants can be identified by selection in appropriate biochemically selective media; b) if the foreign insert expresses a function such as glycoprotein that localizes itself on the membrane of the infected cell, then such a cell harboring the recombinant vaccinia virus can be identified by serological methods and infectious recombinant virus recovered; c) if the recombinant virus harbors a genetic element that is able to utilize chromogenic substrates, then recombinant virus can be identified by simply scoring for color development; d) if the generation of a recombinant virus is linked to a temperature-sensitive function, then recombinant viruses can be identified by growth under permissive temperature, and so on. A generic approach, independent of any assayable genetic function and based solely on the presence of the foreign DNA sequence, has been developed to identify recombinant vaccinia virus and described in detail (10, 12).

EXPRESSION OF FOREIGN GENETIC ELEMENTS IN VACCINIA VIRUS VECTORS

The insertion and expression of a number of foreign genes by vaccinia virus vectors have been reported to date. These include the herpes simplex virus thymidine kinase (5, 10), the influenza virus hemagglutinin (9, 22), the hepatitis B virus surface antigen (11, 19), the herpes simplex virus glycoprotein D (11), the Plasmodium knowlesi sporozoite antigen (18), the rabies virus glycoprotein gene (3, 24), and the chloramphenicol acetyltransferase gene (6). Other inert genetic elements from herpes simplex virus, bacteriophage λ (21) and the entire plasmid sequences of pBR322 (9), have been inserted into infectious vaccinia virus.

We will describe here some of the pertinent published characteristics of recombinant vaccinia viruses expressing foreign genes with particular reference to their use as live recombinant vaccines.

VACCINIA VIRUS RECOMBINANTS EXPRESSING THE INFLUENZA VIRUS HEMAGGLUTININ

The influenza virus hemagglutinin is localized on the surface of the influenza virus particle and on the surface membrane of cells infected with influenza virus. The hemagglutinin is considered to be a critical antigen in immunity to influenza. Antibodies directed against the hemagglutinin can neutralize the infectivity of influenza virus, and the presence of antibodies inhibiting the agglutination of erythrocytes by the influenza hemagglutinin is considered as serological evidence for immunity to influenza. The RNA segment from influenza encoding the hemagglutinin was prepared as cDNA and inserted into vaccinia virus. These recombinants synthesized influenza hemagglutinin with properties similar if not identical to the hemagglutinin synthesized by influenza virus itself (9, 22). The hemagglutinin was shown to be present on the surface membrane of the infected cell by reacting unfixed infected cells with specific antibody followed by radiolabeled Protein A (9) or by the more standard immunofluorescence test (20). When these recombinant vaccinia viruses were inoculated into animals, the animals were shown to produce antibodies that reacted specifically with authentic influenza hemagglutinin (9, 22), neutralized the infectivity of influenza virus in plaque reduction assays (9), and inhibited the agglutination of erythrocytes in appropriate hemagglutination inhibition tests (9, 16, 22). Furthermore, animals immunized with these recombinant viruses were shown to be resistant to subsequent intranasal challenge with live influenza virus (22). The elaboration of the hemagglutinin as a surface-associated protein has been demonstrated to elicit cellular immune responses (2).

VACCINIA VIRUS RECOMBINANTS EXPRESSING THE HEPATITIS B VIRUS SURFACE ANTIGEN

The hepatitis B virus surface antigen is localized on the surface of the infectious hepatitis "Dane" particle. It can be localized in the plasma of chronically infected individuals where the surface antigen is elaborated as a 22 nm particle as a cellular secretion product. The HBsAg is considered a critical antigen in eliciting immunity to hepatitis. In fact, the use of purified plasma-derived HBsAg as an immunogen results in efficacious immunization. Vaccinia virus recombinants expressing the hepatitis B surface antigen have been constructed and their properties reported (11, 13, 19, 20). Cells infected with these recombinants secrete material serologically reactive with anti-HBsAg serum. The secreted material has a characteristic polypeptide composition, a density of 1.2 g/cc on CsCl gradients, and a morphology characteristic of the 22 nm particles when examined as negatively stained preparations by electron microscopy. By all criteria thus far applied, the HBsAg synthesized under the regulation of vaccinia virus is indistinguishable from the HBsAg elaborated under native conditions.

When laboratory animals were inoculated either intravenously or intradermally with vaccinia virus recombinants expressing the HBsAg, the animals responded by elaborating anti-HBsAg antibodies at very high titers (11, 19). Chimpanzees vaccinated with these vaccinia recombinants were shown to be protected when challenged with infectious hepatitis B virus (7).

VACCINIA VIRUS RECOMBINANTS EXPRESSING THE HERPES SIMPLEX GLYCOPROTEIN D

Herpes simplex viruses elaborate a number of glycoproteins. Of these glycoproteins, glycoprotein D appears to be an excellent candidate for use in a potential vaccine (as noted in (4, 12)). The glycoprotein D is located on the surface of the infectious herpes simplex virion and on the surface membrane of cells infected with herpes simplex virus. The glycoprotein D of herpes simplex type 1, the oral herpes, and herpes simplex type 2, the genital herpes, share some antigenic determinants which are known to cross-react with the two virus types in ways pertinent to vaccine production strategies. The DNA coding sequences from herpes simplex virus were isolated and inserted into vaccinia virus recombinants (11). These engineered vaccinia viruses expressed the HSV-gD in in vitro-infected tissue culture cells as membrane-bound glycoprotein. Rabbits inoculated with these recombinant vaccinia viruses made antibodies which neutralized the infectivity of both type 1 (11) and type 2 HSV (12) as determined in plaque reduction assays. When these vaccinia recombinants were used to immunize laboratory mice it was demonstrated that the mice were protected against a subsequent intraperitoneal challenge with infectious herpes simplex virus (11, 12). Better protection against HSV challenge was obtained when the mice were challenged with the homologous HSV serotype than when challenged with the heterologous serotype. Vaccinia virus recombinants into which two genes have been engineered encoding the glycoprotein D, one deriving from HSV type 1 and the other deriving from HSV type 2, have been constructed (14). These recombinants should afford better levels of protection against HSV infections.

RECOMBINANT VACCINIA VIRUSES EXPRESSING MALARIAL GENES

Evidence for the expression of a malarial gene derived from Plasmodium knowlesi in mammalian cells, as regulated by vaccinia virus recombinants, has been reported (18). Animals inoculated with these recombinants produced appropriate antibodies. These studies illustrate the potential of live vaccinia virus recombinants as malarial vaccines. Because of the different stages involved in the life cycle of these organisms, an optimal vaccine may require a number of antigens derived from the various stages of the organism. That a number of genes can be inserted and expressed in a single vaccinia virus recombinant has been shown elsewhere (15), demonstrating the feasibility of constructing polyvalent vaccinia virus vectors.

VACCINIA VIRUS RECOMBINANTS EXPRESSING THE RABIES VIRUS GLYCOPROTEIN GENE

The rabies virus glycoprotein cDNA was expressed under vaccinia virus regulation in vitro (3, 24). On inoculation of rabbits or mice, anti-rabies glycoprotein antibodies were rapidly induced to significant levels. Laboratory animals vaccinated with these vaccinia virus recombinants were shown to be protected against subsequent intracerebral challenge using several strains of rabies virus (24). Furthermore, expression of cytotoxic T lymphocytes elicited by the recombinant vaccinia virus has been detailed (24). Cellular extracts prepared in vitro after infection with the recombinant vaccinia virus expressing the rabies glycoprotein were shown to be effective in immunizing animals after inactivation of the virus with Beta propiolactone (24).

PROSPECTS AND PROBLEMS

The desire to have a safe and effective vaccinia virus vaccine vector is paramount. Examination of available vaccine strains should provide preferable candidates for immediate use as vectors as well as starting material which could be further manipulated genetically by recombinant DNA technology to provide even safer vaccinia vaccine vectors.

Although the biological properties of the recombinant vaccinia viruses in eliciting responses to the foreign antigens are dramatic, it is clear that these responses can be even further enhanced by manipulating the expression levels of these antigens to induce an even stronger response. This involves identifying regulatory sequences that increase expression of the foreign genes. Progress in this area is slow due to the very nature of the problem, but increased levels of expression of foreign antigens have already been obtained (14).

The possibility exists of generating live recombinant vaccines directed against a variety of infectious diseases. These polyvalent vaccines would express a number of different antigens derived from either viral, bacterial, or parasitic pathogens. Already, three antigens have been expressed when inserted into a single vaccinia genome (14, 15). Inoculation of laboratory animals has demonstrated serological reactivity directed against all three foreign antigens. It is possible that additional genes, perhaps as many as two or three dozen or more, may be vectored by vaccinia virus due to the large capacity of the virus in stably retaining exogenous DNA. Integration of 20-25Kb of foreign DNA has already been reported (9, 21), and even more room for packaging foreign DNA can be made available by deleting the virus of unnecessary endogenous DNA. The question we have to address in this regard is whether the immunological system can be saturated. Furthermore, in constructing polyvalent vaccines one has to monitor that the expression of one antigen does not suppress the immunological reactivity of other antigens. The potential can also be explored for engineering into these recombinant vaccine viruses elements that result in immunostimulation. Furthermore, serial multiple vaccination with recombinant vaccinia virus expressing novel antigens is not excluded.

A point of concern is that the foreign antigens not be incorporated into the viral envelope. The conjecture is that this kind of viral envelope modification might result in altered tissue tropism of the virus with altered infectious characteristics. No evidence for such integration of foreign glycoproteins into the viral envelope has been reported. Neither altered infectivity nor altered responses to inoculation of these recombinant vaccinia viruses has been observed in laboratory animals. If such a problem were to arise, then one may bypass it by additional manipulation of that particular antigen such as by piggybacking it as a fusion protein to an antigen that does not react with the viral membrane.

Because of the broad host range of the virus it is possible to prepare vaccines not only for human use but also for veterinary applications. The potential for manipulating other members of the poxvirus family has yet to be tapped. For example, it should be genetically possible to engineer fowlpox for use as a vaccine in the poultry industry. Vaccine strains of fowlpox already exist and are used in the industry. Furthermore, fowlpox is known not to be infectious for humans.

A great deal of work remains to be done and a large number of questions remain to be answered; nevertheless, it is clear from progress to date that this approach holds considerable promise for the vaccine approach to controlling infectious diseases.

Acknowledgement. We would like to thank K. Dombrowski for preparing the manuscript.

REFERENCES

(1) Baxby, D. 1981. Jenner's Smallpox Vaccine: The Riddle of Vaccinia Virus and Its Origin. London: Heinemann Educational Books Ltd.

(2) Bennink, J.R.; Yewdell, J.W.; Smith, G.L.; Moller, C.; and Moss, B. 1984. Recombinant vaccinia virus primes and stimulates influenza haemagglutinin-specific cytotoxic T cells. Nature **311:** 578-579.

(3) Kieny, M.P.; Lathe, R.; Drillien, D.; Spehner, D.; Skory, S.; Schmitt, D.; Wiktor, T.; Koprowski, H.; and Lecocq, J.P. 1984. Expression of rabies virus glycoprotein from a recombinant vaccinia virus. Nature **312:** 163-166.

(4) Long, D.; Madara, T.J.; Ponce de Leon, M.; Cohen, G.H.; Montgomery, P.C.; and Eisenberg, R.J. 1984. Glycoprotein D protects mice against

lethal challenge with herpes simplex virus types 1 and 2. Infect. Immun. **43:** 761-764.

(5) Mackett, M.; Smith, G.L.; and Moss, B. 1982. Vaccinia virus: A selectable eukaryotic cloning and expression vector. Proc. Natl. Acad. Sci. USA **79:** 7415-7419.

(6) Mackett, M.; Smith, G.L.; and Moss, B. 1984. General method for production and selection of infectious vaccinia virus recombinants expressing foreign genes. J. Virol. **49:** 857-864.

(7) Moss, B.; Smith, G.L.; Gerin, J.L.; and Purcell, R.H. 1984. Live recombinant vaccinia virus protects chimpanzees against hepatitis B. Nature **311:** 67-69.

(8) Nakano, E.; Panicali, D.; and Paoletti, E. 1982. Molecular genetics of vaccinia virus: Demonstration of marker rescue. Proc. Natl. Acad. Sci. USA **79:** 1593-1596.

(9) Panicali, D.; Davis, S.W.; Weinberg, R.L.; and Paoletti, E. 1983. Construction of live vaccines by using genetically engineered poxviruses: Biological activity of recombinant vaccinia virus expressing influenza virus hemagglutinin. Proc. Natl. Acad. Sci. USA **80:** 5364-5368.

(10) Panicali, D., and Paoletti, E. 1982. Construction of poxviruses as cloning vectors: Insertion of the thymidine kinase gene from herpes simplex virus into the DNA of infectious vaccinia virus. Proc. Natl. Acad. Sci. USA **79:** 4927-4931.

(11) Paoletti, E.; Lipinskas, B.R.; Samsonoff, C.; Mercer, S.; and Panicali, D. 1984. Construction of live vaccines using genetically engineered poxviruses: Biological activity of vaccinia virus recombinants expressing the hepatitis B virus surface antigen and the herpes simplex virus glycoprotein D. Proc. Natl. Acad. Sci. USA **81:** 193-197.

(12) Paoletti, E.; Lipinskas, B.R.; Woolhiser, S.; and Flaherty, L. 1984. Live recombinant poxvirus vaccine directed against herpes simplex. **In** Herpesvirus: UCLA Symposia on Molecular and Cellular Biology, New Series, ed. F. Rapp, vol. 21, pp. 663-676. New York: Alan R. Liss.

(13) Paoletti, E.; Panicali, D.; Lipinskas, B.R.; Mercer, S.; Wright, M.; and Samsonoff, C. 1984. Construction of live recombinant vaccines using genetically-engineered poxviruses. **In** Modern Approaches to Vaccines: Molecular and Chemical Basis of Virus Virulence and Immunogenicity, eds. R.M. Chanock and R.A. Lerner, pp. 295-299. New York: Cold Spring Harbor Laboratory.

(14) Paoletti, E.; Perkus, M.E.; Piccini, A.; Wos, S.; and Lipinskas, B.R. 1984. A novel approach to the preparation of live vaccines: Genetically engineered poxviruses. **In** The 1984 International Symposium on Medical Virology, ed. L.M. de la Maza. Philadelphia: The Franklin Institute Press.

(15) Paoletti, E.; Perkus, M.E.; Piccini, A.; Wos, S.; Lipinskas, B.R.; and Mercer, S.R. 1985. Genetically engineered poxviruses expressing multiple foreign genes. **In** Modern Approaches to Vaccines, eds. R.M. Chanock and R.A. Lerner, pp. 147-150. New York: Cold Spring Harbor Laboratory.

(16) Paoletti, E.; Weinberg, R.L.; Davis, S.W.; and Davis, M. 1984. Genetically engineered poxviruses: A novel approach to the construction of live vaccines. Vaccine **2:** 204-208.

(17) Sam, C.K., and Dumbell, K.R. 1981. Expression of poxvirus DNA in co-infected cells and marker rescue of thermosensitive mutants by subgenomic fragments of DNA. Ann. Virol. (Inst. Pasteur) **132E:** 135-150.

(18) Smith, G.L.; Godson, G.N.; Nussenzweig, V.; Nussenzweig, R.S.; Barnwell, J.; and Moss, B. 1984. Plasmodium knowlesi sporozoite antigen: Expression by infectious recombinant vaccinia virus. Science **224:** 397-399.

(19) Smith, G.L.; Mackett, M.; and Moss, B. 1983. Infectious vaccinia virus recombinants that express hepatitis B virus surface antigen. Nature **302:** 490-495.

(20) Smith, G.L.; Mackett, M.; Murphy, B.R.; and Moss, B. 1984. Vaccinia virus recombinants expressing genes from pathogenic agents have potential as live vaccines. In Modern Approaches to Vaccines, eds. R.M. Chanock and R.A. Lerner, pp. 313-317. New York: Cold Spring Harbor Laboratory.

(21) Smith, G.L., and Moss, B. 1983. Infectious poxvirus vectors have capacity for at least 25 000 base pairs of foreign DNA. Gene **25:** 21-28.

(22) Smith, G.L.; Murphy, B.R.; and Moss, B. 1983. Construction and characterization of an infectious vaccinia virus recombinant that expresses the influenza hemagglutinin gene and induces resistance to influenza virus infection in hamsters. Proc. Natl. Acad. Sci. USA **80:** 7155-7159.

(23) Weir, J.P.; Bajszar, G.; and Moss, B. 1982. Mapping of the vaccinia virus thymidine kinase gene by marker rescue and by cell-free translation of selected mRNA. Proc. Natl. Acad. Sci. USA **79:** 1210-1214.

(24) Wiktor, T.J.; Macfarland, R.I.; Reagan, K.J.; Dietzschold, B.; Curtis, P.J.; Wunner, W.H.; Kieny, M.P.; Lathe, R.; Lecocq, J.P.; Mackett, M.; Moss, B.; and Koprowski, H. 1984. Protection from rabies by a vaccinia virus recombinant containing the rabies virus glycoprotein gene. Proc. Natl. Acad. Sci. USA **81:** 7194-7198.

Biotechnology: Potentials and Limitations, ed. S. Silver, pp. 165-177. Dahlem Konferenzen 1986. Berlin, Heidelberg, New York, Tokyo: Springer-Verlag.

The Future Role in Medicine of Proteins Made by Genetic Engineering

W. Fiers
Laboratory of Molecular Biology, State University of Ghent
9000 Ghent, Belgium

INTRODUCTION

Most of the efforts in the pharmaceutical industry are directed toward human medicine, but quite often these concepts and breakthroughs also find parallel applications in veterinary medicine. However, we will not discuss here those possible applications to veterinary science which are not primarily directed towards alleviation of disease such as the possible use of growth hormones in animal husbandry. The term "genetic engineering" will be used to designate the cloning of genes in a heterologous genetic context, and especially the efficient expression of these cloned genes in heterologous cells. Therefore, the use of chemically synthesized peptides in medicine will not be further considered here. It is generally accepted that the cutoff for economical chemical synthesis of peptides is about 20-30 amino acids; above that, genetic engineering methodology becomes the method of choice.

CLONING AND EXPRESSION OF GENES

We will first discuss some technical problems. Although methodological aspects are treated elsewhere (Smith et al., this volume), it is important to summarize here the approaches which are used for making proteins (and nucleic acid sequences) useful in medicine, as some of these approaches are quite specific.

Where Do We Get the Gene That Codes for a Product That We Want to Make?

Chemical DNA synthesis. This is, of course, the most direct approach, and with the considerable progress which has been made, it is now no longer an impossible task to synthesize, within a relatively short time, the coding information for a polypeptide of 200 amino acids or more. Chemical synthesis of the gene for proinsulin, interferon, somatostatin, etc., has been reported (2, 6, 8). This approach has the considerable advantage that one can adapt the codon usage to the host organism such as E. coli or yeast.

However, the exact amino acid sequence of the protein must be known and - incidentally - it may be noted that the protein sequences which have been published in the past often contained serious errors as in, for example, the alignment of different peptides.

Reverse transcription of an mRNA preparation (a polyA$^+$ RNA size-fractionated on a sucrose gradient). A cDNA copy can be made which is then inserted in an appropriate plasmid. This is the approach most often used. It requires, of course, an active mRNA preparation to start with. This means that the mRNA must be taken from a tissue or from a cell line which is actively making the protein of interest. Sometimes, however, a protein is known to be formed in vivo in an experimental animal or a patient, but the tissue or the induction conditions are not known. This used to be a problem with erythropoietin or with coagulation Factor VIII.

Genomic DNA. This is obviously a sure source as the genetic information for the protein of interest must be present somewhere in the genome. However, there are several major problems (which can all be overcome). The first obstacle is that the relative abundance of the sequence of interest is usually considerably lower than in the case of an RNA preparation (only about 5% of the mammalian DNA genome consists of coding sequence and only a small fraction of all genes are expressed in any cell at a given time). This poses considerable constraints on the screening procedure. A second major disadvantage is that, in the case of all higher eukaryotes, it is very likely that the gene of interest contains intron(s) and this again creates problems when one wants to express the gene in a prokaryote. Nevertheless, if enough reliable protein sequence information is available, it is quite straightforward to isolate a genomic clone.

How Do We Identify the Clone of Interest in a Large Library?
Once a large library has been constructed (either a cDNA or a genomic library), one still faces the most difficult problem, namely, the identification of the clone of interest. Although a great many approaches have been published, four approaches are of particular interest because of their widespread use or their promise.

Hybrid selection-translation. This method requires an mRNA preparation (preferably in abundant quantity) which can be translated into a product which can be identified either immunologically or by a biological assay. Translation is possible either in an in vitro system (possibly enriched with a processing cocktail: see (1)) or translation in vivo such as injection into oocytes of Xenopus laevis. Undoubtedly, the latter approach is the method most often used. The general procedure, then, is to screen groups of clones by preparing plasmid (or phage) DNA from the group, immobilize this DNA, and hybridize to this the mRNA preparation. The hybridized fraction is then eluted and translated in the aforementioned system. When a positive group of clones has been identified, subgroups can be made up and the procedure repeated until an individual hybridizing clone can be identified. This approach has been very widely used and has led to the cloning of such important genes as human interferon-α (11), -β (3, 17), -γ (4, 7), interleukin-2 (5, 16), etc. Nevertheless, this procedure is not without potential pitfalls such as cross-hybridization due to, for example, the presence of homologous sequences. Therefore, further confirmation is always necessary based, for example, on expression of biological activity or on a correlation between the amino acid and the nucleotide sequence.

Identification by hybridization with a synthetic oligonucleotide probe. This fairly straightforward procedure requires that the protein of interest first be obtained in a chemically pure form such that part of its amino acid sequence can be determined. On this basis, chemically synthesized probes can then be used for selective hybridization. Previously, highly redundant sets of probes having a length of 15 to 20 nucleotides were often used. More recently, with the methodology for chemical synthesis of oligonucleotides becoming more powerful, longer probes have been used (40 to 60 nucleotides in length). In this case, no full homology is required and, therefore, the probe may contain little or no degeneracy. This procedure can be used for cDNA screening as well as for genomic screening. It has been successfully used for the cloning of human blood Factor VIII, TNF, and others.

Identification on the basis of direct immunological expression. In this approach, one is not attempting to obtain a full length clone, but rather a partial sequence, fused in phase to an efficiently translated gene. The system best worked out for this purpose is λgt11. It is necessary to screen a large number of clones, as only one out of six of each particular cDNA will score (two orientations and three reading phases). The method requires the availability of a good antiserum: either a monoclonal or a polyclonal, provided that the latter was raised with highly purified protein.

Identification based on direct biological expression. The protocol of choice here makes use of the Okayama and Berg-type vector (12). This procedure is very elegant and requires the availability of an mRNA source to make the gene bank; the latter can then be scored groupwise. All that is further needed is a very sensitive biological assay. In order to be successful, the cloning should be nearly optimal as only full-length clones can potentially be scored.

Expression of Cloned Genes in a Heterologous Cell

It is obviously impossible within the framework of a general background paper to review all aspects governing expression of a cloned gene in a heterologous cell. We will restrict ourselves to giving a general outline. As the large majority of proteins used in medicine, or biotechnology for that matter, act extracellularly, they are nearly all produced via secretory mechanisms. With a few exceptions (e.g., ovalbumin), all extracellular proteins are made via a precursor which starts with a hydrophobic signal sequence guiding secretion outside the cell and which is subsequently processed away, giving rise, in the extracellular medium, to the fully active, mature protein. By genetic engineering, it is possible either to make the mature protein directly from a properly manipulated gene or to use the gene for the preprotein either in a homologous form, i.e., with its original signal sequence, or else provided with a heterologous signal sequence.

Expression as a mature protein. Expression in the bacterium Escherichia coli is still the method of choice as the latter is so well understood genetically. It is usually possible to make perfect constructions, which means replacement of the signal sequence by a single initiating methionine codon. Many studies have been devoted to the optimalization of expression on the basis of increased gene copy number, better transcription efficiency, and especially, better initiation of translation. Although the latter aspect is still not understood in great detail and there is still a great deal of "black magic" and "trial and error" involved, it is usually possible to obtain a fairly high level of foreign gene expression in E. coli. In some examples, up to 30% of the total bacterial protein can be the gene product of interest such as human interferon-γ (15). The most critical factor is, in fact, the efficiency of initiation of translation. This is influenced not only by the Shine-Dalgarno sequence and the sequence between the latter and the initiating AUG, but also by the nucleotide sequence before the Shine-Dalgarno sequence and that after the AUG; undoubtedly these sequences play a role because they determine the overall folding (secondary and tertiary structure), and it is this folding which determines the presentation of the initiating AUG in the proper context to the translation machinery. It would also, however, not be correct to give the impression that everything has been solved. Some very serious problems remain. For example, very few long heterologous polypeptides have been made in E. coli. It is possible that there are intrinsic problems such as sequences which easily lead to frame shifting resulting in premature termination. Another problem is that although for many proteins the initiating methionine is subsequently removed, in some cases it remains present and this can cause problems as in the case of human growth hormone. Another major difficulty is that many heterologous proteins, when they are abundantly synthesized in E. coli, accumulate in the cell as an amorphous protein mass which subsequently has to be solubilized with denaturing agents and finally refolded. Although this insolubility facilitates the initial purification, the costs relating to this solubilization and renaturation are considerable. Other organisms or cells have rarely been used for expression of mature heterologous genes as they usually cannot compete in efficiency with expression in E. coli, their other advantages being negligible.

Expression in a secreted form. This presents many advantages as the product is released in a mature form in the extracellular medium and, hence, it does not contain an additional N-terminal sequence or methionine. Usually the extracellular medium is of low complexity and, thus, purification is fairly straightforward; most important of all, the protein is in a native configuration. Bacteria would be the organisms of choice as they are growing rapidly on inexpensive media. E. coli does secrete some proteins into the periplasmic space (e.g., β-lactamase, bacterial alkaline phosphatase) or insert them into the outer membrane, and also mutants have been isolated which release the proteins from the periplasm into the medium. Nevertheless, the expression of heterologous genes, in a secreted form in E. coli, has been most disappointing to date. Gram-positive organisms such as Bacillus subtilis hold much promise. They can secrete proteins in considerable amounts (e.g., 8 g/l of α-amylase has been reported (13)) and, furthermore, industry is quite familiar with their fermentation. Nevertheless, expectations using this system have not been fulfilled and high-yield production of foreign proteins is difficult to obtain. One of the obstacles is ubiquitous proteases though this is presumably not the only limitation. Much genetic engineering work is also being done with various Streptomyces species. Of course, many of these studies are aimed at improving production of various antibiotics or their precursors. But it is quite possible that some Streptomyces systems may also be useful for expression of foreign genes in a secreted form. The familiar brewing yeast Saccharomyces cerevisiae has a rather elaborate cell wall and does not secrete many proteins into the medium. Nevertheless, it is this microorganism which has been most useful for synthesis of valuable products. For example, this is currently considered to be the best method for production of the hepatitis surface antigen for preparing a vaccine; although the product is not released into the medium, it is nevertheless associated with membrane components and fairly closely mimics structures which are formed by the native virus. It has also often been mentioned that yeast should offer an advantage over bacteria in that it can glycosylate secreted proteins. This has not been observed so far in proteins released into the medium, but the genetically engineered rabies surface protein is glycosylated and this again is a promising vaccine. The most interesting system for obtaining products in the medium is to make use of the fairly sophisticated machinery for secretion which, for example, allows the yeast cell to release a mating factor into the outside medium. Fusion of a mature gene to these secretory elements allows release of valuable proteins such as human proinsulin or human interleukin-2 in the outside medium.

Various cDNA or genomic genes, especially from mammalian cells, can be expressed at moderate levels in mammalian cells. This has the considerable advantage that proper processing and modification can occur. Usually, the gene of interest is introduced together with a selection marker such that the transformed mammalian cells can be readily identified. A variety of selection systems have been worked out such as the thymidine kinase gene introduced in TK^- cells, dominant selection by means of the antibiotic G418, and several others. But undoubtedly, the system most often preferred is a Chinese Hamster Ovary (CHO) cell line which is dihydrofolate reductase ($DHFR^-$)-negative. Therefore, the $DHFR^+$ transformants can be obtained by introduction of a heterologous gene (e.g., derived from mouse) put under an appropriate promoter such as the late adeno transcription system (9). The passenger gene can be introduced either on the same plasmid or in an appropriate molar ratio on a second plasmid (14). It is obviously advantageous to put the latter gene under a strong promoter such as the SV40 early promoter, preceded by its enhancer sequences. The elegance of this system is that it is subsequently possible to put methotrexate selection pressure on the transformants, and the surviving clones have an increased expression level of the DHFR gene as well as of the passenger gene, which had been introduced at the same time. Indeed, very often the latter is amplified as well and, in this way, considerable amplification of the gene copy number and, hence, of the expression level is achievable. In the case of human interferon gamma, the yield of secreted product is about 5-10 µg/ml (this amount of secreted protein is about the same level as a monoclonal antibody produced by a hybridoma line). Although the glycosylation in these cells has not yet been studied in great detail, the product obtained from the CHO cells is very similar or identical to the natural product in this respect.

WHAT NEW PROTEINS ARE NEEDED IN MEDICINE?

A limited number of proteins have been identified as being useful for prophylaxis, diagnosis, and therapy in medicine. The genes for many of these proteins have been cloned and expressed and the list of useful proteins, well-known and characterized but not yet provided by genetic engineering, is rapidly dwindling. However, a rich lore of useful products undoubtedly remains to be discovered and studied. It seems likely, then, that progress in the future will not so much be limited by the ability to clone genes and express them in appropriate organisms but rather by the difficult and slow pace of further medical and fundamental, biological research. Below, a non-exhaustive list is given of the main classes of possibly useful proteins.

Vaccines

The need for effective immunization against parasites, viruses, and bacteria can hardly be overemphasized. It is also obvious - but it must be stressed again and again - that vaccines made by genetic engineering are much safer and can be prepared in abundant amounts. However, their efficacy is often a problem and these aspects will be the object of considerable research. Nevertheless, one should not overlook the considerable progress that has been made with genetically engineered vaccines. Promising results have been obtained with some antibacterial vaccines (colitis, cholera), but we should be aware that immunity against many bacteria is directed against polysaccharides and mucopolymers, and it is not known whether the same bacteria also present appropriate protein targets which then can be mimicked by genetically engineered products. Viral vaccines are extremely important and some of these have almost reached the market stage, such as the surface hepatitis B vaccine and, further down the road, rabies and perhaps herpes and the antiparasite vaccine against malaria.

Proteins Required As a Substitute Because Their Natural Counterpart Is Missing Due to a Genetic Defect or a Similar Mishap

Some of these proteins have been identified for quite some time; the absence of a single protein offers a clear indication for therapy and the provision of the missing substance gives immediate relief. These proteins are medically interesting and fill a need in a well-defined, quantifiable market. The most obvious example is, of course, human insulin, although here the economics are such that the recombinant product is not necessarily advantageous over the substitutes obtained from pigs or cattle. Factors VIII and IX are, perhaps, more striking examples because here only the natural product obtained from human blood constitutes an effective alternative. But the high risk connected with blood fractions (AIDS is the latest addition to the list) and the very low concentration of the two aforementioned factors in the blood means that the recombinant product, when it becomes available, is clearly the way to go. Its cloning, recently reported by two groups, is one of the highlights of genetic engineering considering its size (9 kb cDNA; about 200 kb for the genomic clone (18, 19). It may be noted that in the Caucasian population there are about 10 to 20 hemophiliacs/100,000 males, and until 1950 the average life span of these patients was only about 20 years. It may also be added that these cloned genes and further research on these diseases will undoubtedly lead to better prenatal diagnosis and, therefore, may ultimately reduce the number of live births affected with hemophilia. Human growth hormone is perhaps the clearest example of a protein whose function has already been known for quite some time and which was not available in adequate amounts as it shows a strict species specificity. Its efficacy in treating a common form of dwarfism is well established and, therefore, its future use in medicine seems guaranteed. Moreover, there is some evidence that growth hormone can also be used for such indications as wound healing.

Proteins Required in Greater Than Physiological Quantities for Short- or Long-term Treatment

The first example which comes to mind here is tissue plasminogen activator, a sophisticated protease which is very promising, indeed, for treatment of thrombosis as it dissolves the clot without doing much harm anywhere else. Considerable effort has already been spent on its cloning and expression and the product has reached the clinical evaluation stage.

Also, the interferons (both leukocyte or -α and fibroblast or -β) can be considered in this class, at least as far as their antiviral action is concerned. However, due to the fact that they prevent or slow down viral infection rather than curing an infected cell, their use - as antiviral agents - is mainly restricted to prophylactic treatment (e.g., in high risk patients) or else to treatment in the case of slow viruses or diseases where a viral etiology is suspected.

Erythropoietin, which accelerates the rate of erythropoiesis, may have a promising role after a trauma or major surgery in addition to its potential in certain hematological disorders.

Proteins Required for Interference in (Disturbed) Regulatory Networks:

Immune system. a) Biological adjuvants - It is tragically well-known that a child can only survive in a world full of microbes, viruses, and other aggressions if it has a well functioning immune system. Some deficiencies might perhaps be complemented by lymphokines such as interleukin-2, B cell growth factor (BCGF), or B cell differentiation factor (BCDF). Furthermore, we also should not forget that inactivated viruses and subunits, although innocuous, are often not very effective for vaccination and must be used in larger quantities. Proteins, which can play a proper adjuvant role, may supply here a welcome answer.

b) Anticancer activity of the immune system - Arguments for or against immunosurveillance as a major defense mechanism against cancer lead to a somewhat complex and confused picture. However, it is clear that, at least in experimental animals, deterioration of the immune system is often noted and there are indications that tumor growth is responsible for deterioration and/or exhaustion of the immune system. In view of the complexity of this network, treatment with proteins or immune factors, which clearly play a very important role, often cannot be interpreted in terms of a single simple mechanism. For example, human interferon-γ, which was first detected on the basis of its antiviral activity but is undoubtedly an immunomodulator, has a multitude of effects. It is cytostatic on many cell lines and even cytotoxic on some tumor lines and, therefore, treatment with interferon-γ might have a direct effect on the tumor. Moreover, interferon-γ has various other potent effects. For example, it dramatically increases the expression of class II HLA antigens, not only on cells which normally express it to a certain extent such as macrophages, but also on cells which are normally DR^- such as endothelial cells and fibroblasts. This induction of HLA antigens, also in tumor cells, may render them more immunogenic and, therefore, boost the immunological response against these cells. On the other hand, there is also evidence that IFN-γ-treated cells are worse targets for natural killer cells. The latter, which at least in some types of cancer or leukemia may play a role in immunosurveillance, can be activated by interferon-γ but also by interferons -α or -β.

Interleukin-2, a product of T helper cells, plays a very central role in the immune system. It causes proliferation of antigen-activated T helper cells as well as cytotoxic cells. Amplifying the level of CTLs may then, in turn, exert an antitumor activity. However, the role of IL-2 is manyfold: it is now also known to stimulate activated B cells and even to induce secretion of immunoglobulin. Furthermore, interleukin-2 also activates NK cells and there is evidence that it may be a key stimulator of interferon-γ release.

Interleukin-1, which is a macrophage product, again plays a very important role in the network as it acts not only on T helper cells and on B cells, but moreover on some cells outside the immune system such as fibroblasts and chondrocytes. As IL-1 is presumably identical to endogenous pyrogen, its use in medicine may seem limited. However, it may have interesting activities at concentrations below the unacceptable pyrogenic level and, furthermore, its use as a diagnostic indicator remains to be evaluated.

A number of novel proteins, made by various cells of the immune system, remain to be studied with respect to their potential role in antitumor biology. Some of these, such as Tumor Necrosis Factor and Lymphotoxin, have been cloned and look very promising. Others, such as Macrophage Activating Factor, remain elusive (is there an MAF different from interferon-γ?).

c) In autoimmune diseases, we are dealing with subdomains of the immune system which for one reason or another have deviated from their normal physiological mechanisms. Presumably, some minor but essential actor in an autoregulatory system is deficient and responsible for causing the imbalance. As more lymphokines and monokines become cloned and available as probes, and since more subpopulations of the immune cells can be differentiated by means of monoclonal antibodies, it is likely that we will gain a better understanding of diseases such as lupus, multiple sclerosis, allergy, etc. On the basis of this, we will perhaps be able to design an adequate treatment.

Cell growth and control. There is much evidence that in multicellular organisms the rate of cell growth and differentiation is determined by a set of external factors. A few of these have been well characterized and some have even been cloned such as epidermal growth factor, nerve cell growth factor, platelet-derived growth factor (PDGF), and others. Also, insulin and some insulin-like factors such as somatomedins may belong to this category. We are still only at the beginning in our understanding of how these factors work and what the molecular basis is of endocrine, autocrine, and paracrine control. Presumably, some of these factors may have a role in tissue regeneration after injury or other activities such as stimulation of hair follicle growth, etc. But perhaps most expectations are in the anticancer field. For example, it is well-known that epidermal growth factor receptors are enhanced on a number of tumor cells and that the PDGF gene has reemerged in evolution as an oncogene, etc. Furthermore, a tumor growth factor has been identified which presumably interacts with the epidermal growth factor receptor, and natural antagonists, if these can be obtained in sufficient amounts, would be of interest to test for possible therapy.

Growth after injury or surgery. In addition to wound healing (Human Growth Hormone), various factors may have a possible application for faster or better regeneration of tissues (some of these were discussed in the previous section). Other factors can be useful for preventing allograft rejection or preventing graft-versus-host disease, etc. One can also include serum albumin in this list. This is the human protein used in highest amounts. Its production by genetic engineering on a scale which is economically attractive remains a great challenge.

Proteins (and Nucleic Acid Sequences) Needed As Diagnostic Reagents and for Immunotherapy

An increasing list of reagents can now be better prepared by genetic engineering than by classical approaches. This is true for certain antigens which can be used as a reference or for immunization to raise antibodies needed in immunoassays or as a labelled competitor in competition-type assays. Also, there is a rapidly increasing list of nucleotide sequences, synthesized either chemically or biosynthetically, which are essential as hybridization probes. These allow the identification of viral, parasitic, and bacterial infections, detection of oncogene expression in tumor cells, prenatal diagnosis, etc.

Second Generation Products: Do They Have a Future?

These are gene products which are derived from their natural counterpart genes and which have improved properties. Our technological skills are so improved that we can now create almost any mutation at any site in a cloned gene. However, it is much more difficult to decide what we actually want. One can hope for a more efficient product, a better active site, a longer half-life in circulation, a better tolerance and less toxicity, better thermostability, etc. But methods of protein engineering which would allow us to do this without trial and error are still in their infancy. Some progress has been made as, for example, by the Cetus company and by Biogen; they have replaced cysteine in position 17 of human interferon-β by serine (10). This resulted in a product which was easier to purify, still biologically active, and had less tendency to aggregate. Similar results have been obtained with human interleukin-2. Whether these products will actually be admitted for widespread use in humans is still an open question. Although the change of an -SH into an -OH group is minor, one will still have to prove that the modified protein does not induce specific antibodies and, if so, then the latter certainly should not interact with the authentic homologous protein. One may believe that the changes should only be made in the interior of the molecules, such that, e.g., thermostability could be enhanced without changing the exposed immunogenic epitopes. This is deceptive as there is more and more evidence that macrophages, for example, can digest the antigenic proteins before exposing certain subregions and some of the latter may very well be buried in the original antigenic protein. On the other hand, rapid progress has also been made in immunology, and the better understanding of these networks may reveal how one can make proteins nonimmunogenic or may tell us that a low level antigenicity is not harmful at all due to the waves of anti-idiotype interactions.

Non-autologous Proteins

Some proteins, even of bacterial origin, have found a use in internal medicine such as, for example, the treatment of certain leukemias with bacterial arginase. Of course, such a treatment will almost inevitably result in an immune response, but the advantages may far outweigh the disadvantages. Also, it is sometimes possible to identify an equivalent protein which is less immunogenic in man. Some enzymes can be useful for therapy and simply be nonexistent in the mammalian species. There is an interesting market if somebody could come up with an enzyme which dissolves gallstones or which prevents the graying of hair, etc.

Impact of Genetic Engineering on Public Health

One should not overlook the fact that the introduction of refrigerators has done more to increase the average life span of the Western population than such dramatic breakthroughs as the widespread use of antibiotics. Therefore, the introduction of new procedures and new products made possible by genetic engineering into the food industry and agriculture can also have a considerable impact on the public well-being. The effects can be fairly direct, such as better foodstuffs, on which basis adequate diets can be developed. For example, treatment with genetically engineered enzymes which lower cholesterol or saturated fatty acids can have a very real impact on cardiovascular diseases. In terms of the nutrition problems of the Third World, one can think of enriching staple foods such that the amino acid composition becomes more balanced, for example, by the introduction of S-amino acids-rich proteins into soya, or lysine-rich proteins into cereals by plant genetic engineering. Other approaches can improve public health through indirect ways such as a reduction of chemically synthesized herbicides and pesticides released into the environment by replacement with genetically engineered alternatives which are more biocompatible. Cheaper ways to manufacture ethanol may help in the introduction of lead-free gasoline, etc.

GENERAL CONCLUSIONS

It cannot be stressed enough that the main impact of genetic engineering is its enormous contribution to biological research including medicine and agriculture. A better understanding can only help medicine's slow transition from an art into a science. But one of the more immediate benefits of genetic engineering methodology is that it has allowed the preparation, in abundant quantities, of various proteins which previously had only an elusive existence. At present, it is often sufficient to have only a reliable biological test for a supposedly new factor (or else a monoclonal antibody which reacts with a particular protein) or even an open reading frame in a sequenced genome, and these hints may be sufficient to detect a new protein, clone its gene, and express it such that the protein becomes available in large quantities. Genetic engineering in itself is not a discipline but a powerful tool, just like organic synthesis used to be for the pharmaceutical industry in previous decades: in order to be useful, it must be applied and made use of. This goes hand in hand with progress made in biological science, cell biology, clinical medicine, pharmacology, and so on. The various products made by genetic engineering are gradually entering the hospitals and the clinical laboratories. Even yesterday's harshest critics of recombinant DNA will have to concede the major impact of these new products on the medicine of tomorrow.

Acknowledgements. Research in the author's laboratory was supported by grants from the "Fonds voor Geneeskundig Wetenschappelijk Onderzoek," the "Gekoncerteerde Onderzoeksakties" of the Belgian Ministry of Science, the A.S.L.K., and Biogen, N.V.

REFERENCES

(1) Content, J.; De Wit, L.; Derynck, R.; De Clercq, E.; and Fiers, W. 1982. In vitro cotranslation processing of human pre interferon β_1 enhances its biological activity. Virology **122:** 466-470.

(2) Crea, R.; Kraszewski, A.; Hirose, T.; and Itakura, K. 1978. Chemical synthesis of genes for human insulin. Proc. Natl. Acad. Sci. USA **75:** 5765-5769.

(3) Derynck, R.; Content, J.; De Clercq, E.; Volckaert, G.; Tavernier, J.; Devos, R.; and Fiers, W. 1980. Isolation and structure of a human fibroblast interferon gene. Nature **285:** 542-546.

(4) Devos, R.; Cheroutre, H.; Taya, Y.; Degrave, W.; Van Heuverswyn, H.; and Fiers, W. 1982. Molecular cloning of human immune interferon cDNA and its expression in eukaryotic cells. Nucl. Acids Res. **10:** 2487- 2501.

(5) Devos, R.; Plaetinck, G.; Cheroutre, H.; Simons, G.; Degrave, W.; Tavernier, J.; Remaut, E.; and Fiers, W. 1983. Molecular cloning of human interleukin 2 cDNA and its expression in E. coli. Nucl. Acids Res. **11:** 4307-4323.

(6) Edge, M.D.; Greene, S.R.; Heathcliffe, G.R.; Meacock, P.A.; Schuch, W.; Scanlon, D.B.; Atkinson, T.C.; Newton, C.R.; and Markham, S.F. 1981. Total synthesis of a human leucocyte interferon gene. Nature **292:** 756-762.

(7) Gray, P., et al. 1982. Expression of human immune interferon cDNA in E. coli and monkey cells. Nature **295:** 503-508.

(8) Itakura, K.; Hirose, T.; Crea, R.; Riggs, A.D.; Heyneker, H.L.; Bolivar, F.; and Boyer, H.W. 1977. Expression in Escherichia coli of a chemically synthesized gene for the hormone somatostatin. Science **198:** 1056-1063.

(9) Kaufman, R.J., and Sharp, P.A. 1982. Construction of a modular dihydrofolate reductase cDNA gene: Analysis of signals utilized for efficient expression. Molec. Cell. Biol. **2:** 1304-1319.

(10) Mark, D.V.; Ru, S.D.; Creasey, A.; Yaniamoto, R.; and Lin, L. 1984. Site-specific mutagenesis of the Human Fibroblast Interferon Gene: A structure-function analysis of the protein. Proc. Natl. Acad. Sci. USA **81:** 5662-5666.

(11) Nagata, S.; Taira, H.; Hall, A.; Johnsrud, L.; Streuli, M.; Ecsodi, J.; Boll, W.; Cantell, K.; and Weissmann, C. 1980. Synthesis in E. coli of a polypeptide with human leukocyte interferon activity. Nature **284:** 316- 320.

(12) Okayama, H., and Berg, P. 1982. High-efficiency cloning of full length cDNA. Molec. Cell. Biol. **2:** 161-170.

(13) Palva, I.; Petterson, R.F.; Kalkkinen, N.; Lehtovaara, P.; Sarvas, M.; Söderland, H.; Takkinen, K.; and Kääriäinnen, L. 1981. Nucleotide sequence of the promoter and NH_2-terminal signal peptide region of the α-amylase gene from Bacillus amyloliquefaciens. Gene **15:** 43-51.

(14) Scahill, S.J.; Devos, R.; Van der Heyden, J.; and Fiers, W. 1983. Expression and characterization of the product of a human immune

interferon cDNA gene in Chinese hamster ovary cells. Proc. Natl. Acad. Sci. USA **80:** 4654-4658.

(15) Simons, G.; Remaut, E.; Allet, B.; and Fiers, W. 1984. High-level expression of human interferon gamma in Escherichia coli under control of the p_L promoter of bacteriophage lambda. Gene **28:** 55-64.

(16) Taniguchi, T.; Matsui, H.; Fujita, T.; Takaoka, C.; Kashima, N.; Yoshimoto, R.; and Hamuro, J. 1983. Structure and expression of a cloned cDNA for human interleukin-2. Nature **302:** 305-310.

(17) Taniguchi, T.; Sakai, M.; Fujii-Kuriyama, Y.; Muramatsu, M.; Kobayashi, S.; and Sudo, T. 1979. Construction and identification of a bacterial plasmid containing the human fibroblast interferon gene sequence. Proc. Jpn. Acad. **55(B):** 464-469.

(18) Toole, J.J., et al. 1984. Molecular cloning of a cDNA encoding human antihaemophilic factor. Nature **312:** 342-347.

(19) Wood, W.I., et al. 1984. Expression of active human factor VIII from recombinant DNA clones. Nature **312:** 330-337.

Biotechnology: Potentials and Limitations, ed. S. Silver, pp. 179-183. Dahlem Konferenzen 1986. Berlin, Heidelberg, New York, Tokyo: Springer-Verlag.

Some Applications of Modern Immunological Techniques

U. Krawinkel
Institut für Genetik der Universität zu Köln
5000 Köln 41, F.R. Germany

Abstract. There are many innovative applications for immunological techniques in a wide variety of fields. To cover them all is far beyond the scope of this brief summary. I rather concentrate on the description of some techniques that have become important for clinical and biotechnological research in recent years: a) hybridomas producing antibodies for the identification and purification of molecules and cells of interest; b) genetically engineered antibodies; c) cell sorting; and d) bone marrow transplantation. These techniques are interrelated in that the latter three are dependent on the achievements made in the monoclonal antibody field.

OVERVIEW

Monoclonal antibodies produced by hybridomas are now routinely used in biotechnology and clinical diagnostics (see (2) for review). The range of available specificities grows rapidly. The application of monoclonal antibodies will become routine in clinical therapy, for instance, to inhibit specifically neoplastic growth. Successful treatment of patients has not yet been achieved with antibodies prepared against specific surface structures of tumor cells, but the growth of murine and human tumor cells transplanted into immune-deficient mice has been inhibited by treatment with antitumor antibodies. Krolick et al. (9) covalently coupled the cytotoxic peptide of the plant toxin ricin to antibodies binding tumor-specific surface antigens of the murine lymphoma BCl_1. This lymphoma normally kills a recipient of tumor cells within 3-5 months. After treatment with antitumor antibodies, spleen cells from tumor-bearing mice did not develop lymphomas in syngeneic hosts. Even lethally irradiated hosts which have no functional immune system can be successfully repopulated with tumor-depleted spleen cells from BCl_1-bearing donor mice.

This last experiment has implications for autologous bone marrow transplantation following supralethal tumor therapy in man. Bone marrow is removed before the patient is treated with supralethal radiotherapy and/or chemotherapy to kill all tumor cells. The autologous bone marrow is then injected into the patient to reconstitute the hemopoietic system. One major problem with this approach is that the marrow may and usually does contain tumor cells. The depletion of these cells from the marrow by ricin-tagged antibodies may circumvent this problem.

Herlyn and Koprowski (5) suppressed the growth of human colorectal carcinoma cells transplanted into thymus-deficient nude mice by injection of monoclonal antitumor antibodies. They found that specific inhibition of tumor growth can only be achieved with murine antibodies of the IgG2a isotype. This indicates that attachment of antibodies to isotype-specific Fc receptors of the macrophage is essential to effect tumor destruction.

The Herlyn-Koprowski experiment shows a possible way of achieving tumor therapy in man utilizing tumor-specific antibodies as clinical tools. The potential of monoclonal antibodies in this field is obvious and, generally speaking, hybridoma technology will become an even more powerful tool when combined with other techniques such as production of chimeric antibodies and fluorescence-activated cell sorting (see below).

Passive immunization is an effective therapy for acute bacterial and viral infections. One problem with this therapy is the source of the passively immunizing antibody which is usually produced from sera of immune animals: application of xenogeneic antibodies may cause severe allergic complications. Genetic engineering of human antibodies is a way of circumventing this problem. The genes for the variable portions of heavy and light chains of a murine antibody carrying a desired specificity can be combined with human immunoglobulin constant region genes (3, 10, 12). The recombinant antibody genes are then transfected into a lymphoid cell line and a chimeric antibody can be isolated. This antibody minimizes the undesired side effects of passive immunization across the species barrier and provides the proper effector functions of human immunoglobulins.

A hybrid mouse/human antibody carrying a specificity suitable for application in passive immunization has not yet been constructed but work is in progress in many laboratories. Antibodies against viral (i.e., hepatitis) and bacterial (i.e., Pseudomonas) antigens certainly will be available soon.

The fluorescence-activated cell sorter (FACS) allows the analysis and isolation of cells of interest by utilizing fluorescence markers, usually antibodies that have directly or indirectly been coupled to fluorescent dyes. It is desirable for many reasons to isolate cells producing proteins which have undergone defined structural changes.

Fluorescence-activated cell sorting has been used for the isolation of mutant myeloma and hybridoma cells that express immunoglobulins with altered constant regions (8, 14). Variants with altered heavy chain constant regions occur at a low frequency (10^{-5} - 10^{-6}) in many hybridomas and have been isolated by staining the variant cells with fluorescent antibodies and separating the bright fluorescent cells from the unstained ones. Enrichment of stained - i.e., variant - cells is 100 - 1000-fold in one sorting step. Isolation of a variant which occurs at a low frequency thus requires 2-3 sorting cycles. Isolation of isotype switch variants from hybridomas may be needed when a particular isotype (for instance, IgG2a; see section on Monoclonal antibodies produced by hybridomas) is associated with a desired immunological function. Cell sorting has also been successfully used for the isolation of genes encoding the transferrin receptor (6). Living transfected cells were separated and isolated according to the surface expression of proteins encoded by the transfected genes.

Another field of FACS application is the isolation of cell lines which are high producers of a particular product. The selection of cells expressing high amounts of target surface molecules in a stable fashion is technically important for experiments with transfected cell lines. FACS has successfully been used for the isolation of stably transfected high producers of human Leu2 antigen (7). A 100-fold increase in staining intensity reflecting a 30-fold increase in gene copy number could be achieved.

The limitation of FACS is the impossibility of applying fluorochromated antibodies to secreted proteins or to intracellular proteins, because external antibodies cannot penetrate the membrane of living cells. Sequential sublining (11) is a method particularly applicable to the isolation of any myeloma or hybridoma variant whose secreted product has acquired a new predefined antigenic determinant. By this method hybridoma cells are grown in culture wells and the supernatant fluids of individual wells are tested in a solid phase radioimmunoassay for the presence of antibodies expressing an isotype other than the one expressed by the wild type. Cells from wells scoring positive for a variant immunoglobulin isotype are then transferred to new culture wells and expanded. Cell populations in which variants are enriched are subjected to the next round of sublining by using 100-fold fewer cells per well than in the previous round. The main advantage of sequential sublining over the FACS lies in its simplicity: variants can be selected from large populations of cells without complicated and expensive equipment.

The examples described here represent only a small fraction of the numerous applications of FACS. A broad review and rationale for FACS application has been given by Shapiro (15).

Functional deficiency of an enzyme is the cause of many inborn metabolic diseases. The logical therapy for treating the resulting clinical abnormalities should involve the functional enzyme. However, problems arise when the enzyme is administered from an external source: apart from finding an adequate source of the enzyme, purifying the enzyme at a reasonable cost, and delivery of the enzyme to the appropriate tissue, the most serious problem is that treatment is transient at best. In order to achieve permanent and more physiological enzyme replacement therapy, transplantation of cells, tissues, or whole organs and gene therapy of somatic tissue will soon be the treatment of choice for some inborn metabolic errors (for review see (13)).

Bone marrow transplantation will become particularly important, since it is now standardized and reasonably effective and has the advantage of being rather universal since a variety of stem cells capable of performing many different metabolic activities exist in marrow (see (1) for review). Mature bone marrow cells circulate to nearly all organs of the body, although they do not pass across the blood-brain barrier. Bone marrow is relatively easy to obtain and bone marrow cells can be subjected to treatment and analysis with antibodies. For instance, they can be separated from undesired cells by using the fluorescence-activated cell sorter (see sections on Monoclonal antibodies produced by hybridomas and on The fluorescence-activated cell sorter (FACS)).

An example of curing an inborn metabolic disorder by bone marrow transplantation has been presented by Gasper et al. (4) in an animal model system. Mucopolysaccharidosis VI is the result of arylsulfatase B deficiency and causes severe cardiac, skeletal, and pulmonary complications in man and cat. Soon after bone marrow transplantation an arylsulfatase-deficient cat showed signs of improvement. Leucocyte arylsulfatase B activity persisted for the length of the study (250 days), and the cat's ability to walk and chew along with its coarse facies and demeanor improved rapidly. Such animal models of inborn errors of metabolism could lead the way to a much wider use of marrow transplantation.

SUMMARY

The potential of some immunological methods has been described. Monoclonal antibodies secreted by hybridomas are already widely used in clinical diagnostics and will be important for therapy as well. Antibodies against tumor-specific antigens can be used for the targeting and destruction of tumor cells. Hybrid antibodies carrying human C-regions and murine V-regions with a pre-chosen specificity will be useful for passive immunization devoid of undesired side effects. Fluorescence-activated cell sorting allows the analysis and isolation of cells producing proteins of clinical or industrial interest. Bone marrow transplantation appears to be a powerful therapy for inborn metabolic diseases.

The examples presented here do not represent a complete overview but may give the reader an idea about current trends in immunological techniques.

REFERENCES

(1) Bayever, E., and Feig, S.A. 1983. Bone marrow transplantation for inherited metabolic disorders. In Recent Advances in Bone Marrow Transplantation, ed. R.P. Gale, pp. 185-195. New York: Alan Liss, Inc.

(2) Bernard, A.; Boumsell, L.; Dausset, J.; Milstein, C.; and Schlossman, S.F., eds. 1984. Leucocyte Typing. Berlin, Heidelberg, New York, Tokyo: Springer-Verlag.

(3) Boulianne, G.L.; Hozumi, N.; and Shulman, M.J. 1984. Production of functional chimaeric mouse/human antibody. Nature **312:** 643-646.

(4) Gasper, P.W.; Thrall, M.A.; Wenger, D.A.; Macy, D.W.; Ham, L.; Dornsife, R.E.; McBiles, K.; Quackenbush, S.L.; Kesel, M.L.; Gillette, E.L.; and Hoover, E.A. 1984. Correction of feline arylsulphatase B deficiency (mucopolysaccharidosis VI) by bone marrow transplantation. Nature **312:** 467-468.

(5) Herlyn, D., and Koprowski, H. 1982. IgG2a monoclonal antibodies inhibit human tumor growth through interaction with effector cells. Proc. Natl. Acad. Sci. USA **79:** 4761-4765.

(6) Kamarck, M.E.; Barbosa, J.A.; Kühn, L.; Messer-Peters, P.G.; Schulman, L.; and Ruddle, F.H. 1983. Somatic cell genetics and flow cytometry. Cytometry **4:** 99-108.

(7) Kavathas, P., and Herzenberg, L.A. 1983. Stable transformation of mouse L cells for human membrane T-cell differentiation antigens, HLA and β_2-microglobulin: Selection by fluorescence-activated cell sorting. Proc. Natl. Acad. Sci. USA **80:** 524-528.

(8) Kipps, T.J., and Herzenberg, L.A. 1985. Hybridoma immunoglobulin isotype switch variant selection using the fluorescence activated cell sorter. In The Handbook of Experimental Immunology, 4th ed., eds. D.M. Weir, L.A. Herzenberg, and C.C. Blackwell. Edinburgh: Blackwell Scientific Publishers, Ltd.

(9) Krolick, K.A.; Uhr, J.W.; and Vitetta, E.S. 1982. Selective killing of leukaemia cells by antibody-toxin conjugates: implications for autologous bone marrow transplantation. Nature **295:** 604-605.

(10) Morrison, S.L.; Johnson, M.J.; Herzenberg, L.A.; and Oi, V.T. 1984. Chimeric human antibody molecules: Mouse antigen-binding domains with human constant region domains. Proc. Natl. Acad. Sci. USA **81:** 6851-6855.

(11) Müller, C.E., and Rajewsky, K. 1983. Isolation of immunoglobulin class switch variants from hybridoma lines secreting anti-idiotope antibodies by sequential sublining. J. Immunol. **131:** 877-881.

(12) Neuberger, M.S.; Williams, G.T.; and Fox, R.O. 1984. Recombinant antibodies possessing novel effector functions. Nature **312:** 604-608.

(13) Pyeritz, R.E. 1984. Treatment of inborn errors of metabolism by transplantation. Nature **312:** 405-406.

(14) Radbruch, A. 1985. Isotype switch variants. In The Handbook of Experimental Immunology, 4th ed., eds. D.M. Weir, L.A. Herzenberg, and C.C. Blackwell. Edinburgh: Blackwell Scientific Publishers, Ltd.

(15) Shapiro, H.M. 1983. Multistation multiparameter flow cytometry: A critical review and rationale. Cytometry **3:** 227-234.

Biotechnology: Potentials and Limitations, ed. S. Silver, pp. 185-196. Dahlem Konferenzen 1986. Berlin, Heidelberg, New York, Tokyo: Springer-Verlag.

Gene Transfer into Mouse Stem Cells

E.F. Wagner, U. Ruether, and C.L. Stewart
European Molecular Biology Laboratory
6900 Heidelberg, F.R. Germany

Abstract. The various approaches that can be used to introduce genes into embryos and stem cells will be reviewed. In particular, the results obtained using DNA injection and retroviral vectors will be summarized and the potential of these systems discussed.

INTRODUCTION

The transfer of recombinant genes into fertilized mouse eggs and various murine stem cells offers an excellent tool to analyze the regulation of gene expression. In such an experimental system, the introduced gene is exposed to all possible factors that would act during embryonic development and in the process of differentiation of a particular cell lineage. The correct expression of an exogenous gene in the context of the whole animal will provide us with new information concerning mechanisms controlling gene expression in development and differentiation.

Qualitative and quantitative correct expression of introduced genes, which are thought to be expressed in a particular cell at a given time during development, will allow us to elucidate factors that control cell-type specific gene expression. The transfer of certain genes, such as onc genes, into embryos should help to unravel the function of their products. In addition, such an approach could provide us with new information regarding the control of cell growth and the induction of malignant transformation. The feasibility of using gene transfer in somatic gene therapy for humans will certainly first be addressed by experiments with murine hemopoietic stem cells.

In this paper we would like to summarize recent results obtained in three experimental systems: a) gene transfer studies into mice by egg injection; b) introducing genes into embryonic stem cell lines and into mice; and c) the transfer of genes into murine hemopoietic cells using retroviral vectors.

Following a brief introduction and summary describing the current state of the field, we will discuss the future aspects and potentials of the particular experimental systems for basic and applied research.

GENE TRANSFER INTO MICE BY EGG INJECTION

Summary of Current Data

The first attempts to transfer foreign DNA into mice were made about ten years ago, when SV40 viral DNA was injected into early mouse embryos and DNA sequences were detected in the adult animals. Within the last four years, microinjection of cloned genes into one of the two pronuclei of the fertilized egg has become a standard technique and one of the most promising systems for studying the expression of exogenous genes in the context of the whole organism. To date, a number of cloned genes have been stably introduced into mice (see Table 1). In most cases the injected genes are integrated into the host cell DNA and are transmitted through the germ line to offspring, whereby new strains of mice homozygous for the acquired genotype can be obtained. Such animals carrying foreign genes are nowadays commonly called "transgenic."

In the first series of experiments using thymidine kinase genes, β-globin genes or growth hormone genes, expression of the introduced gene either was poorly investigated or was not found (Table 1). In no case was it possible to detect tissue-specific expression of the introduced gene. The best results concerning a regulated expression were obtained when the genes, such as thymidine kinase or growth hormone, were linked to the mouse metallothionein promoter region. Expression was often found in response to heavy metal induction, but the level of expression varied to a large extent between individual animals as well as between parent and offspring. In addition, the observed level of expression was not in the normal range, nor was it controllable from the external inducers.

First indications of a possible cell-type specific expression came from transgenic mice carrying the chicken transferrin or mouse immunoglobulin genes. The introduced genes were preferentially expressed in the right tissues, where they are normally synthesized, e.g., transferrin in the liver. The immunoglobulin genes were found to be expressed exclusively in the lymphoid cells; whether expression is restricted to B cells is still under investigation.

Transgenic mice expressing the rat elastase I gene in the pancreas was the first example where a foreign gene was expressed strictly in a cell-type specific way, both qualitatively and quantitatively. Three additional genes, the rat myosin light chain II gene, a truncated version of the α-fetoprotein gene, and a β-globin hybrid gene were found to be expressed in the appropriate target cells; however, the level of expression did not correlate with the normal level of expression. No clear picture is yet emerging as to why certain genes can be expressed quite specifically, whereas others show a very poor functional activity. Recently it has been suggested that the

TABLE 1 - DNA sequences injected into mouse eggs.

DNA	Modes of expression			Ref.
	None	Unspecific	Specific	
SV40	+			14
TK-HSV[1)]	ND			11
TK-HSV		+		36
MLV[2)]		+		13, 30
MT-TK[3)]		+		3
MT-GH[4)], rat		+		22
GH, human		+		35
β-globin, rabbit		+		7, 18
β-globin, human	+			33
Transferrin, chicken			+)	19
Ig kappa, mouse			+)	3
Ig mu, mouse			+)	12
Ig kappa & mu, mouse			+	25
Elastase I, rat			+	21, 34
Myosin light chain II, rat			+	28
AFP[5)], mouse			+	17
β-globin, hybrid, mouse & human			+	6
SV40-T-antigen			+)	2
MMTV-<u>myc</u>[6)]			+)	2

Not all data concerning DNA injections into eggs are listed in this table.

Abbreviations: 1) TK-HSV, thymidine kinase from herpes simplex virus; 2) MLV, Moloney murine leukemia virus; 3) MT-TK, metallothionein-thymidine kinase fusion gene; 4) MT-GH, metallothionein-growth hormone fusion gene; 5) AFP, alpha-fetoprotein; 6) MMTV-<u>myc</u>, mouse mammary tumor virus-<u>myc</u> fusion gene; ND, not determined; +) only preference for specific expression.

presence of the vector sequences may inhibit the expression of the foreign gene (6).

The aim of a different set of experiments was to introduce genes into the germ line of mice and ask for a biological effect in the animal as a consequence of expression of the introduced gene(s). The most dramatic effect was obtained when eggs were injected with a metallothionein-growth hormone fusion gene leading to increased growth of the resulting mice. Expression of the transferred growth hormone genes was not found in the pituitary cells, where growth hormones are normally synthesized (22). Another aspect directed toward identifying the function of genes and their role in malignant transformation was addressed in experiments where oncogenes were introduced into the germ line of mice. In two cases, transgenic mice expressing either SV40 T-antigen genes or a modified version of the myc oncogene displayed an inheritable tendency to develop cancers (2, 32). The correlation between cell-type specific oncogene expression and the onset of a particular malignancy is incomplete. However, these experiments provide a way of studying in living animals the effects of genes that have been linked to the development of murine and human cancers.

One last aspect concerning the consequence of foreign DNA integration into the host cell genome should be mentioned. In at least two published cases, the integration of foreign DNA into the germ line of transgenic mice using DNA injection led to mutagenic effects, described as insertional mutations. In one case, prenatal lethals were obtained in mice homozygous for the foreign DNA, and an unusual transmission pattern was observed in the other transgenic mouse strain (23, 35). The identification of the gene(s), which had been disrupted by the foreign DNA insertion and which seems to be responsible for the biological effect, turns out to be rather difficult. The integration pattern appears to be much more complex than the insertion of a retroviral genome in the α1-collagen gene, which also caused homozygous lethals (26).

Potentials and Future Aspects

In the following we would like to list several applications where transgenic mice may help to answer still unsolved biological problems.

Analysis of cell-type specific gene expression. In combination with in vitro tissue culture experiments, it will be feasible to study cis-acting elements responsible for tissue specificity by using deletion mutants of the gene of interest. Possible trans factors controlling gene expression might also be identifiable by this approach.

Identification of the functions of isolated genes. With inducible or constitutive expression vectors it will be possible to overexpress a certain gene product of unknown function in a specific cell within the organism. The consequence of gene expression of certain onc genes or growth factor genes and their role in cell growth, malignant transformation, as well as the inheritance of cancer(s) can be studied. With "anti-sense" constructions specific gene products may be inactivated and thereby their functional role assessed.

Induction of insertional mutants and isolation of "developmental" genes. As a consequence of DNA integration, essential genes might be disrupted and recessive lethals induced. The gene(s) acting at a certain stage during pre- or postnatal development can be isolated and their function possibly studied by complementation analysis. This can already be tested with existing developmental mutants, e.g., from the T complex.

Correction of a genetic defect. Gene replacement experiments at the germ and somatic cell level can be performed on some already existing mouse models for human genetic diseases, e.g., the β-thalassemic mouse.

To optimize the chances of a successful somatic gene therapy experiment, directed gene insertion by homologous recombination should be attempted. In addition, conditions for single gene integration have to be established. A germ-line therapy experiment in humans is presently not considered.

Study of aspects of immunology. Questions concerning the mechanism of immunoglobulin gene rearrangements during the ontogeny of B cell differentiation can be addressed in the context of transgenic mice. Furthermore, aspects of T cell function, H-2 restriction, and the study of autoimmune responses will be accessible to molecular analysis through genetically altered strains of mice.

Commercial aspects. The use of transgenic mice for applied research may be relevant for obtaining substances such as hormones or vaccines which are synthesized in small amounts and which require a mammalian cell for optimal biological activity. In addition, "animal farming," the improvement of domestic animals by genetic manipulation, is another area that can first be tested in the mouse before a bigger and less understood host system will be applied.

Finally, we would like to point out that gene transfer into eggs by DNA injection can serve as a powerful tool for numerous aspects of basic and applied research, provided that the derivation of transgenic mice can be made routine (at least 10-50% of animals born carry injected DNA sequences) and that sufficient quantitative expression of the injected gene can be obtained, e.g., 10-20% of the "transgenic" animals are positive for expression at levels close to normal.

GENE TRANSFER INTO MOUSE EMBRYONIC STEM CELLS

Embryonic carcinoma (EC) or embryonic stem cells can be induced to differentiate into a wide variety of cell types both in vitro and in vivo. These cells were originally produced by transferring embryos to extrauterine sites where they would form tumors. The stem cells or EC cells could be isolated from these tumors and established in vitro. More recently, cells very similar to EC cells have been produced directly from mouse embryos explanted in vitro (8). This latter form is usually called ES cells. EC and ES cells have attracted much interest for a number of reasons which are outlined below.

Studies on the Control of Gene Expression in Vitro
ES and EC cells have been shown to share many morphological, biochemical, immunological, and developmental features in common with certain cell types of early mouse embryos (9). Indeed, the control of gene expression during differentiation of the stem cells appears to parallel the differentiation of the mouse embryo. Results derived from the introduction of cloned genes, e.g., H-2 class 1 antigens and α-fetoprotein, into these stem cells have shown that these genes are only expressed after the onset of differentiation (24, 27). In addition, studies using viruses such as SV40, polyoma, and murine retroviruses as model genes have shown that the full expression of viral genes is dependent on the state of differentiation of the cells. Thus these viruses are unable to replicate and express correctly viral genes in the stem cells, although they can do so in the differentiated derivatives (16). These results correlated well with studies performed on infecting embryos with these viruses. In general, virus expression was only found in the differentiated cells of the embryos.

A different approach has been to analyze what effect the expression of exogenous genes in these stem cells would have on differentiation. It was shown that transfection of the c-fos protooncogene into F9 EC cells resulted in the isolation of many clones that had a diffentiated phenotype and were expressing the fos protein (20). These results further support the hypothesis that the expression of protooncogenes has some role in stem cell differentiation and development.

The Control of Gene Expression in EC/ES Cells in Vivo
A second reason why EC and ES cells have attracted interest has arisen from the demonstration that these cells could be reintroduced back into embryos where they participate in normal development in vivo (29). Furthermore, in some individuals, gametes were produced from the EC and ES cells. This approach has therefore offered the possibility of manipulating the mouse genome by any technique available to the somatic cell geneticist. Various attempts have been made to apply this approach, such as using mutant EC cell clones lacking the thymidine kinase or HGPRT enzymes (29). None of these attempts were successful in producing chimaeras which could have transmitted the mutant genotype to their progeny. However, enthusiasm has recently been reawakened by the observation that many ES cell lines can be extremely efficient in colonizing the germ line (1), and genetically altered ES cell clones are now being tested for germ line transmission. Here, we would like to summarize our results that demonstrate some of the advantages of this system.

Our particular approach has been to introduce genes into EC and ES cells with retroviral vectors, since they overcome many of the problems associated with DNA-mediated gene transfer such as the difficulty of transfecting the cells as well as the stability of expression of the introduced DNA. There is one disadvantage with these vectors in that it has been previously shown that early embryos as well as EC and ES cells possess an efficient mechanism that suppresses retroviral gene expression (10, 31). However, retroviral vectors containing selectable marker genes under the control of the long terminal repeat (LTR) or an internal promoter can be used to select clones of EC and ES cells in vitro (37). Some of these clones have been reintroduced into embryos and a number of chimaeric mice were obtained. It has been found that the introduced vector carrying the neomycin gene under the control of the LTR can be expressed in most of the adult tissues analyzed (Stewart and Wagner, in preparation). From a different series of experiments chimaeras were produced with selected ES cells carrying in addition the v-myc oncogene. Presently these animals are being bred to see if they will transmit the vector to their offspring. They are now being analyzed for the expression of the introduced sequences.

Potentials and Future Aspects

It is still true today that teratocarcinomas continue to attract interest largely because of their potential rather than for what they have revealed about how the development of the mouse embryo proceeds. Here again we will list some of the potential areas for future research using ES and EC cells.

Gene expression and early mouse development. It has been argued that there appear to be close similarities between the control of gene expression in these cells and in embryos. Thus, further study of gene control mechanisms may be of particular use since it is possible that in these stem cells the molecular mechanisms exist by which the major developmental decisions are made, i.e., the molecular events that would lead to the formation of the major organs and tissues of the body.

Alternative route to manipulating the mouse genome. The most encouraging development has been the discovery that ES cells can be reproducibly isolated from blastocysts explanted in vitro (8). Prior to the discovery of these cells, the prospect of being able to manipulate the mouse genome in vitro looked grim, largely due to the inefficiency of established EC cells in colonizing the germ line. However, the demonstration that some ES cell lines can colonize the germ line extremely efficiently (1) has done much to stimulate an interest in showing that this in vitro approach may provide an alternative to DNA injection of fertilized eggs (see section on GENE TRANSFER INTO MICE BY EGG INJECTION). It is, however, yet to be shown that ES cells whose genome has been manipulated in vitro are still capable of forming functional gametes. However, their potential will probably be restricted as an experimental system for introducing genes into mouse embryos and studying the control of gene expression in stem cells, and they will not easily offer a means to human gene therapy.

Animal models for human genetic diseases. Following the points made above, we feel that the derivation of mutant ES cells carrying enzyme defects (for instance, lacking a functional HGPRT enzyme) should be attempted again to obtain an animal model for Lesch-Nyhan disease.

Malignancy and normal differentiation. It still remains an unsolved question as to why ES and EC cells form tumors when injected into adult mice, whereas when introduced into embryos they form chimaeras. It will be of interest to discover if introduced onc genes that are constitutively expressed in these stem cells will perturb the interaction between EC/ES cells and embryos.

Studies with human ES cells. What the approach of establishing ES cells in vitro does offer is a potential method for establishing equivalent human cell lines from the excess embryos that are produced from couples who require in vitro fertilization. Such cell lines would be of enormous use in uncovering many aspects of the biology, growth control, and molecular genetics of human development, provided that the experiments were strictly confined to in vitro studies.

GENE TRANSFER INTO BONE MARROW CELLS BY RETROVIRUS INFECTION

Introduction and Summary of Current Data

Pluripotent stem cells of the hemopoietic system are another population of cells well suited for gene transfer experiments. These cells, found in adult bone marrow, can be manipulated in vitro and then transferred to hemopoietically deficient hosts where they are able to generate cells of both the myeloid (e.g., red cells, granulocytes, macrophages) and lymphoid lineages. The stem cells are capable of producing progeny which remain pluripotent ("self-renewal") and cells which are more restricted in their differentiation potential ("commitment"). Although the primitive stem cells can be assayed only in long-term reconstitution experiments, their descendents, the more restricted pluripotent and terminally differentiated cells, can be detected in in vivo and in vitro clonal assays (5).

Gene transfer into bone marrow cells by conventional techniques such as DNA transfection, microinjection, etc., has not succeeded. Initial experiments in the field do not seem to be reproducible. The first successful attempts to introduce and express a selectable marker gene conferring neomycin (neo) resistance have only recently become possible through the use of retroviral vectors. In one report, expression of the transferred selectable neo gene was assayed by in vitro colony assay following selection of infected bone marrow cultures (15). A different study combined the infection of bone marrow cultures in vitro with the reintroduction of the infected cells into lethally irradiated recipients. In these experiments, vector DNA was found in the reconstituted mice. However, expression of the introduced genes was not analyzed (38). In collaboration with G. Keller from the Basel Institute of Immunology, we were able to introduce and express the selectable neo gene as well as a nonselectable gene, the v-myc oncogene in infected bone marrow cultures. In addition, expression of neo RNA and protein was found in various hemopoietic cell types of reconstituted mice, and the clonal origin of myeloid and lymphoid precursors can also be traced from DNA restriction patterns (G. Keller et al., in preparation). It appears that many new possibilities are now available for a molecular analysis of hemopoietic stem cell differentiation.

Potentials and Future Aspects

Since gene transfer studies into hemopoietic stem cells combined with a molecular analysis of the introduced genes is a very new research area, we do not want to speculate too much on the potential and will only briefly outline a few possible outstanding applications (some are already mentioned in the section on the Control of Gene Expression in Vitro).

Use in gene therapy experiments. This approach seems at present to be the only possible way for a correction of a genetic defect at the somatic cell level and will certainly have future applications in human gene therapy. The first experiments will certainly be done using the β-thalassemic mouse as a model for human β-thalassemia.

Characterization of pluripotent stem cells and differentiation studies. It is now necessary to determine whether or not the most primitive stem cell, the cell capable of giving rise to all myeloid and lymphoid cells, can be infected with a recombinant virus. If this can be shown satisfactorily, then viral vectors containing transforming genes will be used to immortalize the pluripotent stem cells so as to expand these cells to large homogenous populations. If, in addition, the transforming principle could be inactivated at any time so that their normal physiology is restored, or vice versa, if transformation could be induced at a given time during stem cell differentiation, then many requirements for a molecular approach to their function in differentiation could be fulfilled.

Cell-type specific expression. Since the hemopoietic system is well characterized from a cell biological viewpoint, studies can be performed directed towards analyzing cell-type specific gene expression using genes that are only expressed in one cell of a given lineage. In addition, it may become possible to express regulatory genes which have yet to be defined in stem cells, thereby inducing a distinct differentiation event.

It should be clear that many future potentials of the hemopoietic stem cell system have medical importance for alleviating disease symptoms in the treatment of human cancers and in treating human genetic diseases.

Acknowledgements. We thank J. Collins and S. Silver for critically reading the manuscript and I. Benner for typing it. U. Ruether is a recipient of a long-term EMBO fellowship.

REFERENCES

(1) Bradley, A.; Evans, M.; Kaufman, M.H.; and Robertson, E. 1984. Formation of germ-line chimaeras from embryo-derived teratocarcinoma cell lines. Nature **309:** 255-256.

(2) Brinster, R.L.; Chen, H.Y.; Messing, A.; van Dyke, T.; Levine, A.J.; and Palmiter, R.D. 1984. Transgenic mice harboring SV40 T-antigen genes develop characteristic brain tumors. Cell **37:** 367-379.

(3) Brinster, R.L.; Chen, H.Y.; Trumbauer, M.; Senear, A.W.; Warren, R.; and Palmiter, R.D. 1981. Somatic expression of herpes thymidine kinase in mice following injection of a fusion gene into eggs. Cell **27:** 223-231.

(4) Brinster, R.L.; Ritchie, K.A.; Hammer, R.E.; O'Brien, R.L.; Arp, B.; and Storb, U. 1983. Expression of a microinjected immunoglobulin gene in the spleen of transgenic mice. Nature **306:** 332-336.

(5) Burgess, A., and Nicola, N. 1983. Growth Factors and Stem Cells. New York: Academic Press.

(6) Chada, K.; Magran, J.; Raphael, K.; Radice, G.; Lacy, E.; and Costantini, F. 1985. Specific expression of a foreign β-globin gene in erythroid cells of transgenic mice. Nature **314:** 377-380.

(7) Costantini, F., and Lacy, E. 1981. Introduction of a rabbit β-globin gene into the mouse germ line. Nature **294:** 92-94.

(8) Evans, M.J., and Kaufman, M.H. 1981. Establishment in culture of pluripotential cells from mouse embryos. Nature **292:** 154-156.

(9) Gardner, R.L. 1983. Embryonic and germ cell tumours in man and animals. Cancer Surv. **2(1).** Oxford: University Press Oxford.

(10) Gautsch, J.W., and Wilson, M.C. 1983. Delayed de novo methylation in teratocarcinoma suggests additional tissue-specific mechanisms for controlling gene expression. Nature **301:** 32-37.

(11) Gordon, J.W.; Scargos, G.A.; Plotkin, D.J.; Barbosa, J.A.; and Ruddle, F.H. 1980. Genetic transformation of mouse embryos by microinjection of purified DNA. Proc. Natl. Acad. Sci. USA **77:** 7380-7384.

(12) Grosschedl, R.; Weaver, D.; Baltimore, D.; and Costantini, F. 1984. Introduction of a μ immunoglobulin gene into the mouse germ line: specific expression in lymphoid cells and synthesis of functional antibody. Cell **38:** 647-658.

(13) Harbers, K.; Jähner, D.; and Jaenisch, R. 1981. Microinjection of cloned retroviral genomes into mouse zygotes: integration and expression in the animal. Nature **293:** 540-542.

(14) Jaenisch, R., and Mintz, B. 1974. Simian virus 40 DNA sequences in DNA of healthy adult mice derived from preimplantation blastocysts injected with viral DNA. Proc. Natl. Acad. Sci. USA **71:** 1250-1254.

(15) Joyner, A.; Keller, G.; Phillips, R.A.; and Bernstein, A. 1983. Retrovirus transfer of a bacterial gene into mouse haematopoietic progenitor cells. Nature **305:** 556-558.

(16) Kelly, F., and Condamine, H. 1982. Tumor viruses and early mouse embryos. Biochim. Biophys. Acta **651:** 105-141.

(17) Krumlauf, R.; Hammer, R.; Tilghman, S.; and Brinster, R.L. 1985. Developmental regulation of alphafoeto protein genes in transgenic mice. Molec. Cell. Biol. **5:** 1639-1648.

(18) Lacy, E.; Roberts, S.; Evans, E.P.; Burtenshaw, M.D.; and Costantini, F. 1983. A foreign β-globin gene in transgenic mice: integration at abnormal chromosomal positions and expression in inappropriate tissues. Cell **34:** 343-358.

(19) McKnight, G.S.; Hammer, R.E.; Kuenzel, E.A.; and Brinster, R.L. 1983. Expression of the chicken transferrin gene in transgenic mice. Cell **34:** 335-341.

(20) Müller, R., and Wagner, E.F. 1984. Differentiation of F9 teratocarcinoma stem cells after transfer of c-fos protooncogenes. Nature **311:** 438-442.

(21) Ornitz, D.M.; Palmiter, R.D.; Hammer, R.E.; Brinster, R.L.; Swift, G.H.; and McDonald, J.R. 1985. Specific expression of an elastase-human growth hormone fusion gene in pancreatic acinar cells of transgenic mice. Nature **313:** 600-602.

(22) Palmiter, R.D.; Brinster, R.L.; Hammer, R.E.; Trumbauer, M.E.; Rosenfeld, M.G.; Birnberg, N.C.; and Evans, R.M. 1982. Dramatic growth of mice that develop from eggs microinjected with metallothionein growth hormone fusion genes. Nature **300:** 611-615.

(23) Palmiter, R.D.; Wilkie, T.M.; Chen, H.Y.; and Brinster, R.L. 1984. Transmission distortion and mosaicism in an unusual transgenic mouse pedigree. Cell **36:** 869-877.

(24) Rosenthal, A.; Wright, S.; Cedar, H.; Flavell, R.; and Grosveld, F. 1984. Regulated expression of an introduced MHC H-2K^{bm1} gene in murine embryonal carcinoma cells. Nature **310:** 415-417.

(25) Rusconi, S., and Köhler, G. 1985. Transmission and expression of a specific pair of rearranged immunoglobulin μ and K genes in a transgenic mouse line. Nature **314:** 330-334.

(26) Schnieke, A.; Harbers, K.; and Jaenisch, R. 1983. Embryonic lethal mutation in mice induced by retrovirus insertion into the α1(I)collagen gene. Nature **304:** 315-320.

(27) Scott, R.W.; Vogt, T.F.; Croke, M.E.; and Tilghman, S.M. 1984. Tissue-specific activation of a cloned α-fetoprotein gene during differentiation of a transfected embryonal carcinoma cell line. Nature **310:** 562-566.

(28) Shani, M. 1985. Tissue-specific expression of rat myosin light-chain 2 gene in transgenic mice. Nature **314:** 283-286.

(29) Stewart, C.L. 1984. Teratocarcinoma chimaeras and gene expression. **In** Chimaeras in Developmental Biology, eds. N. Le Douarin and A. McLaren, pp. 409-427. London: Academic Press.

(30) Stewart, C.L.; Harbers, K.; Jähner, D.; and Jaenisch, R. 1983. X chromosome-linked transmission and expression of retroviral genomes microinjected into mouse zygotes. Science **221:** 760-762.

(31) Stewart, C.L.; Stuhlmann, H.; Jähner, D.; and Jaenisch, R. 1982. De novo methylation, expression and infectivity of retroviral genomes introduced into embryonal carcinoma cells. Proc. Natl. Acad. Sci. USA **79:** 4098-4102.

(32) Stewart, T.A.; Pattengale, P.K.; and Leder, P. 1984. Spontaneous mammary adenocarcinomas in transgenic mice that carry and express MTV/myc fusion genes. Cell **38:** 627-637.

(33) Stewart, T.A.; Wagner, E.F.; and Mintz, B. 1982. Human β-globin gene sequences injected into mouse eggs, retained in adults, and transmitted to progeny. Science **217:** 1046-1048.

(34) Swift, G.H.; Hammer, R.E.; McDonald, R.J.; and Brinster, R.L. 1984. Tissue-specific expression of the rat pancreatic elastase I gene in transgenic mice. Cell **38:** 639-646.

(35) Wagner, E.F.; Covarrubias, L.; Stewart, T.A.; and Mintz, B. 1983. Prenatal lethalities in mice homozygous for human growth hormone gene sequences integrated in the germ line. Cell **35:** 647-655.

(36) Wagner, E.F.; Stewart, T.A.; and Mintz, B. 1981. The human β-globin gene and a functional viral thymidine kinase gene in developing mice. Proc. Natl. Acad. Sci. USA **78:** 5016-5020.

(37) Wagner, E.F.; Vanek, M.; and Vennström, B. 1985. Transfer of genes into embryonal carcinoma cells by retrovirus infection: efficient expression from an internal promoter. EMBO J. **4:** 663-666.

(38) Williams, D.A.; Lemischka, I.R.; Nathan, D.G.; and Mulligan, R. 1984. Introduction of new genetic material into pluripotent haematopoietic stem cells of the mouse. Nature **310:** 476-480.

Top row, left to right:
Peter Starlinger, Francesco Salamini, Klaus Hahlbrock, Elmar Weiler, Dick Flavell, Meinhart Zenk

Middle row, left to right:
Wolfgang Barz, Regine Kahmann, Philip Abelson, Heinz Saedler

Bottom row, left to right:
Jim Peacock, John Beringer, Paul Broda, Doug Eveleigh

Biotechnology: Potentials and Limitations, ed. S. Silver, pp. 199-221. Dahlem Konferenzen 1986. Berlin, Heidelberg, New York, Tokyo: Springer-Verlag.

Plants and Agriculture

Group Report

R.B. Flavell, Rapporteur

W. Barz, J.E. Beringer, P. Broda, D.E. Eveleigh, K. Hahlbrock, R. Kahmann, W.J. Peacock, H. Saedler, F. Salamini, P. Starlinger, E. Weiler, M. Zenk

INTRODUCTION

The selection of "improved" plants by man and the practice of agriculture are thousands of years old. For many species, selection and careful culture have produced remarkable increases in harvestable products to provide man with food, fuel for cooking, paper, fiber, energy, and a range of special chemicals including medicinal compounds and wood for housing, construction, etc. The need for continuation of the improvements in many species, for adaptation of plants to new industrial processes and for manipulating agriculture to the changing patterns of societies and economies is greater than ever before.

Improvement in productivity in agriculture often involves soil microorganism modifications as well as changes in the germplasm of plants and in agronomy and husbandry. The central involvement of plant and microbial genes in determining agricultural productivity assures that the application of the powerful new techniques of gene engineering and cell culture will without question result in enhanced plant productivity. But what is needed for this to happen? This question provided the background for the deliberations of our discussion group. It was assumed that these techniques will supplement and should not replace existing plant improvement programs. Indeed, they will be dependent upon the continuation and expansion of the conventional programs if they are to reach the commercial marketplace.

It is worth noting at the beginning of this report that the application of the techniques of molecular biology to the genetic improvement of most of the plants and microorganisms important in agriculture has not been initiated. Molecular biological studies have been established for a few species by only relatively few people for less than ten years. Therefore, it is too early to see most of the opportunities that will emerge for the application of this technology as well as many of the problems that will arise in those instances where a relatively straightforward solution to a problem appears possible. It is necessary to point out that the genetic complexity of most of the properties that need to be changed and improved is unknown. This is a serious deficiency when applying the new powerful techniques of molecular biology for plant and microbe improvement, but at the same time it is the development of these techniques that is providing the opportunities to discover much of the missing information. This conclusion can be amply supported by the dramatic progress made in the last few years in those areas where research effort has been concentrated.

The goals for biotechnology in agricultural and plant cell research are in most cases the same as those of more conventional programs. In virtually all crop improvement programs the principal goals are to improve yield potential, resistance to diseases, pests, and environmental stresses, and the quality of the harvested product. Where economic efficiency is important, changes to crops or cells which can reduce chemical, energy, or labor inputs without proportional losses in yield are desirable. However, biotechnology also offers the possibility of new kinds of solutions to problems. For example, the biochemical pathways of specific secondary products may be genetically engineered so that high concentrations of products can be synthesized in and extracted from cells in culture more cheaply than from specialized cells of natural plants grown in fields. Also, completely new tools can be developed through molecular biology to make conventional plant improvement programs more efficient.

The range of agriculturally important species that man has selected to satisfy his various needs and preferences is extremely diverse. Therefore, the opportunities for biotechnological intervention in future improvement programs are extensive and varied. The enormous size of this target for research and application is perhaps bewildering for any group which sets out to consider the opportunities and priorities. The diversity is being created and sustained by many different interest groups, public and private, in every country in the world. Each will have its own specific needs and preferences. The group therefore spent much of its time addressing general opportunities and needs rather than the details of specific experimental approaches.

In this report the plants and microorganisms important in agriculture are reviewed from the point of view of their current technical suitability for the application of biotechnology. From this some needs and opportunities are apparent. Some of the general applications and developments of biotechnology to crops, plant cells, and microorganisms that received special consideration in our discussions are also presented. We also considered some other issues associated with the application of biotechnology to agriculture, and these are summarized at the end of the report.

IMPORTANT PLANT/MICROBES AND THEIR CURRENT TECHNICAL SUITABILITY FOR THE APPLICATION OF BIOTECHNOLOGY

Plant species important in agriculture, horticulture, biomass production, and other industries are included in Tables 1 to 5. All of the species have probably been subjected to breeding and selection, but the effort devoted to improvement of some is ongoing and substantial (e.g., Triticum aestivum (wheat) and Zea mays (maize)) while that to most is small or virtually non-existent. The quality and depth of breeding programs of a particular species will influence strongly the ease of success of the application of biotechnology, especially in the near and medium-term future.

The insertion of defined single genes into a plant species requires that a whole plant be regenerated from the single cell receiving the new gene. This is possible via tissue culture for some species, especially the Solanaceae, e.g., Nicotiana (tobacco), Solanum tuberosum (potato), and Petunia hybrida (petunia). For some species it is possible to regenerate at least some genotypes from protoplasts, while for others it is possible only from collections of cells - in callus or suspension culture. Sometimes it is possible to regenerate only from cells which have been derived from embryos and which have not lost their capacity to proliferate embryo-like structures in culture. The current status of the ability of cells of different species to regenerate plants is therefore included in Tables 1-5.

Genes can be efficiently inserted into cells of some plant species from the Ti plasmid of Agrobacterium tumefaciens and so the ability of a plant to stimulate this complex, gene transfer process is a very important property for genetic engineers. Research is advancing both on extending the number of suitable recipient plant species capable of receiving genes via Agrobacterium and on how to regenerate the modified cells into whole plants. Some of the species into which engineered genes have been introduced via Agrobacterium or direct uptake of DNA into protoplasts are listed in Table 6.

Many (but by no means all) of the important genes to be isolated, manipulated, and reinserted will first have to be recognized in a plant species. For this reason those plants which have been studied by geneticists and those processes whose genetic bases are understood are likely to be manipulated earlier. It is anticipated that important genes regulating development, the production of defined end products and, for example, disease resistance will be isolated after being tagged by a molecularly characterized, transposable element. If such elements are not available for use within a species, then the isolation of such genes may be extremely difficult and time-consuming. Therefore, information as to whether or not genetically defined transposable elements have been recognized and isolated is included in the tables.

Microorganisms which are appropriate candidates for genetic improvement such as Rhizobium species and mycorrhizal fungi are shown in Table 7. A large inoculum industry already exists for Rhizobium. If any of these organisms is to be modified to improve its host range, competition with endogenous strains, N-fixing ability, or interaction with appropriate plant cells, then advanced genetic knowledge will be required such as is now available for the "fast growing" Rhizobium strains.

From perusal of the tables several key conclusions are easily drawn which are important to note if biotechnology is to make a major impact on plant improvement. The crop plant species which have most of the desirable features for immediate biotechnology application are very few. For Nicotiana tabacum (tobacco), Lycopersicum esculentum (tomato), Solanum tuberosum (potato), and some Brassica species there are good breeding programs, genetic markers, the ability to introduce genes by Agrobacterium, and efficient regeneration of plants from single cells. For the most important legumes such as Glycine max and Vicia faba and for cereals (e.g., wheat, barley, rice, and maize) there remain important deficiencies for the development of plants with newly inserted genes. Similar deficiencies remain for the important tree species. For most species it is not known how easy or difficult it will be to insert new genes into them.

With such a broad range of organisms (especially plants) needing improvements to keep up with the world production needs, which species should receive the most attention from a limited number of biotechnologists? Undoubtedly scientists will choose those organisms and biological systems where progress can be made - often research funding agencies demand annual (i.e., rapid) progress. However, it can be argued that the crops which have received the most attention will be harder to improve dramatically and that in any case the existing commitment to improvement of these plants will suffice. Instead, many important crops whose rate of improvement is "in the doldrums" should receive the special attention of both conventional breeders and biotechnologists. This debate is likely to emerge more strongly as more species are successfully manipulated by biotechnological methods. Opportunities are also likely to change with time. For example, crop production in the tropics may become much more important than at present as more investment is made in crop production systems.

BIOTECHNOLOGY APPROACHES THAT CAN BENEFIT PLANT IMPROVEMENT

From our discussions, some of the many perspectives as to how biotechnology can help improve agricultural and plant product yields are summarized below.

Micropropagation of Plants from Meristems and Callus

The propagation of plants from meristems and to a lesser extent from callus is already being exploited commercially on a considerable scale, particularly with ornamental species and trees. The advantage to this is that successful heterozygous genotypes can be propagated rapidly without loss of the successful gene combinations that would accompany meiosis and propagation by seed. However, the application of this technology should and can be expanded considerably, especially for trees (for oil, timber, paper, pulp, fiber, and energy) where the development and multiplication of improved germplasm by seed are very slow and the needs for yield improvements urgent. Local assessments of successful genotypes can be made throughout the world and meristems of cells from those individuals bulked up, where the technology for the species is available.

Such propagation programs inevitably produce populations of individuals which are virtually identical (although plants regenerated from callus often carry mutations of various sorts). Genetic uniformity increases the vulnerability of the crop to specific diseases, but the replacement by resistant genotypes can also be achieved efficiently by micropropagation if disease-resistant forms can be recognized.

Propagation of plants from meristems or cells usually eliminates viruses and viroids, and so this technology is also playing an important role in providing disease-free stocks for agriculture. This is especially valuable in vegetatively propagated crops, such as potato and cassava, in which the presence of these pathogens causes considerable yield losses throughout the world.

Molecular Diagnostic Methods to Improve Breeding Programs

In many advanced breeding programs the breeder is not short of variation for crop improvement, but analyzing it in large populations is time-consuming, inefficient, and expensive. Rapid methods that can facilitate the selection or screening of plants with the products of desirable genes or of genes tightly linked to desirable genes would be very helpful. Biochemistry and molecular genetics can provide such methods. If it is possible to detect easily the protein or metabolic variant associated with a phenotypic trait which is difficult to measure easily and reliably (e.g., disease resistance, cold tolerance), then detection of the protein or metabolite in samples of the seed or seedling will on many occasions provide a real improvement in efficiency. Therefore, the molecular basis of important phenotypic traits should be sought. Also, the genetic location of easily measured markers in relation to the locations of desirable genes should be established so that convenient, linked markers can be used wherever possible. Biochemistry and molecular genetics offer new ways of producing generalized linkage maps much more rapidly than the methods of classical genetics.

Considerable variation that is not manifest in the phenotype exists in the DNA of all crop species. Thus, if a series of DNA probes can be developed which are capable of detecting this variation at known sites throughout the chromosome complement, then these probes will enable single (or combinations of) genes to be followed by genetic linkage in segregating populations. This approach is technically simplest where many copies of the same marker sequence are clustered near the gene of interest - as is sometimes the case when segments of chromosomes are transferred from one species to another. The presence or absence of the species-specific, repeated marker sequences can be assayed under these circumstances in the DNA extruded when root tip or other tissue segments are squashed. To detect variation in DNA sequences common to segregating plants it is necessary to purify the DNA from each plant and perform several manipulations before investigating the variant sequences with DNA probes. Here, fewer samples can be handled but many more chromosomal sites can be surveyed. The development of libraries of DNA probes which can facilitate the simultaneous detection, by linkage, of several alleles in a large number of individuals could and should be developed for all major crop species. Where selectable or easily screenable markers do not exist in elite germplasm, they can be inserted from the test tube into species where gene insertion is possible. This offers the opportunity for very considerable improvements in the efficiency of breeding programs in the long term.

Molecular Diagnostics in Plant Pathology

The susceptibility or resistance of plants to specific viruses, viroids, and virusoids is still determined in many breeding programs by scoring visual symptoms. This is time-consuming and expensive, and sometimes difficult and unreliable. The use of DNA probes (for all three) and/or monoclonal antibodies (for viruses) for detecting extremely low levels of these pathogens is more efficient, and often only small samples of tissue are necessary. Susceptible genotypes which allow the proliferation of a virus are easily distinguished from those which do not. All breeding programs in which resistance to such pathogens is sought should use these new procedures if maximum efficiency of time, space, and labor is required. Similar diagnostics may be developed for fungi and bacteria. The use of these diagnostics should simplify decision making in plant quarantine laboratories and in the certification of disease-free seed or plant material. Currently, use of the DNA probes requires labelling the probes with ^{32}P and detecting the probe by X-ray film autoradiography. Nonradioactive enzyme-linked detection systems are being developed and should be adapted to plant systems so that the pathogens can be detected rapidly in field laboratories.

The Isolation of Useful Mutants and Genes

Plant improvement by biotechnological as opposed to most conventional methods depends in part on inducing and selecting new favorable mutations and isolating important genes. It may be possible under certain circumstances, using cells in culture, to select for resistance to pathogens' toxins and other environmental toxins; such resistance will be retained through plant development. These possibilities should be explored for many species, including trees.

It is now well-known that the frequency of many sorts of mutations rises when plant cells are maintained in culture. Some (but only very few) of the mutations may be favorable, thus making cell culture an attractive means of introducing limited useful variation into crop plants. This approach to crop improvement should be explored more fully and the molecular basis of the various mutational processes explored. Research devoted to the means of controlling mutation frequency should also be undertaken. Recent results suggest that the mutation frequency is higher in protoplasts and is related to the length of time the cells are maintained in culture.

As noted earlier, many important genes that are required for the modification of important traits such as disease resistance, flowering time, and vernalization requirement are unknown at the biochemical or molecular levels. However, they could be isolated if they could be inactivated by insertion of a molecularly characterized transposon and the phenotypic consequences observed in subsequent generations. This transposon is then reisolated, together with the neighboring gene. A piece of the gene is then used to isolate allelic functional variants from other varieties. The development of this approach is urgently needed in the important crop species.

Another approach to the molecular recognition of important genes is currently being tested. Segments of the antisense strands of genes are inserted into plants via transformation methods such that their transcription is ensured. If the antisense messenger RNA complexes with the natural messenger RNA, the gene product may be in a significantly lower concentration. If this affects the phenotype, such plants might be recognized easily. The gene homologous with the antisense RNA can then be isolated.

Modification of Plants by Insertion of New Genes

Because all plant species can be improved or modified for new purposes by the insertion of single genes, the methods required to achieve gene insertion, stable expression, and inheritance should be developed for crop species, as discussed earlier. It will be necessary to regulate the expression of the introduced genes. Part of the regulation will be determined by the structure of the DNA inserted. Therefore more information on gene control mechanisms should be gathered, together with pieces of DNA that will provide the appropriate control for each of the recipient plant species.

When these systems are available, the opportunity to improve plants by single gene insertion will emerge. However, it will be necessary to identify and isolate favorable genes or alleles to enhance harvested yield, resistances to disease and environmental stresses, and product quality. This will be the major preoccupation of future plant biotechnology research together with learning how to alter gene control mechanisms to optimize gene action in plant development. The identification and exploitation of genes to control breeding systems, especially hybrid seed production, is also an important objective.

Some features of our current kowledge of plant molecular genetics were discussed because of their potential relevance to the genetic modification of plants. Some genes are present in only one copy per haploid genome while others are found in many copies. Why? No simple answer can be given because nature has probably exploited many different solutions to evolve adequate gene control mechanisms. However, there are some clues to suggest that levels of gene product may be manipulated by varying the number of genes inserted. Also, the differing expression of genes in different tissues may require the insertion of different gene variants.

Many of the genes required to influence plant phenotype will be regulatory genes. We know so little about these and how they work that much new research is needed to isolate and characterize such genes. However, such genes are likely to be vital in coordinating the activities of the many genes required to modify the phenotype in defined, desired ways.

Some discussion developed on the extent to which deficiencies in crop or cell performance were due to physiological barriers rather than genetic deficiencies. Although physiological barriers do exist, it is likely that only rarely do they currently limit the accumulation of valuable metabolites. Even when they occur, solutions to the problem can often be developed by breeding.

Many desirable genes for insertion into plants were nominated in our discussions, but few are yet characterized or indeed have been looked for. However, some new genes which are likely to lead eventually to new, commercially valuable germplasm have already been inserted into plants. These include genes for resistance to herbicides and insect pathogens.

Strains of bacteria and yeast resistant to specific established herbicides have been selected and the single genes responsible for the resistance have been isolated and inserted into tobacco plants in such a form that the enzyme product, resistant to inactivation by the herbicide, is synthesized (3). The rapid progress in these experiments was due to the easy recognition (selection) of the desired genes in microorganisms. Further research should be oriented towards determining how these genes may be regulated during plant development for optimum value. It was noted that creation of new herbicide resistances might lead to increased agricultural efficiency by reducing competition from weeds and also by reducing the labor required to remove weeds from emerging crops. However, there is the risk that with more crops resistant to herbicides, more herbicides will be used with adverse consequences for the environment.

Genes have been isolated from Bacillus thuringiensis which specify synthesis of a spore protein which is toxic to lepidopterous larvae. This gene has been inserted into tobacco plants to see if it can serve as a substitute for chemical insecticides in other crops without causing major reduction in plant performance. For this example, and particularly for other examples where the insertion of new genes for resistance to pathogens was discussed, it was stressed that the resistance (toxicity to pathogen and plant cells) might involve compounds toxic to man or to animals if consumed regularly. Plant cells often produce toxic secondary products only in response to pathogens or in specialized cells. Genes inserted for toxin production by the genetic engineer should be similarly regulated where possible and potential problems emanating from higher levels of toxic chemicals should not be forgotten.

The genes already known to bestow specific resistance to pathogens in crop plants are not understood in biochemical or molecular terms. There is an urgent need for intensive research here - on both plants and pathogens. If we had a better appreciation of the molecular basis of the different ways plants defend themselves and microbes attack plants, then molecular solutions for host resistance might be more obvious. It was also noted in our discussion that single gene solutions to disease susceptibility often present a selection pressure to the pathogen that results in the pathogen population changing rapidly to a more virulent form. This should be kept in mind in genetic engineering for disease resistance, but perhaps batteries of alternative alleles could be isolated/created by molecular geneticists to distribute to different cultivars in concurrent use and so reduce the opportunity for selection of virulent mutants.

Genetic solutions to relieve susceptibility to heavy metal toxicity in soils, salinity, and drought stress were discussed by the group. More research is needed on the biochemical bases of these problems, but it was noted that genes have already been isolated from animal and microbial cells whose products bind heavy metals and permit growth in high concentrations of the metals. The use of these or similar plant genes for plant improvement should be explored.

Nematodes are particularly troublesome to some crops in certain soils. The group discussed the possibility of inserting new genes into plants to produce and release nematocides from root cells. If the nematocide was a general toxin, care had to be taken not to destroy desirable microorganisms. An alternative approach is genetically to engineer root-associated microorganisms - rhizobium or mycorrhiza - with the capacity to produce a nematocide and be stable in the soil. The chemical could be released directly into the soil or initially into the root of the plant.

Improvement of Harvested Crop Quality

There is tremendous scope for modification of crop quality by biotechnology methods. Furthermore, as industrial uses and processing of crop products increase, the demand for more sophisticated seeds, etc., will increase. For example, oil crops need to be further modified in diverse ways to create different kinds of oils. The genetic control of the biosynthetic pathways needs to be understood and gene activities modified. The amino acid balance in protein of cereal foods is not optimal for animal consumption. This could be altered by inserting additional genes into the plants which will specify the synthesis of proteins with higher concentrations of the limiting amino acids. Since the synthesis of seed proteins from newly inserted genes has been achieved in tobacco (7), this goal appears readily achievable when gene insertion techniques into cereal plants are developed. It may also be possible to increase the total amount of protein produced in potatoes to create a much better balanced diet food and a source of protein for extraction by modifying the activity of defined tuber storage protein genes, although it is not known if the protein production will be limited by the availability of nitrogen. Crop quality is a very diverse and complex field but it should receive special attention from biotechnologists. Useful things may be achieved by single gene manipulations in this area. Biotechnological approaches may also help reduce food spoilage and enable food wastes to be exploited more efficiently.

Production and Utilization of Lignocellulose

Biomass is likely to become an increasingly important source of energy and chemicals throughout the world, as utilizable fossil fuels decline. Biotechnology is likely to play a vital role in exploiting biomass more efficiently by contributing to improvement in plant productivity (as discussed above) and also by modifying microorganisms to enable them to break down and hence separate hemicellulose, cellulose, and lignin more efficiently. Most work on cellulose biodegradation has been done with the filamentous fungus Trichoderma reesei (Broda, this volume). It contains many different degradative enzymes which are regulated in a complex manner. The separate genes should be cloned and studied and their control altered so that they are produced in higher (and appropriate) amounts during fungal growth, and not subject to end product repression and inhibition. Such changes might make the economics of cellulose breakdown much more attractive. Other potentially important organisms such as Thermonospora and Zymomonas mobilis should be engineered, if possible, to be more efficient lignin and cellulose degraders, respectively.

Lignin solubilization by biological processes needs to be explored intensively. Organisms such as the hymenophyte fungus Phanerochaete chrysosporium degrade lignin only slowly and only when depleted of alternative carbon and nitrogen sources. The genes regulating the extracellular ligninases need to be altered to uncouple production from cessation of growth. This may be achieved by combined mutagenesis and the reinsertion of reconstructed genes into fast-growing strains. Mesophilic and thermophilic actinomycetes may also provide useful sources of ligninases and should be explored.

Environmental lobbyists as well as problems due to soil erosion are demanding more efficient solutions to the rapid breakdown of straw after cereal harvests to avoid on-site burning. Where it is too expensive to harvest and transport the straw, on-site degradation is desirable. Microorganisms which can degrade the straw have been identified and could probably be improved by genetic manipulation (Broda, this volume). Alternatively, on-the-farm conversion to an upgraded animal feed is a possibility.

IMPROVEMENTS IN PLANT-MICROBE INTERACTIONS

Nodulated legume crops are not generally nitrogen-limited. Inoculation with Rhizobium is needed where appropriate species are not already present in the soil. Improvements in the symbiotic proficiency of Rhizobium strains and host plants may lead to increased yields. Similar improvements may be obtainable with other nitrogen-fixing interactions, though until we know more about the cost of nitrogen fixation and assimilation to plants it is difficult to point to specific objectives.

Frankia species nodulate a wide range of plant species. A better understanding of these symbioses should lead to information about the potential for extending N-fixing symbioses to a wider range of crop plants. Little molecular genetic research has been done with Frankia species to date, though future prospects look promising.

Mycorrhizal fungi have an important role in facilitating the uptake of phosphate. They are involved in water relationships and also possibly in the control of some diseases. Opportunities exist to exploit these functions, though commercialization will require effective methods for large-scale production.

The nature of symbiotic associations is such that it should be feasible to utilize the microorganisms concerned to carry into and express foreign genes in host plants. The exploitation of any such microorganisms in agriculture will depend upon finding methods for providing manipulated strains with attributes that will enable them to compete realistically with indigenous microorganisms.

Important progress in controlling pathogens may be made using microbes. For example, there are reports that microbial inoculants can make wheat less susceptible to take-all disease. However, to achieve good control high doses of inoculum may have to be sustained in the soil. Also, the control of whitefly and other insects by fungal pathogens has been developed in the glasshouse industry. Many difficulties exist in developing successful biological control agents, but little genetic manipulation has been applied to make the agents more effective. There is considerable scope for novel research here.

IMPROVEMENT OF THE PRODUCTION OF PLANT SECONDARY METABOLITES

Plants are the source of many commercially important compounds (pharmaceuticals, fragrances, etc.) whose chemical synthesis is impossible or too expensive. However, yields are often low. To improve production many different biotechnological approaches can and should be explored. More knowledge about the pathways of biosynthesis and the genes involved is vital - especially for the novel and/or rate-limiting enzymes. Enormous potential exists for modifying secondary metabolite pathways. It may be possible to isolate and alter the regulation of the rate-limiting gene before inserting it back into the plant cell. Alternatively, the genes for the novel enzyme(s) may be inserted into other plants or microorganisms to achieve more efficient production of the end product. Because immobilization of cultured cells may lead to enhanced product yields, more experience is needed with this technology.

Plants are also good sources of industrially important enzymes, e.g., specific lytic enzymes, hydroxylases, epoxidases, and glycosyl transferases. Glutathione transferases and malonyl transferases appear to be the major enzymes for the detoxification of xenobiotics in plant cells (Hahlbrock, this volume). Production of these might be improved by adoption of some of the genetic modifications outlined above, and their production might also be usefully transferred to a microorganism from which the enzymes could be efficiently excreted. Opportunities may also exist for combining organic syntheses with isolated plant enzymes or competent immobilized cells to overcome the conversion steps which are difficult to achieve chemically.

PERSPECTIVES

The survival of mankind is dependent on agriculture. So central is agriculture to human life that it is not surprising that a high proportion of the population is directly employed in it or because of it. Future progress in crop improvement through all the various ways, genetic as well as agronomic, will impinge on us all by influencing either the amount of food we have to eat, the diversity of food or plant products available to us, or the standard of life available to us. Biotechnology will play an increasingly important part in the future progress of agriculture through the next century, although the extent to which it will contribute to the world food problem will depend on economic, political, and social factors. The group recognized that agricultural progress has in the past and can in the future be to the detriment as well as to the benefit of societies, sections of societies, and individuals. These possibilities should be monitored by politicians, scientists, and biotechnologists alike.

Because of the long time scales of plant breeding, the slow adoption of new agricultural systems in some countries, and the biological complexities of plants, the more dramatic financial returns in biotechnology investment will, in general, take longer in agriculture than in the pharmaceutical industry. However, they will emerge and it is essential that governments and international funding agencies sustain the necessary scientists, breeders, agriculturalists, and agronomists to enable the necessary technological advances to occur.

We are considering the application of gene technology to agriculturally important species today because of the investment of governments in basic research in bacterial genetics during the 1940s to 1970s. We may shortly find ourselves inserting genes into maize and wheat because of work on a soil pathogen of tobacco and other dicot plants. Thus the biotechnology research essential for future agricultural production should not be confined to crops of economic interest in the developed countries today - or transferred from governments to industry. Sustained high quality basic research is required to realize many of the goals outlined in this report.

A country's economy is often closely intertwined with its agricultural economy. This can lead to calls for rapid changes in priorities for agricultural production. Fluctuations in priorities should not, however, undermine the sustained long-term research which will be necessary to achieve our ultimate aims.

We, like all scientists, recognize that we have a responsibility to consider any problems that could result from adopting the technical recommendations in this report. One of the issues we discussed was the release of plants with newly inserted genes into the agricultural system. On reflection, the group agreed that the kinds of problems that could be generated by modifying crop plants with foreign genes were not of such a novel kind that existing legislation, toxicological testing, and quarantine procedures would not cover them. Societies can and should allow genetically manipulated plants to be released, subject to the legislation and checks already in existence. If it were ever necessary to withdraw a particular variety from agriculture, this would be straightforward in most cases, in contrast to the situation which would exist if a microbe had been distributed in soils. This view, of course, does not imply that no potentially harmful effects will emerge from breeding new varieties of crop plants, especially those resistant to pests and diseases. It is well-known that many wild plants produce potent toxins. Only where there is a risk significantly greater than that in conventional breeding will different testing practices need to be adopted.

TABLE 1 - Status of manipulation of some Gramineae species (from (4, 5) and from volumes 1(1981) through 5(1985) of the journal Plant, Cell, Tissue and Organ Culture).

Plant	Nature of regeneration			Status of the art						Well developed breeding programs?
	SE	SH	PR	callus →plant	cell susp.→ plant	p → callus	p → plant	haploid through ant. cult.	estab-lished cell lines	
Agropyron	X		X	X						
Agrostis		X	X	X						
Alopecuros	X		X	X						
Avena sativa*	X	X	X	X						Yes
Bromus inermis	X		X	X						
Dactylis glomerata	X	X	X	X				X		
Festuca arundinacea	X	X	X	X				X		
F. ovina										
F. rubra	X		X	X						
Hordeum vulgare*		X	X	X				X	X	Yes
Lolium perenne	X	X	X	X				X	X	Yes
L. multiflorum	X	X	X	X		X				Yes
L. ridigum		X	X	X						
Oryza sativa	X	X	X	X	X	X		X	X	Yes
Panicum maximum	X	X	X	X	X			X		

TABLE 1 (cont.)

Plant	Nature of regeneration			Status of the art						Well developed breeding programs?
	SE	SH	PR	callus → plant	cell susp. → plant	p → callus	p → plant	haploid through ant. cult.	estab-lished cell lines	
P. miliaceum										
P. miliare										
Pennisetum americanum	X		X	X	X	X			X	Yes
P. purpureum	X		X	X	X		X	X	X	
P. typhoides								X		
Phleum	X		X	X						
Saccharum officinalis	X	X	X	X	X	X			X	Yes
Secale cerale	X	X	X	X				X		Yes
Sorghum	X	X	X	X						Yes
Stipa	X		X	X						
Triticale	X	X	X	X				X	X	Yes
Triticum aestivum*	X	X	X	X				X		Yes
T. durum		X	X	X						Yes
T. monococcum									X	
Zea mays* +	X	X	X	X	X	X		X	X	Yes

Legend: SE = somatic embryogenesis; SH = production of shoots from callus; PR = plant regeneration from callus; p → = from protoplast to . . . ; * = chromosomal gene mapping established; + = transposable element isolated.

TABLE 2 - Status of manipulation of some Leguminosae species (from (1, 2, 4, 5) and from volumes 1(1981) through 5(1985) of the journal Plant, Cell, Tissue and Organ Culture).

Plant	Nature of regeneration			Status of the art						Well developed breeding programs?
	SE	SH	PR	callus →plant	cell susp. → plant	p → callus	p → plant	haploid through ant. cult.	estab-lished cell lines	
Acacia boa (from (2))			X							
Albizzia lebbek (from (2))	X[1]									
Arachis hypogen		X	X	X						
Cajunus cajan		X	X	X						
Cicer arietinum		X	X	X						
Coronilla varia	X		X	X						Yes
Cyamopsis tertagono-loba	X	X	X	X	X				X	
Glycine soja	X	X	X		X					
G. max* '	X	X	X	X	X				X	Yes
G. canascens	X	X	X	X						
G. tabascina	X	X	X			X	X		X	
Indigofera enneaphylla		X	X	X						
Lathyrus sativus		X	X	X						
L. odoratus						X				Yes
Lotus caucasicus		X	X	X						
L. corniculatus	X	X	X	X		X	X		X	
Medicago sativa	X	X	X	X	X	X	X		X	Yes
M. coerulea						X	X		X	
M. glutinosa						X	X		X	
M. arborea	X		X	X						

Table 2 (cont.)

Plant	Nature of regeneration			Status of the art						Well developed breeding programs?
	SE	SH	PR	callus →plant	cell susp. → plant	p → callus	p → plant	haploid through ant. cult.	estab-lished cell lines	
Melilotus alba				X					X	
Onobrychis viciifolia						X				
Phaseolus vulgaris		X	X	X					X	Yes
P. aureus				X					X	
Pisum sativum*	X	X	X	X						Yes
Psophocarpus tetra-gonolobus		X	X	X		X				
Stylosanthes guyamensis (from (2))	X(1)	X	X	X						
S. hamata		X	X	X						
Trifolium alexandrinum		X	X	X						
T. pratense		X	X	X						
T. repens		X	X	X		X			X	
T. subterraneum				X						
T. hybridum				X						
Trigonella Toenumgraecum	X		X	X		X	X			
T. corniculata	X		X				X			
Vicia faba				X		X		X	X	Yes
V. hajastana						X				
Vigna unguiculata		X	X			X				
V. aconitifolia		X	X	X		X	X			

Legend: SE = somatic embryogenesis; SH = production of shoots from callus; PR = plant regeneration from callus; p → = from protoplast to . . . ; * = chromosomal gene mapping established; + = transposable element isolated.

TABLE 3 - Status of manipulation of other plants of agricultural, horticultural, or pharmaceutical value (from (4,5) and from volumes 1(1981) through 5(1985) of the journal Plant, Cell, Tissue and Organ Culture).

Plant	Nature of regeneration SE	SH	PR	MP	Well developed breeding programs?
Allium cepa		X	X		Yes
Allium sativum		X	X		Yes
Anethum graveolens	X		X		
Apium graveolens	X				
Asparagus officinalis	X	X	X		Yes
Atropa belladonna	X	X	X		
Beta vulgaris		X	X	X	Yes
Brassica oleracea					Yes
var. gemmifera		X	X		
var. capitata		X	X		
B. napus		X	X		Yes
Carum carvi	X				
Cichorium endivia	X	X	X		
C. intybus		X	X		
Conium maculatum	X				
Coptis japonica		X	X		
Coriandrum sativum	X				
Cucumis sativus		X	X		Yes
Cucurbita pepo	X	X	X		Yes
Cuminum cyminum		X	X		
Datura innoxia		X	X		
Daucus carota	X		X		Yes
Dioscorea sansibarensis			X		
D. bulbifera		X			
Fagopyrum esculentum		X	X		
Foeniculum vulgare	X				
Hedychium roxburghii		X	X		
Hypoxis rooperi				X	
Ipomea batatas	X	X	X	X	
Lactuca sativa		X	X		Yes
Lycopersicon esculentum		X	X		Yes
Nicotiana sp.		X	X		Yes
Petroselium hortense	X		X		
Pergularia minor	X	X	X		
Physalis ixocarpa	X		X		
Pogostemon cabli		X	X		
Rauwolfia serpentina	X		X		
Solanum khasianum		X	X		
Solanum melongena	X		X		
Solanum tuberosum		X	X	X	Yes
Solanum xanthocarpum		X			
Symphytum officinale		X	X		
Tylophora indica	X	X			
Urginea indica		X	X	X	
Yucca gloriosa		X	X		

Legend: SE = somatic embryogenesis; SH = shoot production from callus; PR = plant regeneration; MP = micropropagation.

TABLE 4 - Status of manipulation of some ornamental species (from (4-6) and from volumes 1(1981) through 5(1985) of the journal Plant, Cell, Tissue and Organ Culture).

Plant	Nature of regeneration SE	SH	PR	MP
Aloe pretoriensis		X	X	
Amaryllis				X
Alstroemeria				X
Adiantum cuneatum				X
Anagallis arvensis	X	X		
Alsophila auistralis				X
Antirrhinum majus* +	X			
Aranda (see (6))				X
Aranthera (see (6))				X
Asclepias curassavica	X			
Ascotinetia (see (6))				X
Ananas comosus				X
Aechmea fasciata				X
Anthurium andreanum				X
Begonia		X	X	
Begonia herbácea		X	X	
Billbergia				X
Bougainvillea glabra		X	X	X
Cactacea	X	X	X	X
Caladium				X
Calathea				X
Codiaeum variegatum		X		
Consolida orientalis		X		
Chrysanthemum		X	X	
Crepis capillaris		X	X	
Convolvulus arvensis		X	X	
Cyclamen persicum		X		
Cryptanthus bivittatus				X
Cryptbergia				X
Convallaria				X
Clivia				X
Cymbidium				X
Dampiera diversifolia		X		X
Deudrobium (see (6))				X
Dianthus		X	X	
Dieffenbachia				X
Doritaenopsis (see (6))				X
Doritis (see (6))				X
Dracaena				X
Dyckia sulphuria				X
Epidendrum			X	
Eschscholzia californica	X			
Euphorbia pulcherrima				X
Freesia		X	X	X
Gasteria verrucosa		X	X	X
Gerbera jamesonii		X	X	X
Gladiolus		X	X	
Gloxinia				X
Haworthia fasciata	X	X	X	X

TABLE 4 (cont.)

Plant	Nature of regeneration SE	SH	PR	MP
Haworthia variegata		X	X	
Helianthus annuus		X		
Hippeastrum		X		
Ipheion				X
Hyacinthus				X
Kalanchoe pinnata	X	X	X	
Lilium		X	X	X
Lunaria annua		X		
Laeliocattleya		X		
Microlepia strigosa				X
Montbretia crocosmaeflora		X	X	
Mesembryanthemum floribundum	X	X		
Macleaya cordata	X			
Mazus pumilio		X		
Nigella damascena	X	X		
Narcissus				X
Neostylis (see (6))				X
Nephrolepis				X
Nerine				X
Nidularium				X
Neoregelia				X
Pteris cretica				X
Philodendron				X
Passiflora coerulea	X	X		
Pelargonium hortorum		X	X	X
Petunia hybrida* +	X	X	X	
Petunia inflata	X	X	X	X
Petunia axillaris		X	X	
Phalaenopsis (see (6))				X
Phlox drummondii		X		X
Prostanthera rotundifolia		X	X	X
Ranunculus sceleratus	X		X	
Rosa		X		
Scilla				X
Sparaxis				X
Schizostylis				X
Streptocarpus				X
Salpiglossis sinuata		X	X	
Saintpaulia ionantha		X	X	X
Sinapis alba	X	X		
Scopolia parviflora		X		
Tulipa				X
Torenia fournieri		X	X	
Vanda (see (6))				X
Vascostylis (see (6))				X
Woodwardia fimbriata				X

Legend: SE = somatic embryogenesis; SH = shoot production from callus; PR = plant regeneration; MP = micropropagation; * = chromosomal mapping of genes established; + = transposable element isolated.

TABLE 5 - Status of manipulation of some tree species (from (4,5) and from Volumes 1(1981) through 5(1985) of the journal Plant, Cell, Tissue and Organ Culture).

Plant	Nature of regeneration SE	SH	PR	MP	Well developed breeding programs?
Albizzia lebbeck		X	X		
Alnus sp.		X		X	
Annona squamosa		X	X		
Azadirachta indica		X			
Biota orientalis	X				
Carica papaya	X		X		
Citrus	X		X		Yes
Coffea arabica	X		X		
Coffea canephora	X		X		
Coffea liberica	X		X		
Cryptomeria japonica		X			
Dalbergia sissoo		X	X	X	
Elaeis guineensis	X		X		
Eucalyptus citriodora		X	X		
E. viminalis		X	X		
E. obliqua		X	X		
Eremocitrus glauca	X				
Fortunella crassifolia	X				
Hevea brasiliensis	X	X			
Ilex aquifolium	X				
Malus domestica		X	X	X	Yes
Mangifera indica	X		X		
Manihot esculenta		X	X	X	
Microcitrus australasica	X		X		
Microcitrus warburgiana	X				
Musa textilis		X		X	
Panax ginseng	X		X		
Picea sp.	X	X	X		
Pinus banksiana	X				
P. cembra		X			
P. lambertiana		X			
P. palustris	X		X		
P. taeda	X				
P. strobus	X				
P. elliotti	X		X		
P. echinata	X				
P. ponderosa		X			
P. contorta	X	X	X		
Poncirus trifoliata	X				
Populus sp.		X	X	X	
Prunus amygdalis		X	X		
Pseudotsuga menziesii	X				
Sapium saebiferum		X	X		
Vitis vinifera		X	X	X	Yes

Legend: SE = somatic embryogenesis; SH = shoot production from callus; PR = plant regeneration; MP = micropropagation.

TABLE 6 - Plant species into which engineered genes have been inserted.*

Plant	Mode of Transformation	Regenerated Plants (Yes - No)
Nicotiana tabacum	in vivo infection	Yes
	in vitro co-cultivation of protoplasts with Agrobacterium tumefaciens	Yes
Nicotiana plumbaginifolia	in vitro co-cultivation with A. tumefaciens	Yes
Kalanchoe dagremontiana	in vivo infection with A. tumefaciens	No
Petunia hybrida	in vitro co-cultivation with A. tumefaciens	Yes
Hyosciamus muticus	in vitro co-cultivation with A. tumefaciens	No (?)
Brassica napus	in vivo infection and in vitro co-cultivation with A. tumefaciens	Yes
Asparagus officinalis	in vivo infection with A.tumefaciens	established cell lines
Narcissus cv. Paperwhite	" "	(?)
Chlorophytum capense	" "	(?)
Triticum monococcum	DNA - feeding	established cell lines
Zea mays	DNA - electroporation	No
Lolium multiflorum	DNA - feeding	No
Solanum tuberosum	in vivo infection and in vitro co-cultivation with A. tumefaciens	Yes

* Results obtained from screening the journals Theoretical and Applied Genetics, Molecular and General Genetics, Cell, Nature, Science, and EMBO Journal from 1978 to 1985.

TABLE 7 - Microorganisms important for agricultural biotechnology.

Organism	Major value in agriculture	State of genetic research
Rhizobium	Symbiotic nitrogen fixation with legumes	Good
Leuconostoc Lactobacillus Pediococcus Streptococcus	Used in silage making and the dairy industry	Moderate in some species
Azotobacter Azospirillum Cyanobacteria Photosynthetic bacteria	Low-level nitrogen fixation and probably also benefits to plants through hormone production	Moderate Moderate Moderate Moderate
Bacillus thuringiensis	Control of lepidopterous insects	Good
Verticillum	Control of aphids, whitefly, nematodes	Poor
Agrobacterium radiobacter	Control of crown gall	Good
Mycorrhizal fungi	Nutrient uptake by plants	Poor

Acknowledgements. Thanks are due to B. Lupotto who contributed substantially to the compiliation of Tables 1-6.

REFERENCES

(1) Ammirato, P.V. 1983. Embryogenesis. In Handbook of Plant Cell Culture, vol. 1, pp. 82-123. New York: MacMillan.

(2) Flich, C.E.; Evans, D.A.; and Sharp, W.R. 1983. Organogenesis. In Handbook of Plant Cell Culture, vol. 1, pp. 13-81. New York: MacMillan.

(3) Hardy, R.W.F., and Giaguinta, R.T. 1984. Molecular biology of herbicides. Bioassays **1:** 152-156.

(4) Razdau, M.K., and Cocking, E.C. 1981. Improvement of legumes by exploring extra-specific genetics variations. Euphytica **30:** 819-833.

(5) Reinert, J., and Bajaj, Y.P.S., eds. 1977. Applied and Fundamental Aspects of Plant, Cell, Tissue and Organ Culture. Berlin: Springer-Verlag.

(6) Sagawa, Y., and Kunisaki, J.T. 1984. Clonal propagation of Orchids. In Cell Culture and Somatic Cell Genetics of Plants, vol. 1, pp. 61-67. Orlando: Academic Press.

(7) Sengupta-Gopalan, C.; Reichert, M.A.; Barker, R.F.; Hall, T.C.; and Kemp, J.D. 1985. Developmentally regulated expression of the bean β-phaseolin gene in tobacco seed. Proc. Natl Acad. Sci. USA **82:** 3320-3324.

Biotechnology: Potentials and Limitations, ed. S. Silver, pp. 223-239. Dahlem Konferenzen 1986. Berlin, Heidelberg, New York, Tokyo: Springer-Verlag.

Plant Gene Engineering and Plant Agriculture

W.J. Peacock and E.S. Dennis
CSIRO Div. of Plant Industry
Canberra, ACT 2601, Australia

Abstract. Many plant breeders feel there is little prospect that genetic engineering will make any real contribution to plant improvement programs. We cannot agree, as it is already clear that recombinant DNA technology will make impacts on plant production systems. Diagnostic tools will provide increased power to many selection schemes and increased accuracy in the selection of appropriate parents in breeding programs. Genetic engineering should also make direct contributions to plant improvement by providing additional means of introducing specific genes to the genetic structure of a cultivar. Gene addition should extend the life of many of our most efficient cultivars and enable them to be used in different and more marginal environments. It should enable a plant breeder to respond to yield limitation in a shorter time. It may also significantly reduce the scale of a plant breeding program. Time and scale of operation are probably the principal factors in determining the effectiveness of any plant improvement effort.

All the components of a gene transfer system are presently in place for only one or two commercial crops. However, it should be possible to have effective systems in each of the major crop species in the near future. Gene transfer technologies will increase the range of variation available to a plant breeder for any given crop species and will provide opportunities for quite novel adjustments to the workings of a genome, adjustments which would not have been possible by classical breeding and selection schemes. We feel sure that the analytical power of recombinant DNA technology will also help to dissect the apparent complexity of many of the major agronomic physiological characteristics so fundamental to successful plant production. We may thus be able to provide substantial yield increases in even the most well developed crop species.

GENETIC DEMANDS IN PLANT AGRICULTURE

Throughout the past ten thousand years man has been manipulating wild species by selective breeding. He has produced a handful of crop species with characteristics finely attuned to his needs, which account for the major supply of food in the world. The changes he has made are dramatic in morphology, in physiological tolerances, and, most strikingly, in yield of the primary food source, often the seed or fruit. However, there probably have never been the demands for plant improvement to equal those which presently apply. Increased competition in world markets and increasing costs of inputs have underlined the need for efficiency in plant-based agriculture. Plant production systems are under increasing pressure to maximize yields with fewer inputs. Excessive inputs have led to the demise of many production areas. Irrigation systems are running into trouble with increasing salinity, and frequently the extensive use of fertilizers leads to problems such as increasing soil acidity. Excessive use of herbicides and of pesticides can lead to problems of resistance in plants and pests and to ecological destruction.

Increased intensities of cropping have led to changes in the goals of plant improvement. In the past the primary goal has always been that of maximum production. Now in intensive cropping systems where there is a rotation of crops and where crops have to be fitted together into the production year, selection is frequently for an optimal rather than for a maximal yield. There is an increasing realization that genetic answers to a problem are often the preferred solution. Plants which themselves produce the effectors for disease or pest resistance could lead to significant decreases in the costs of inputs and provide a more balanced and stable production system.

Another major feature of global agriculture is the need to obtain economic yields from more marginal environments. There are ever-increasing demands for plant food production, and the increasing mechanization of agriculture, both in developed and developing countries, has meant that less than ideal environments can be brought into crop production. Developments such as minimum or reduced tillage have increased the opportunism of agriculture, enabling cropping systems to be established in lower rainfall zones and in sloping country previously unavailable to conventional tillage practices. These trends have associated with them a whole range of new problems in the need for plant adaptation and for resistances to different ranges of pests and diseases.

A third major trend has been the move towards greater genetic homogeneity in our cropping species. This trend has been accentuated by increased mechanization and by increased sophistication of management. A farmer needs to have a more predictable and uniform plant response to the set of environmental conditions which apply. This has inevitably meant that the better developed crop and pasture species are resting on a narrower genetic base than previously. It also has the corollary that the crop is more susceptible to a major challenge from an environmental stress condition or from a pest or disease epidemic and the perturbation to the production system can often be costly. Genetic homogeneity has also been favored by the need to have plants better tailored to fit particular ecoclimatic zones and by the increasing sophistication of market forces making more stringent demands on various characteristics of plant products.

All of these major trends in agriculture demand greater control over the characteristics of the plant production system. Any plant production system is a product of the interactions of its genetic makeup, the environment to which it is exposed, and the management system used. Management systems are constantly improving. The genetic improvement of a crop has certain limitations. A breeder, even if he has an appropriate pool of variability at his disposal, has to cope with the two major factors of time and scale of operation in the achievement of his objective. In the most highly developed crops the modern specialized cultivar frequently requires a period of ten years or more for development, evaluation, and release. The rates of change of environments and management systems are frequently faster than this and so have a time scale shorter than the breeding response time scale. Consequently, the efficiency and output of the production system suffer.

The question we address in this paper is whether the new tools of genetic engineering and biotechnology will be valuable in the tailoring of plant production systems. Will they, for example, add precision and novelty to the breeding process? Will they reduce the time scale of the process or can they reduce the magnitude of the enterprise needed for any particular goal, thereby releasing resources to achieve additional objectives? Apart from its application to genome manipulation, we will briefly consider the role of genetic engineering in the production of diagnostic tools which could have a significant impact on plant breeding by enhancing the efficacy of selection procedures. Brief mention will also be made of the opportunities for improving plant production systems not only by genetic manipulation of the plants themselves but by manipulation of the genomes of some of the micro-organisms which are intimate partners of the plant in the production system.

GENE TECHNOLOGY IN DIAGNOSTIC TOOLS

In many plant production systems, in broad acre cropping as well as in intensive horticultural systems, disease organisms frequently limit yield. The extent of yield losses resulting from individual diseases is in general not known. Many diseases are difficult to diagnose and very few surveys have accurately assessed crop loss. The application of gene technology to the detection and accurate diagnosis of disease organisms is one of the most immediate impacts genetic engineering will have on plant-based agriculture. As an example we will consider the detection of virus diseases. All crop, pasture, and horticultural plant species have virus diseases. Some viruses, such as Barley Yellow Dwarf Virus (BYDV), are ubiquitous in cereal cropping systems around the world and BYDV is estimated to cause up to 30% yield losses. The best available data suggest that even where the virus is not recognized as a problem it is in all likelihood accounting for a few percent yield loss, which of course is a significant proportion of gross margins obtained. Diagnosis by symptomology is difficult for many diseases, and accurate confirmation by biological indexing procedures is frequently not practicable in actual cropping situations. The logistics, time scales, and expense of biological indexing have in general caused virus limitations to yield to be ignored.

The production of cloned nucleic acid probes has opened up a whole new concept of plant health evaluation. Viruses, viroids, and virusoids can be detected by such probes using hybridization procedures which can be completed in a few hours (3). Probes can be selected to provide differing levels of specificity for strains of disease organisms. The emergence of these new diagnostic tools has significance for a number of components of a plant production system. It may be important for giving precision to quarantine procedures in germ plasm exchange; it will facilitate quality certification of seed or other crop material; and for some viral diseases, in particular those transmitted by insects, early detection in a crop may allow appropriate management. Perhaps the most important use will be in providing efficient and accurate selection tools in breeding programs designed to introduce resistance genes into a crop. Breeding for Barley Yellow Dwarf resistance in wheat has, for example, been a formidable task. The currently known sources of resistance genes are other cereal or grass species, which means that the breeding strategy is difficult. Direct monitoring with a probe should greatly facilitate the selection of a resistant line. Similar nucleic acid probes should be able to be constructed for the unequivocal detection of other disease organisms, such as bacteria and fungal species.

Another type of diagnostic probe of significance in plant improvement programs is a probe for a particular DNA sequence from a plant genome. The probe may be for a class of repeated sequences or it may be for a specific gene segment. These probes could be of great value when the characteristic being selected requires difficult and lengthy testing procedures. Nucleic acid probes can track the segment containing the gene so that individual plants having the required segment can be identified quickly and used with precision in the breeding strategy. This application of genetic engineering in plant breeding is already used in cereal breeding in Australia (2) and presumably in other countries.

Probes for unique sequences can lead to the development of a linkage map of restriction enzyme polymorphisms (8). This is likely to prove a valuable tool in plant breeding since it can rapidly provide genetic markers in crop and pasture species which otherwise have scanty genetic information. Linkage associations of a desired character with one or more of these markers can provide for effective and precise breeding procedures. We believe that one of the major advantages that this technique offers will be in the identification of the major genetic determinants of yield and for their effective combinatorial selection.

MICROORGANISMS IN PLANT PRODUCTION SYSTEMS

Improvement in efficiency of operation of plant production systems does not necessarily mean that the major crop species are the only targets for improvement. Increasingly we have become aware that microorganisms associated with a crop species can strongly influence the biological output of the system. The manipulation of microorganisms may be a more immediately achievable objective than manipulation of the crop species itself. The most obvious example is where there is an interdependent relation, such as the symbiosis between rhizobium and its legume hosts. The opportunity for utilizing genetic engineering to increase the positive inputs of microorganisms are discussed by Beringer (this volume), but we will mention two examples which illustrate the range of possible manipulations. An infection with a nonvirulent or mildly virulent organism can often give cross-immunity against more pathogenic strains. There are already cases of infective but nonvirulent forms of a disease organism being used in a protective capacity, e.g., in the glasshouse tomato industry (12). The capacity of genetic engineering to identify genes of importance in virulence by techniques such as transposon mutagenesis should enable manipulation of many disease organisms so that they can be used to generate immunity.

An entirely different approach is to think of modifying naturally occurring microorganisms such that they perform specific tasks in a production system more efficiently and extensively than they might otherwise do. Retention of stubble or plant residues in fields is an increasingly common practice in cropping systems. Sometimes the residues are incorporated into the soil and, in these cases, have often been regarded as a nitrogen sink. The microorganisms breaking down the residues compete for available nitrogen with the roots of the next crop. However, in some situations retention and incorporation of crop residues lead to a positive nitrogen balance. In some areas of Australia continuous wheat cropping systems have been developed with stubble retention such that the crops are largely independent of exogenous nitrogen application (24). Microorganisms have been identified which break down the cellulose and other components of wheat straw providing carbon sources for other organisms which fix nitrogen (Halsall and Gibson, submitted). There are even some organisms capable of carrying out both of these tasks. Once the genetics and molecular basis of these processes are understood, there is the distinct possibility of increasing the efficiency of the system by genetic manipulation. Utilization in this way of only 5% of Australia's stubble residues could lead to enormous reductions in the application of exogenous fertilizer or, alternatively, could lead to increased yields for the same inputs of fertilizer. Beringer (this volume) discusses a number of other potential manipulations of microorganisms involved in nutrient acquisition into plants, in pathogen protection, and in modulating effects of environmental extremes such as frost damage.

CURRENT STATE OF PLANT GENE TECHNOLOGY

If genetic engineering is to be an effective tool in plant improvement programs, we need satisfactory techniques for the various components of the transformation process. We need to be able to identify and isolate particular genes, we need to be able to deliver those genes to a target nucleus where they will undergo stable integration, and we need to be able to recover the transformed cell as a regenerated plant to be used in further stages of the plant breeding process.

Gene Identification and Isolation

Although many plant genes have now been isolated and sequenced, they fall into the class of genes which have been easy to isolate. These include genes which have prevalent messenger RNAs in particular tissues, such as the seed storage proteins of legumes and cereals, or genes which are induced by a particular signal permitting cDNA clones to be identified on the basis of mRNAs which are present in induced and absent in uninduced tissues. This has permitted the isolation of genes responding to stresses such as heat shock and anaerobic shock, or to light (6, 9, 25). Another class of genes has been isolated by hybridization to heterologous genes isolated initially from other species or types of organisms, e.g., the actin genes (21). One other class of plant genes which has been studied in some detail includes the genes occurring in the much smaller genomes of chloroplast and mitochondria (27). Advances in protein microsequencing and oligonucleotide synthesis have made it possible to isolate other plant genes for which a protein product can be isolated on a gel system. A partial amino acid sequence can be obtained and the sequence of the corresponding oligonucleotide predicted. This oligonucleotide can be synthesized and used as a probe to identify genes in either cDNA or genomic libraries. We have recently used this technology in isolating genes for hemoglobins in non-legume species.

Gene Delivery

The pathogenic organism Agrobacterium tumefaciens, which causes Crown Gall disease in a number of dicotyledonous plants, has provided the most effective gene delivery system for plants. In the normal infection process a particular segment, T-DNA, of the Ti plasmid of the bacterium is incorporated into chromosomal material of the host plant. This system has now been vastly improved as a gene tranfer system (16). Tumorogenesis and regeneration abnormalities have been avoided by removal of most of the genetic material of the T-DNA segment. Any gene or genes placed between the boundaries of the T-DNA segment can be delivered and integrated into a host chromosome provided that the appropriate transfer factors are present. These are coded by the so-called vir region of the Ti plasmid and may be coupled with a T-DNA plasmid in a binary vector system (14). Separating the two components simplifies the genetic engineering of the T-DNA plasmid which can be manipulated in E. coli and then replicated and delivered by Agrobacterium. Infection of the bacterium has also been improved with concomitant improvements in efficiency of transformation. Rather than using entry through wound sites, Agrobacterium-mediated transformation is now more conveniently achieved by co-cultivation with protoplasts, or more recently by application of the bacteria to leaf discs of the target species. In many dicotyledonous species regenerant transformants can be isolated directly from the leaf disc. Recently the T-DNA system has been reported to function in asparagus and amaryllus, two genera of the Liliales (13, 15). Whether this signals that the T-DNA system will become an effective vector for a number of other monocots, including cereals, is not yet determined. There are many convincing examples of Agrobacterium-mediated transformation, and in some cases at least the integration and subsequent inheritance of the transferred genetic material would be perfectly satisfactory as a component in a breeding program.

Another natural vector system which has been used successfully is the Cauliflower Mosaic Virus (CaMV), one of the few double-stranded DNA viruses of plants. It has been possible to introduce a gene into tobacco by replacing a particular segment of the CaMV genome by a selectable foreign gene (5). The length of the segment that can be introduced using this vector may be a severe limitation. Finally, there have been successful transformations of protoplasts by naked DNA. This has been achieved in tobacco where protoplasts have been able to be regenerated to plants and shown to have a stable, heritable, introduced DNA segment (22). DNA-mediated transformation has also been demonstrated in cereal protoplasts but without regeneration (19). DNA-mediated transformations occur with low frequency, with a probability of between 10^{-5} and 10^{-6}, and this limits its usefulness to genes which are readily selectable. However, if the efficiency of the system could be markedly increased, then DNA-mediated transfer into protoplasts or perhaps into other target cells, such as pollen, could become a technique of major value in plant improvement.

Gene Recovery

In a number of dicotyledonous species transformants have been successfully regenerated from callus tissue, from protoplasts, or from leaf discs, and in some of these cases the transformation event has been shown to result from integration of a gene which is functional and stable. In many plants the regeneration step essential for effective gene recovery is not available and presents a major hurdle to gene manipulation.

RESULTS FROM GENE TRANSFORMATION

One of the principal results of experiments using T-DNA mediated transformation is that promoters may operate in heterologous plant systems. For example, within the dicotyledoneae the light-inducible promoter of the small subunit of RUBP-carboxylase, isolated from pea, is operative in both petunia and tobacco. A similar promoter isolated from wheat is not operative in these species. So it is clear that a promoter will not necessarily work correctly in all heterologous systems. A promoter from a monocotyledonous plant may not work in a dicotyledonous plant (Sengupta- Gopalan et al., submitted), perhaps as a result of the long evolutionary separation of these two major plant groupings.

One of the most important results for plant improvement is the demonstration that when the gene for phaseolin, a seed storage protein gene of French bean, is introduced into the tobacco genome, it is expressed in the correct tissue (seed) at the correct developmental time (18). These two examples of controlled expression, one of an inducible gene and one of a developmentally specific gene, increase the feasibility of gene transfer being a useful tool in plant improvement.

The other point to emerge from many of the transformation studies is that a gene or its promoter may be expressed correctly in many sites in a target genome, i.e., chromosomal location does not have to be specific. It is not clear whether all locations are successful and, if the situation is comparable to that found in Drosophila, insertions into heterochromatin are not likely to be expressed correctly. Another technical development of some importance has been the recent demonstration that a coding sequence can be placed under the control of a specifically inducible promoter and at the same time be equipped with an additional segment coding for a sequence of amino acids at the N-terminus of the polypeptide, enabling the polypeptide to be delivered into a specific cellular location. In the particular example a gene was integrated into the nucleus but its product, manufactured in the cytoplasm, was delivered to a location within the chloroplast (26). As our knowledge of delivery and processing sequences increases we should have the capacity to direct an introduced gene product into a number of specific cellular locations.

Transient Expression Systems

Recent data demonstrate that gene segments can be introduced into cereal protoplasts and expressed. In our lab we have introduced the bacterial-derived chloramphenicol acetyl transferase gene driven by a nopaline synthase promoter and 3' signals into protoplasts of Black Mexican Sweet maize (17). We are now proceeding to use this system to investigate the essential components of the anaerobically induced promoter of the Adhl gene of maize. Transient expression systems may provide a convenient tool for tuning chimeric gene constructions to give desired expression. Another promising expression system is that based on brome mosaic virus RNA (1), with gene segments replacing components of the virus genome. In this case the system can provide information on products translated from the replicated viral genome which acts directly as a messenger RNA. It may be useful for looking at targeting of proteins and at the effect of specific mutations on the activity of enzymes.

Transposable Sequences

In maize and Antirrhinum there are genetic and molecular descriptions of another significant component of the genome, sequences which are capable of transposition around the chromosomes. These sequences were termed controlling elements by McClintock in her early genetic studies (20). In maize, where a number of systems have now been characterized at a molecular level, controlling elements may be important for at least two aspects of gene manipulation in higher plants. First of all, they may be used in transposon mutagenesis studies to locate specific gene sequences in maize (11). Some also have properties which suggest that they will be valuable components of transformation systems. Just as the P transposable element has provided an effective transformation vehicle in Drosophila, so too could the Ac/Ds system in maize. Controlling elements may enable genetic segments to be inserted efficiently without their integrity being challenged. In this regard, controlling elements are similar to T-DNA but may be more manageable, as there are mutant forms which affect frequencies of transposition. The sites of integration of controlling elements are not well documented, but some may have a specificity of target which could be used to advantage. Finally, there is a possibility that controlling elements could be used as in vivo mutagenic agents in a particular gene. Enough is known of the consequences of this excision and reinsertion of controlling elements to explore the possibility of using them for generating sequence variability at specific locations (23).

It seems likely that most plant species, including the major crop species, will have their own suite of controlling element systems. However, it is not clear that it will be a simple task to obtain enough molecular and genetic information in many of the species to make their use probable in the short term. We must determine whether the well characterized systems isolated from maize will function in other plants or plant cells.

PROSPECTS FOR GENE MANIPULATION

Where a crop species is represented by a successful cultivar, breeders will attempt to meet any deficiency, such as the susceptibility to a new disease, by introducing appropriate resistance genes from elite lines closely related to the cultivar. This form of limited back-crossing plant breeding has probably the fastest response time of any plant breeding schema. Basically what the breeder is trying to do is to provide new genetic material with as little disturbance as possible of the favorable genotype he had constructed previously. This method is not always available to a plant breeder because he may have to go to more distant genetic material as a source for the required variability.

Gene Addition in Plant Breeding

It is perhaps by gene addition that genetic engineering will have its first direct applications in plant improvement. The technology that is currently available or under development is such that we can plan to insert a specific gene sequence into a recipient target species, with gene expression being controlled and stable. Situations will arise when a highly tuned and successful plant genotype will be challenged by a new pest, a new disease, or by a new aspect of management such as application of particular herbicides in the cropping system. The plant gene engineer may be able to add the required genetic information to the existing well balanced genotype to overcome the major limitation in yield and shore up the genotype for an extended period of productivity.

In a number of crop species there are many instances of resistance to disease or pests being under the control of a single dominant gene function. This is precisely the situation where the isolation of a gene and its subsequent transfer to the target genotype would provide that genotype with the new function without disturbing its many other favorable attributes. Let me cite two examples from Australian agriculture to illustrate this point. Australian wheat production is dependent on continued plant breeding to provide resistance against stem and leaf rusts. Recently a new rust, the striped (or yellow) rust, entered Australia and quickly became a significant limit to yield in many areas. A number of different cultivars are used over the wide range of ecoclimatic zones of our cereal belt. Now we face the need to find suitable sources of resistance genes and breed them into all these cultivars or to develop entirely new cultivars containing the resistance character. This task has already demanded diversion of major segments of our wheat breeding expertise. Sources of resistance are known and appear to be associated with a single locus. If that locus or gene segment could be isolated and transferred to each of the specialized cultivars it would solve the striped rust problem in the short term.

Our lucerne industry is another example of a situation where genetic engineering could have been of help. This major pasture legume in Australia has been selected for a series of specialized properties in our integrated animal and cropping production systems. It was found to be highly susceptible to aphid species not previously in Australia but accidentally introduced. Aphid-resistant cultivars were bred by a process of limited back-crossing, using related American material which contained appropriate resistance genes (10). Nevertheless, we were not able in the period of 3-4 years to back-cross sufficiently to restore completely the highly adapted genotype of the previously existing cultivar. The aphid resistance was encoded by one or two dominant genes for each of the two aphid pests, and it would have been preferable to add this capability to the existing cultivar by gene transfer rather than to go through the hybridization and subsequent back-crossing procedures used.

Identification of Needed Genes

In the two examples mentioned above, or for the many other situations where gene addition breeding would be useful, one of the major problems is the location and isolation of the needed gene segment. If the function to be added is herbicide resistance, then it may be that this gene function can most easily be obtained from microorganisms, e.g., glyphosate resistance; a coding segment may be isolated and equipped with all of the appropriate plant gene controls to ensure its successful operation in the target crop species. Another much-discussed example of this approach is the attempt to transfer the Bacillus thuringiensis insecticidal protein gene into a recipient plant with controls such that it will be expressed in the appropriate tissues. There are many unknowns, even in a process conceptually as simple as this, but it is worth emphasizing that the planning of such experiments is now possible because of the advances in knowledge of regulation of plant genes and the properties of particular promoters.

Another example of gene addition concerns engineering readily isolatable genes. It may be desirable for a particular objective in plant breeding to reinsert a gene with different control functions such that it adds another capability to the crop species. A gene coding for a seed storage protein with high sulfur content could be equipped with controls which cause it to be expressed in leaf cells of the same or a related plant species. This could dramatically change the nutritional value of the foliage of that species. Another related example is to mutate a plant storage protein gene in vitro and insert it back into the plant as an extra copy or with a promoter that is slightly more efficient so as to provide enough of the changed product to change the amino acid composition of the total storage proteins of the seed.

In a majority of the cases where one would like to isolate and transfer a gene, it is likely that the gene will be without any known product and not have an identifiable messenger RNA. What to do in these cases? One approach would be to use transposon mutagenesis. If it is in a species with characterized controlling elements, of which there few, mutants caused by insertion of the controlling elements could lead to gene isolation. Another possibility is that T-DNA insertions could suffice, as could uncontrolled or random insertions of any other DNA in a transformation system as long as the DNA which has inserted and inactivated the gene function can be located. The principal limit will be frequencies - can one develop the methodologies such that the frequency of transformation will be high enough to use controlling elements, T-DNA, or other DNA insertions in this way? It may be possible, using the restriction enzyme mapping technique mentioned earlier, to detect tight linkage to a particular function and limit the location of the gene to a segment of DNA sufficiently small that it could be cloned by "walking" along the chromosome.

Finally, one other general method of gene isolation is shotgun transformations where a whole genome is tested segment by segment and the appropriate transformant selected. Again, feasibility of this approach depends on the efficiency of the shotgun transformation system but, more importantly, on the efficacy of a selection system. Gene engineering is no different from any other form of plant breeding in that ultimately its efficiency rests upon the availability of suitable selection techniques. If selection can operate at cell culture level, or in any way where large numbers of transformants can be examined, shotgun cloning could turn out to be a useful tool. It may also be possible to select initially using microorganism expression systems.

Gene Replacement in Plant Breeding

There are a number of examples in crop species where particular gene functions have had to be removed in order to fit the crop product to market requirements. A modern example is the removal of erucic acid and glucosinolate from rapeseed to produce the double zero Canola.

Gene function has been removed either by classical mutagenesis or frequently by the identification of the appropriate null variant in the range of existing genetic stocks. There are some examples where appropriate variation has not been identified and where mutagenesis has not been successful. Once again DNA-mediated mutagenesis, either by controlling elements, by

T-DNA insertions, or by heterologous DNA insertions, may be a way of removing specific functions. The probability of removal of a function may be enhanced by a targeted insertion event. If, for example, some controlling elements did insert in specific sequences or if DNA insertion could be targeted to a particular genetic segment by inclusion of sequences from that segment, an efficient gene removal system may be generated. Modification of gene function might be achieved if a similar homologous recombination system could be applied to DNA insertion events. We know of no data at present suggesting that naturally occurring specific target systems are available in plants, but certain specificities are known. For example, the Ac/Ds controlling element system frequently transposes to neighboring sequences; thus the probability of inserts into a particular gene target can be increased if translocations or other arrangements are available which bring the target sequence into close proximity to the location of one of the elements.

Gene technology may be important to convert crop species to alternate uses. For example, and this is discussed in more detail by Hahlbrock (this volume), the removal of certain so-called secondary metabolites might convert a crop that is presently grown for industrial purposes into a crop which would be a valuable food source either for animals or humans. For example, if nicotine were removed from Nicotiana by gene dysfunction, perhaps we could capitalize on the history of breeding and management skills that have been developed for this crop; its value, say, as a leaf protein crop, may be further enhanced by a specific gene insertion or modification to change amino acid composition.

Complex Agronomic or Physiological Characteristics

So far we have only discussed characteristics which can be attributed in large part to single genes. Plant breeders are often disdainful of the contribution that genetic engineering may add to their craft because of the apparent complexity of their objectives, chiefly biological yield. However, we think there is no reason to be despondent when considering complex characters. Let us consider one particular example. The amount of biological production of a crop can be equated in first approximation to the amount of dry matter production by the plant, and this is directly dependent on the amount of leaf material available. The conversion of light energy into assimilate or photosynthate is dependent on leaf area and on the biochemical conversion efficiencies. By far the dominant factor in this equation is leaf area. It is leaf area which determines the sum total of intercepted light and it is leaf area that is generally most sensitive to stress, such as water stress or salt stress. The rate of leaf area production, then, may be the prime biological process that we need to address if we are to modulate yield. One can look at leaf area increase as being a consequence of both cell division and cell expansion, and it has been demonstrated at least for cereals that the limiting process is the rate of cell expansion, not division. If there is a single dominant component or control pathway for cell expansion, then manipulation of this control may enable some sophisticated gene manipulation so that at known critical periods of development the rate of production of leaf area will be increased and lead to subsequent increases in yield.

We feel sure that many complex physiological and agronomic characteristics will be able to be dissected to a small number of major determinants and a series of minor determinants. Recombinant DNA technology will be of tremendous importance in the dissection of such processes. Thus we will in many cases reduce complex processes down to genes of major effect and be in a position to manipulate the processes by manipulating these principal determinants. Earlier in this paper we mentioned that the provision of restriction enzyme polymorphisms was likely to provide a powerful tool in plant breeding for following segments identified as having a major effect on yield, even though the precise biological mechanism or gene product may not be known at all. Normally, in a so-called complex polygenic system, plant breeders are fighting to achieve the average or modal situation with many genes depressing yield against the desired additions of yield by other genes. If one could account for 70-80% of yield by a small number of principal determinants and manipulate these in a known way, then breeding for these major characters would become much more efficient.

Another general area which we think holds particular promise in manipulation of economic plants is our increasing ability to manipulate developmental pathways. If, for example, we could identify the principal genetic components of flowering induction, we may be able, by marrying appropriate coding and promoter sequences, to have control over flowering time. Developmental genes or genes which control major pathways of development have had significant impacts on yield; the few major genes that have wrought the conversion of the female flowering spike of teosinte to that of maize is a good example (4). The addition of the dwarfing genes to wheats and other cereals have been important in increasing harvest index values. Many of the dicotyledonous crop species, including the important oilseeds, have harvest indices much lower than those obtained in cereals. We may be able to obtain dramatic increases in yield in these plants by manipulating harvest indices through changes in development. In some cases it may be possible to transfer particular developmental genes from unrelated species, e.g., genes for flowering time or root structure.

Finally, gene manipulation might be used to alter characteristics of breeding systems. As we dissect the bases of cytoplasmic male sterility systems, we are likely to be able to control these systems and introduce them into many more species than we have been able to in the past through traditional breeding. We may also be able to clone the genes responsible for pollen growth retardation in incompatibility systems and introduce them with good effect into other species to promote production of hybrids.

SOURCE OF GENES

Frequently an appropriate gene for transfer may be obtained from an existing cultivar. That gene may need to be modified in vitro or associated with a different set of controls. Sometimes genes will come from very different sources, even from microorganisms, as has been indicated above. We can expect more and more truly novel objectives to be introduced into plant breeding as we gain knowledge of the basic biological processes involved in plant production systems and of the workings of other organisms which impinge upon the production system. An example is with wheat rust, discussed earlier, which is a major limitation to yield in Australia and in a number of other countries. Recently it has been determined that there are double-stranded RNA molecules, similar to mycoviruses, present in all strains of wheat rusts which have been examined (Pryor, unpublished). Immediately the possibility surfaces that these double-stranded RNA molecules, which are ubiquitous, may be similar to those described in yeast and in the smut fungus of maize where a mycovirus codes for a protein toxin killing nonimmune strains. If similar systems operate in wheat rusts, then the isolation and transfer of the appropriate gene segment and its inclusion into the wheat genome with suitable controls may provide for quite new biological methods of resistance.

Apart from the novel breeding objectives which we can expect to emerge, another major consequence of the addition of gene technology to the methods of plant breeding is the provision of a vastly increased range of genetic variation. Plant breeders, when faced with a problem, look first in the pool of available variability and frequently find what they need in other breeders' lines or in more primitive land races. In some instances where genetic methods have reached sufficient levels of sophistication, breeders are able to tap species which can only be hybridized with extreme difficulty to the target species. There are instances of valuable alien genes being integrated into cereal production systems. But there are cases where we can identify in other species what would be suitable genetic variation but which is unavailable because there are no known techniques for hybridization. It is in these instances that gene transfer would be an important additional tool for the plant improver. One example is the rust disease caused by Phakopsora pachyrhizi in soybean which can result in serious and sometimes total yield losses. In existing soybean genetic material there is no satisfactory source of field resistance to the rust, but species of Glycine, the same genus as soybean, which are indigenous to Australia have a range of resistance genes (7). These Glycine species evolved in the presence of the rust organism with resultant selection for resistance genes. If these genes can be isolated and transferred, they are likely to provide a major contribution to soybean production systems in Southeast Asia and nearby areas.

As an increasing range of crop plants are introduced to an increasing range of environments around the world there will be new pests, disease, and other problems, and frequently the source of suitable genes will be in the species which are native to those regions. This introduces the concept of genetic resource material and poses a question as to what really is the meaningful perimeter of genetic resources for any given crop plant. Do we need to consider entirely new principles of genetic resource conservation? Should we think of the impact that genetic engineering methodologies might have in this process? Should we have gene libraries other than those stored in seeds, the usual way in which a genome is stored in a germ plasm bank?

CONCLUSIONS

At present only crops from the Solanaceae, such as potato and tomato, and Brassica from the Crucifereae have all the necessary components of an effective transformation system available. Most of the major crops are deficient in one or more of the needed components. The major limitation is the regeneration of plants from protoplasts or other target cells. It may be that there should be a greater concentration of effort on developing transformation through pollen or other effective vector systems. It may be that T-DNA will be of use in a variety of monocot species. It should not be too long before we have all of the necessary components in the major crop species.

Genetic engineering is already contributing to plant production systems. Examples are in the use of nucleic acid diagnostic kits and in the preparation of linkage maps for species where no genetic tools previously existed. There are manipulated microorganisms ready to be tested in production systems. Plant modification has been achieved, with particular genetic segments being introduced and inherited in a stable fashion.

Most of the present techniques enable the transfer of only single genes or of small groups of genes, but this will be sufficient to make many contributions to plant improvement. The techniques of genetic engineering at present are having their major impact in teaching us about the basic biological processes that are important in plant production. This knowledge will enable us to design new objectives for plant improvement. We will have new ways of using plant species for the production of an increasing variety of products.

Acknowledgements. Many of the ideas and results discussed in this paper stem from our colleagues in the Division of Plant Industry. We are grateful to J. Begg for his comments.

REFERENCES

(1) Ahlquist, P.; French, R.; Janda, M.; and Loesch-Fries, L.S. 1984. Multicomponent RNA plant virus infection derived from cloned viral cDNA. Proc. Natl. Acad. Sci. USA **81:** 7066-7070.

(2) Appels, R., and Moran, L.B. 1984. The molecular analysis of alien chromatin introduced into wheat. **In** Gene Manipulation and Plant Improvement, 16 Stadler Genetics Symposium, ed. J.P. Gustafson, pp. 529-557.

(3) Barker, J.M.; McInnes, J.L.; Murphy, P.J.; and Symons, R.H. 1985. Dot blot procedure with ^{32}P DNA probes for the sensitive detection of avocado sunblotch and other viroids in plants. J. Virol. Meth. **10:** 87-98.

(4) Beadle, G.W. 1980. The ancestry of corn. Sci. Am. **242:** 96-103.

(5) Brisson, N.; Paszkowski, J.; Penswick, J.R.; Gronenborn, B.; Potrykus, I.; and Hohn, T. 1984. Expression of a bacterial gene in plants by using a viral vestor. Nature **310:** 511-514.

(6) Broglie, R.; Bellemaire, G.; Bartlett, S.G; Chua, N.-H; and Cashmore, A.R. 1981. Cloned DNA sequences complementary to mRNA encoding precursors to the small subunit of ribulose-1,5-biphosphate carboxylase and a chlorophyll a/b binding polypeptide. Proc. Natl. Acad. Sci. USA **78:** 7304-7308.

(7) Burdon, J.J., and Marshall, D.R. 1981. Australian native species of Glycine - a source of resistance to soybean rust. Plant Dis. **65:** 44-45.

(8) Davies, K.E. 1981. The application of DNA recombinant technology to the analysis of the human genome and genetic disease. Hum. Genet. **58:** 351-357.

(9) Dennis, E.S.; Gerlach, W.L.; Pryor, A.J.; Bennetzen, J.L.; Inglis, A.; Llewellyn, D.; Sachs, M.M.; Ferl, R.J.; and Peacock, W.J. 1984. Molecular analysis of the alcohol dehydrogenase Adhl gene of maize. Nucl. Acids Res. **12:** 3983-4002.

(10) Downes, R.W.; Salisbury, P.A.; Healy, F.W.; and Mackay, J.H.E. 1980. Cultivar Siriver. J. Aust. Inst. Ag. Sci. **46:** 200-201.

(11) Fedoroff, N.V.; Furtek, D.B.; and Nelson, O.E. 1985. Cloning of the bronze locus in maize by a simple and generalizable procedure using the transposable controlling element Ac. Proc. Natl. Acad. Sci. USA, in press.

(12) Fletcher, J.T. 1978. The use of avirulent virus strains to protect plants against the effects of virulent strains. Ann. Appl. Biol. **89:** 110-114.

(13) Hernalsteens, J.-P.; Thia-Toong, L.; Schell, J.; and Van Montagu, M. 1984. An Agrobacterium-transformed cell culture from the mono cot Asparagus officinalis. EMBO J. **3(13):** 3039-3041.

(14) Hoekema, A.; Hirsch, P.R.; Yooykass, P.J.J.; and Schilperoort, R.A. 1983. A binary plant vector strategy based on separation of vir- and T-region of the Agrobacterium tumefaciens Ti-plasmid. Nature **303:** 179-180.

(15) Hooykaas-Van Slogteren, G.M.S.; Hooykaas, P.J.J.; and Schilperoort, R.A. 1984. Expression of Ti plasmid genes in monocotyledonous plants infected with Agrobacterium tumefaciens. Nature **311:** 763-764.

(16) Horsch, R.B.; Fry, J.E.; Hoffman, N.L.; Eichholtz, D.; Rogers, S.G.; and Fraley, R.T. 1985. A simple and general method for transferring genes into plants. Science **227:** 1229-1231.

(17) Howard, E.A.; Danna, K.; Dennis, E.S.; and Peacock, W.J. 1985. Transient expression in maize. In Plant Genetics, ARCO-UCLA Symposium, in press.

(18) Llewellyn, D.J.; Dennis, E.S.; and Peacock, W.J. 1985. The alcohol dehydrogenase genes of maize: expression studies in tobacco cells. In Molecular Form and Function of the Plant Genome. Proceedings of the NATO Advanced Studies Institute. Renesse, Netherlands: Plenum Press, in press.

(19) Lörz, H.; Baker, B.; and Schell, J. 1985. Gene transfer to cereal cells mediated by protoplast transformation. Molec. Gen. Genet., in press.

(20) McClintock, B. 1956. Controlling elements and the gene. Cold S.H. Symp. Quant. Biol. **21**: 197-216.

(21) Nagao, R.T.; Shah, D.M.; Eckenrode, V.K.; and Meagher, R.B. 1981. Multigene family of actin-related sequences isolated from a soybean genomic library. DNA **1(1)**: 1-9.

(22) Paszkowski, J.; Shillito, R.D.; Saul, M.; Mandák, V.; Hohn, T.; Hohn, B.; and Potrykus, I. 1984. Direct gene transfer to plants. EMBO J. **3(12)**: 2717-2722.

(23) Peacock, W.J.; Dennis, E.S.; Gerlach, W.L.; Sachs, M.M.; and Schwartz, D. 1985. Insertion and excision of Ds controlling elements in maize. **In** Recombination at the DNA Level. Cold S.H. Symp. Quant. Biol. **49**, in press.

(24) Roper, M.M. 1983. Field measurements of nitrogenase activity in soils amended with wheat straw. Aust. J. Agric. Res **34**: 725-739.

(25) Schöffl, F., and Key, J.L. 1982. An analysis of mRNAs for a group of heat shock proteins of soybean using cloned cDNAs. J. Molec. Appl. Gen. **1**: 301-314.

(26) Van den Broeck, G.; Timko, M.P.; Kausch, A.P.; Cashmore, A.R.; Van Montagu, M.; and Herrera-Estrella, L. 1985. Targeting of a foreign protein to chloroplasts by fusion to the transit peptide from the small subunit of ribulose 1,5-bisphosphate carboxylase. Nature **313**: 358-363.

(27) Whitfeld, P.R., and Bottomley, W. 1983. Organization and structure of chloroplast genes. Ann. Rev. Plant Physiol. **34**: 279-310.

Biotechnology: Potentials and Limitations, ed. S. Silver, pp. 241-257. Dahlem Konferenzen 1986. Berlin, Heidelberg, New York, Tokyo: Springer-Verlag.

Secondary Products

K. Hahlbrock
Max-Planck-Institut für Züchtungsforschung, Abt. Biochemie
5000 Köln 30, F.R. Germany

Abstract. Secondary plant products are of great commercial interest in chemical, pharmaceutical, cosmetic, food, and other industries. For the production of particularly valuable compounds, cell suspension cultures in bioreactors have the potential to replace intact plants and complement chemical syntheses, although only a few industrial applications are so far in or near operation. Another area of interest is the relation of secondary products to nutritional quality, yield, and storage properties of crop plants (e.g., substances involved in disease resistance). Despite an urgent need for a better understanding of the physiology, biochemistry, and genetics of secondary product formation, it is foreseeable that this field will soon benefit from the application of gene transfer and other biotechnological methods.

INTRODUCTION

The term "secondary product," or "secondary metabolite," is usually applied, without rigorous definition, to compounds of low molecular weight which are restricted in their occurrence and not involved in primary metabolism, that is, in the basic reactions common to all cells of all organisms (13, 33). A rigorous definition would inevitably cause confusion, partly because an unequivocal distinction between primary and secondary metabolism is difficult if not impossible, partly because the basic pathway for most life on Earth, photosynthesis, and the biosynthesis of many mass products such as starch, lignin, cellulose, and storage proteins, to name a few, might have to be classified as secondary metabolism, depending on how a workable definition is attempted (different schools advocate different definitions). Here, the term will be used as stated above, with its poorly defined implications and confinements.

In this sense, secondary plant products comprise the whole spectrum of pharmaceutically or otherwise commercially interesting compounds whose chemical synthesis is either too difficult (expensive) or impossible; they also include many of those substances (e.g., flower pigments, odorous compounds, antimicrobial "phytoalexins," etc.) whose presence or absence, localization, timing, and extent of accumulation in certain tissues might be decisive in the survival of a plant or its progeny in a given environment.

With regard to biotechnology and agriculture, two rather different aspects are of major interest: the extraction of useful compounds in high yield and the selection for, or "engineering" of, plants with agricultural advantages. Conventional breeding has been very successful over thousands of years in attempts to reach these goals. Modern biotechnology has added two principally new methods, cell culture techniques and the methods of in vitro recombination and transfer of genes. Both of these methods have a high potential in optimizing secondary product formation.

In the following, a brief summary of the present state of the art and a few, probably low-risk predictions of further progress in the foreseeable future will be contrasted with some areas in which basic knowledge is almost completely lacking. These gaps must be filled before occasional success through trial and error can be replaced by fully controlled methods of approach.

SECONDARY PRODUCTS FROM CULTURED PLANT CELLS

So far, secondary products of industrial interest have been mostly the same from cell cultures and intact plants. A possible new area of research involving compounds found in cultured plant cells but not in whole plant tissue will be discussed briefly below. To date, major arguments for the use of cell cultures instead of intact plants for secondary product isolation (51, 53) have been a) potentially high yields from more or less homogeneous cell populations under extremely specialized and optimized conditions; b) independence of political and economical factors, geographic location, climate, and seasons; c) exclusion of losses to pathogens and herbivores; d) rapid adaptability to changes in needs; and e) production of, for instance, narcotics in strictly controlled areas.

A large variety of commercially interesting and potentially useful secondary products from plant cell cultures has been investigated (21). A few selected categories of potential industrial interest are plant-derived antibiotics, antitumor agents, antispasmodics, flavors and fragrances, dyes, spices, sweeteners, tonics, insecticides, gums and hydrocarbons (16, 43). They reflect the size of the market for such substances in different branches of chemical, pharmaceutical, cosmetic, and food industries, with an estimated total value (worldwide) on the order of many billions of U.S. dollars annually. This market certainly has a large capacity for additional products which might become available through further research in this area and perhaps replace to some extent synthetic chemicals. To give only one example of the importance of secondary plant products, about one fourth (23%) of all medical prescriptions except antibiotics analyzed in the USA a few years ago contained one or more of about 50 different plant-derived compounds. More than half of them were steroids (14.7%), followed by codein (2%), atropine (1.5%), and the alkaloid reserpine (1.4%) (16, 17). Many of these compounds have been found in cultured cells, albeit in most cases in low and variable yield. However, a few exceptions thus far clearly demonstrate that both goals, high yield and stable production rates, can be reached and that secondary products can accumulate in cell cultures in very large amounts.

POTENTIALS AND DRAWBACKS OF CELL CULTURE EXPLOITATION

To date, the only running industrial process of large-scale secondary product formation using plant cell cultures is the extraction of shikonin derivatives (naphthoquinones) from Lithospermum erythrorhizon cells by Mitsui Petrochemical Industries in Japan (19, 20). Shikonin (Fig. 1) and its derivatives have antibacterial activity and are used as topical medication for wounds, burns, hemorrhoids, etc. (19), and as dye components for silk and a best-selling Japanese "bio-lipstick." The yield of shikonin derivatives was given as 1.4 g/l (19).

The highest yield of any secondary product so far has been 5.5 g/l (0.9 g/l per day) of rosmarinic acid (Fig. 1) from Coleus blumei cell cultures propagated under carefully established conditions, including a specially designed spiral-stirred bioreactor containing 32 l of medium. The rosmarinic acid content was close to a physiological barrier of around 25% of the dry weight (46). Rosmarinic acid has antiphlogistic properties and is presently being tested for its pharmacological suitability (53). This example of successful, strictly production-oriented studies was based on strategies initially designed for two model systems, the Coleus blumei (56) and Morinda citrifolia (55) cell cultures. Under appropriate conditions, cell suspension cultures of Morinda citrifolia produce anthraquinone derivatives (Fig. 1) whose intracellular concentrations may reach very high levels. Yields from cell cultures (up to 3 g/l) far exceeding those from roots, on a dry weight basis, have been reported (53). Another interesting compound, the spasmolytically active isoquinoline alkaloid, jatrorrhizin (Fig. 1), has been obtained in similarly high yield from Berberis stolonifera cell cultures (26). The large-scale production of berberine from the culture medium of Thalictrum minus cells (36) might soon become another example of industrial application.

This list can be extended (7, 53) but soon reaches a point where compounds are produced in much lower and often extremely variable yields, far below the threshold of commercial interest (51). It is therefore unlikely that more than a few secondary products from cell cultures will in the next few years replace those extracted from intact plants.

One very promising, related procedure, the bioconversion (or "biotransformation") of relatively inexpensive substrates to valuable products, has been widely exploited with microorganisms but only to a small extent with cultured plant cells. At least one such reaction with a high potential for immediate commercial interest has been studied extensively with plant cells. Cell suspension cultures of Digitalis lanata convert the industrial

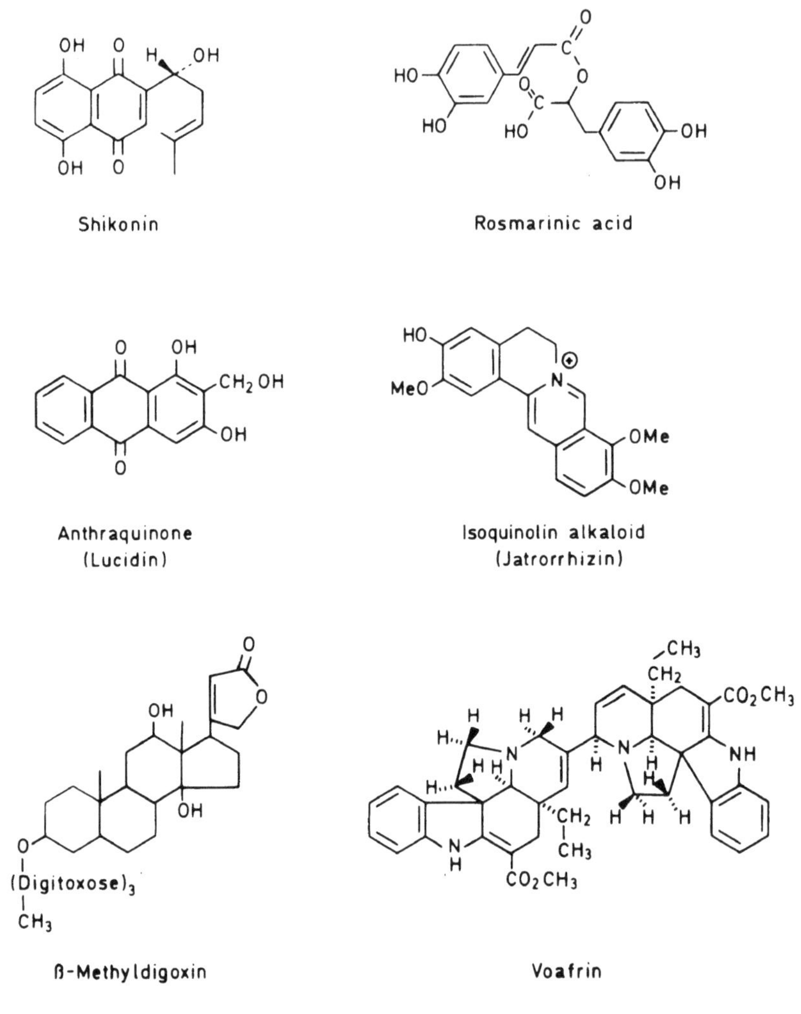

FIG. 1 - Examples of secondary products isolated from plant cell cultures. Upper and center rows: compounds obtained in unusually high yield. Lower row, left: compound produced by bioconversion with immobilized cells; right: compound so far undetected in whole plant tissue (45)

by-product, β-methyldigitoxin, through very efficient and stereoselective 12 β-hydroxylation of the steroid moiety to the widely used cardiac glycoside, β-methyldigoxin (1). An industrial application of this procedure should be feasible, especially when coupled with immobilization of the cells (35). Further reactions of commercial interest involve the conversion of certain lipophilic to hydrophilic compounds.

Immobilization of cultured plant cells (10, 31) or of protoplasts derived either from cultured cells or from whole plant tissues (32) offers the advantages of more or less continuous cultivation and of the potential conversion of a large variety of exogenous, natural as well as synthetic, substrates in chemically inaccessible reactions. Moreover, secondary products may be formed at much higher rates by immobilized than by freely suspended cells (31).

Another potentially useful application of plant cell cultures might result from the fact that several compounds have been obtained which are either not produced or undetectable in intact plants (7, 50, 53). More sensitive assay procedures (48) will probably reveal the presence of small amounts of many previously undetected compounds in plants, but it is possible that some secondary products will be available in appreciable quantities only through plant cells cultured under appropriate conditions. This might include the application of unusual or unnatural substrates to generate novel compounds, the use of inhibitors to cause the accumulation of otherwise rapidly metabolized intermediates (2), and in certain cases the use of photoautotrophic cultures (4). Finally, an interesting suggestion is the conversion of one central metabolite, such as the indole alkaloid precursor, strictosidin, to various products by appropriately mixed and immobilized enzymes, for which in this case cell cultures from selected Catharanthus roseus strains would be a convenient source (53).

Major drawbacks of plant cell culture techniques are the frequent instability of product formation and the unpredictable requirements for growth and production, which often vary from species to species as well as from small to large dimensions of a culture. High production rates for all of the compounds mentioned above were obtained by trial and error at all individual steps (establishment of cell cultures, development of growth media, selection of high yielding strains, establishment of stable strains, development of production media), and only few of the many failures have been reported. A particularly thorough investigation of the instability of indole alkaloid production in Catharanthus roseus cell suspension cultures (14) might serve as an example. Six selected cell lines were monitored over a period of eight years and lost their productivity within a few months. Spontaneous recovery of the initial high production rate was not observed, although high yielding strains could be obtained at any time by repetition of the clonal selection procedure. Many other examples of instability (14) or inability to produce secondary products (8) exist. By contrast, suspension cultures originating from the same Petroselinum crispum cells have yielded, upon UV irradiation, similar amounts of flavonoid glycosides over a period of fifteen years (approximately 750 subcultures) without any selection or other deliberate manipulation (K. Hahlbrock, unpublished results). Many more such cases have been observed (14).

Somewhat less ambiguity is usually involved in the development of growth and production media (54). Increases in sugar and decreases in phosphate and nitrogenous substrate concentrations, as well as a few other changes in medium composition as prerequisites for reduced growth rates and increased rates of secondary product formation, are routinely tested first and are frequently successful. On the other hand, the scale-up from small to large dimensions without loss of production rate is often again an insurmountable obstacle, so far mostly for unknown reasons.

Thus, great potentials still have to be contrasted with serious drawbacks. Despite some remarkable recent progress, most of these drawbacks are no doubt due to large gaps in our knowledge of the physiology, biochemistry, and genetics of secondary product formation. However, this should be a matter of time and effort, and the above-mentioned potentials, together with the manifold demonstrated totipotency of plant cells, the possibilities of somatic hybridization and of using haploid cells (28, 34), as well as the prospects of gene transfer in vitro will certainly have a large impact on future progress in this field.

SHUNT ONTO MICROORGANISMS?

Plant secondary products, by the confined definition used here, are plant-specific substances and therefore not normally produced by microorganisms. However, some possibilities of still exploiting low-cost fermentation of microorganisms are worth considering. Certain "plant products" might be obtained by microbial biotransformation or by using artificial substrates for microbial pathways, either in vivo or in vitro, in analogy to present or suggested future work with cultured plant cells. A particularly exciting area of future research could be the transfer in vitro of specifically engineered plant genes to microorganisms. On the other hand, plant cell cultures will no doubt continue to have a large impact in this field and successfully compete with microorganisms and chemical syntheses, due to the large complexity of most secondary metabolic pathways and their regulatory mechanisms.

SECONDARY PRODUCTS IN PLANT ECOLOGY

The second area of major interest in secondary plant products is their presence or absence and function in intact plants. Man has for thousands of years, deliberately or not, selected for plants with high or low levels of secondary products. For instance, high contents of alkaloids, aromatic compounds, terpenoids, etc., in medicinal and spice plants, carotenoids in carrot or tomato, sucrose in sugar beet, or flower pigments in ornamental plants, and low contents of certain fatty acids in soybean, glucosinolates in rape seed, or phenolics in brewing barley have long been goals of plant breeding and selection. Other than the high-priced or otherwise valuable products to which the list of potentially interesting compounds from plant cell cultures is largely confined, secondary plant products falling under the large category of substances affected directly or indirectly by plant breeding are innumerable.

The functions of most secondary products in plants are unknown, or at least not known for certain. It is quite possible that many of them have multiple functions in such seemingly unrelated areas as disease resistance, sexual reproduction, herbivore repelling, cold hardiness, allelopathy, and many others (6, 18). However, unequivocal proof of a definite function would require that loss of this function be caused by the lack of formation of the very compound and not by any other change. Such hard proof is very scarce for secondary plant products. Besides, it is possible not only that one given product has two or more different functions, but also that two or more different products have one common function. For example, flavonols are almost ubiquitously occurring yellow pigments whose putative functions, possibly among others, are to act as UV-protective substances in leaves, as insect attractants in flowers, and as insect repellents in roots. In flowers, they often occur together with other yellow pigments, carotenoids, which might have, again among others, the same function as flavonols. Another example would be two functionally related antimicrobial substances (phytoalexins) originating from different biochemical pathways, the terpenoid rishitin, and the steroid glycoalkaloid solanine (Fig. 2), which are concomitantly induced by infection of potato tuber tissue with fungal or bacterial pathogens.

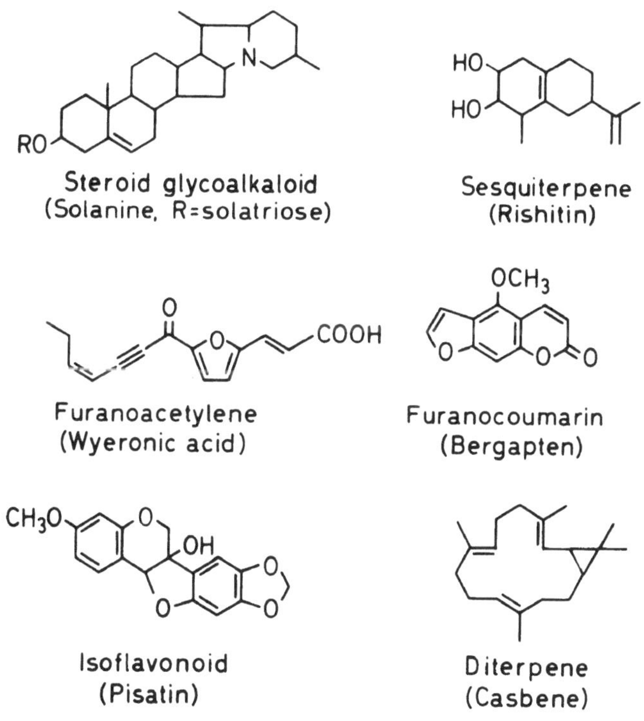

FIG. 2 - Selected phytoalexins from different plants indicating possible large differences in biosynthetic origin.

However unlikely, it cannot be excluded with certainty that some secondary products have no function at all or function merely as metabolic sinks, or as excretion or storage material for superfluous intermediates. At the present state of knowledge, it is useless to enter into the various speculations and controversies about possible functions of such a diversity of substances. The following brief discussion will be confined to a few structurally related groups of compounds, mainly because their biosynthetic pathways and their possible functions are relatively well-known, and genes involved in these pathways have been studied and are available for transfer experiments in vitro (see below).

The putative multifunctional flavonols constitute but one group of flavonoid compounds. They frequently co-occur with other flavonoids (e.g., flavones, anthocyanins, isoflavonoids) in various combinations, depending on plant species, tissue, and developmental stage (25). Their biosynthetic routes all have a chalcone intermediate in common (15, 23) and are further interconnected at an earlier branch point with various other phenylpropanoid pathways (Fig. 3). The three enzymes indicated in Fig. 3 (PAL, 4CL, CHS) are coded for by the first secondary product-related plant genes which have been isolated (see below).

Flavones often co-occur with flavonols and might have similar functions. Anthocyanins are typical red, purple, blue, orange, or yellow pigments occurring in many flowers (e.g., rose, hibiscus, petunia, snapdragon), fruits (e.g., cherry, grape, apple skin, colored maize kernels), and other plant tissues (25). Their putative function is to attract animals for sexual reproduction and geographical distribution. One established function of isoflavonoids is to act as phytoalexins, mainly in leguminous plants. Each plant species has its own characteristic set of secondary products acting as phytoalexins (3). For example, the major phytoalexins in the three legumes, pea, French bean, and soybean, are isoflavonoids (Fig. 2), whereas furanocoumarin phytoalexins in parsley, celery, dill, and related species (Fig. 2) are also of phenylpropanoid origin but products of a different pathway (Fig. 3). The specific role of phytoalexins in disease resistance is not always clear. Their activity might in certain cases contribute only slightly, or not significantly at all, to the concerted defense reactions of a plant but can in other cases be the decisive mechanism. The latter case is exemplified by the dependence of virulence of the fungus, Nectria haematococca, on pea (Pisum sativum) on the presence of an enzyme capable of demethylating the pea phytoalexin, pisatin (29). Genetical studies with crosses between different isolates of the fungus confirmed this relation (30).

POTENTIALS AND DRAWBACKS IN AGRICULTURE

Crop yield, storage properties, and nutritional quality are usually of utmost concern in agriculture, although many other, more specific interests exist. The following considerations apply to many compounds of interest, even if those whose functions in plants are unknown are not mentioned explicitly. As far as secondary plant products are concerned, crop yield and storage

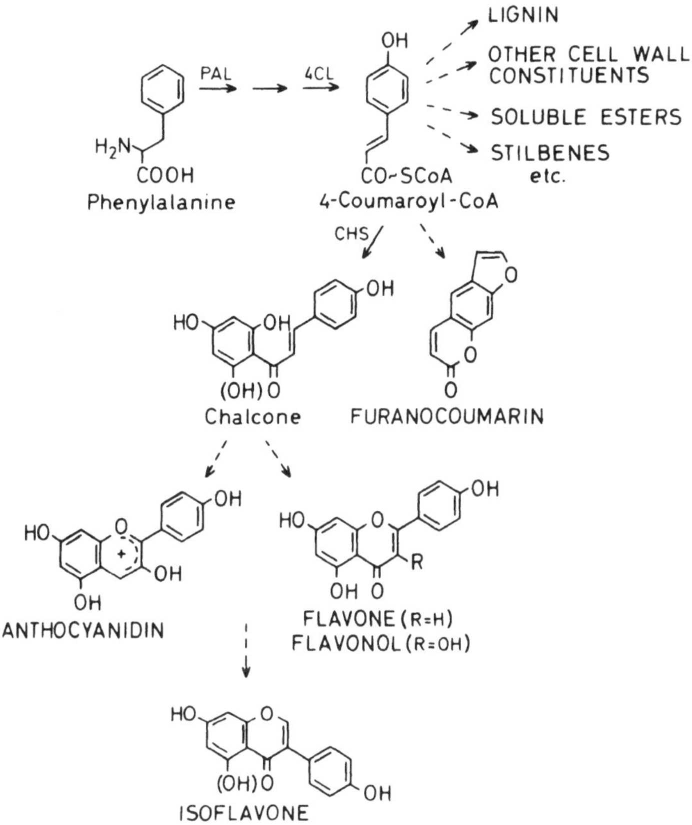

FIG 3 - Biosynthetic origin of various phenylpropanoid derivatives. Enzymes mentioned in the text are phenylalanine ammonia-lyase (PAL), 4 coumarate:CoA ligase (4CL), and chalcone synthase (CHS).

properties are usually associated with compounds involved in pathogen and herbivore resistance. Phytoalexins, as far as a significant role in disease resistance is apparent, might serve as an example.

By formal definition, phytoalexins are low-molecular-weight antimicrobial compounds that are both synthesized by and accumulated in plants which have been exposed to microorganisms (38). This definition is unfortunate in that plants can well contain these compounds without previous exposure to microorganisms. In fact, this might appear to be a crucial point, because many attempted invasions by potential pathogens might fail when phytoalexins are already present at the site of infection as a determining factor in resistance. Thus, plants constitutively containing high levels of phytoalexins would seem desirable. However, disadvantages can be seen immediately: the broad-spectrum toxicity of these compounds often includes higher animals and man; their continuous presence would be a waste of biosynthetic potential and might have negative influences on other properties of the plant; the rate of microbial adaptation might be enhanced.

A more efficient and sophisticated mechanism is probably the specific induction of phytoalexins upon infection. In particular, their rapid and massive induction is almost invariably associated with the hypersensitive response, that is, with the rapid formation of small local lesions at the sites of infection in host as well as non-host resistance. Especially non-host resistance, which by definition is not overcome by simple mutations of an avirulent microorganism, is an extremely desirable phenomenon for plant protection. Its possible causal relation with the timing, site, and extent of phytoalexin accumulation is poorly understood and deserves much attention.

In more general terms, specific induction mechanisms, whether triggered by infections, in association with differentiation and development, or otherwise, are certainly of special interest in studies of secondary product formation. Furthermore, decreased degradation rates might have the same end effect as increased production rates in cases where turnover has a significant role.

OPEN QUESTIONS IN PHYSIOLOGY, BIOCHEMISTRY, AND GENETICS OF SECONDARY PRODUCT FORMATION

Many pieces of information, though often fragmentary, are available on the taxonomical distribution, site of occurrence, and developmental timing of formation of secondary plant products (40, 49). By comparison, relatively little is known about their function, possible translocation, turnover (5), or effects on other pathways. Only in a few selected cases have enzymology and genetics been studied in great detail. Numerous methods have been reported for the induction of certain classes of compounds, notably flavonoids and other phenylpropanoid derivatives. However, hardly anything is known about the molecular mechanisms of signal-reaction chains. Among the few exceptions are the frequent involvement of the photoreceptor phytochrome when light is the inducing agent (39), a possible but unexplained role of ethylene (11), and recent promising studies of receptors of plant hormones, such as abscisic acid (27), which might at least indirectly be involved in secondary product formation. However, neither other receptors nor any molecules possibly acting as "second messengers" have so far been identified. Reports that cinnamic acid, the product of phenylalanine ammonia-lyase, the first enzyme in the biosynthesis of phenylpropanoid derivatives, has a regulatory function in vivo (42) must be regarded with caution (47). End products are often stored in the vacuole and are therefore unlikely to have a major regulatory role in their synthesis.

Important but so far barely touched areas are the mechanisms of interaction, coaction, and synergism of regulatory molecules, as well as the complexity of mechanisms underlying differentiation and "dedifferentiation".

Although "dedifferentiation" is a problematic term, the reversibility of "terminal differentiation" and the principle totipotency of plant cells are the basis for their large potential in cell culture techniques, including secondary product formation and whole plant regeneration.

Only a few complex biosynthetic pathways leading to secondary plant products have been studied completely. Examples are the indole alkaloid, ajmalicin (44, 52), the cyanogenic glycoside, dhurrin (12), and a number of structurally related flavones and flavonols (9, 15, 22, 23). A remarkable observation with possible general implications was made with the enzymes of the dhurrin pathway. Almost the entire sequence is tightly associated with membranes, resulting in the "metabolic channeling" of substrates through all steps with little or no exchange of intermediates with the environment, in vivo as well as in vitro (12). The following possible consequences might be worth considering in the context of this discussion: an increase in the amount of one of the enzymes must not necessarily lead to an increased rate of product formation, whereas a decrease might still result in a decreased product yield; artificial intermediates (not necessarily precursors) might be accepted poorly or not at all; alterations of the protein structure must be regarded with respect to both the kinetics of substrate conversion and the association properties with neighboring enzymes; and finally and perhaps most importantly, various other pathways might be arranged in vivo in a similar manner with similar consequences, although this might go undetected whenever enzymes are "soluble" in vitro (distinctions between "soluble," "microsomal," "particle-associated," etc., are usually based on operational criteria and are therefore often extremely variable and not necessarily reflective of the actual situation in vivo).

The genetics of secondary product formation in plants give an extremely unbalanced picture. On the one hand, the two epochal studies of Mendel and McClintock, which laid the foundations of classical and modern genetics, were largely based on anthocyanin (flavonoid) formation in pea, Pisum sativum, and corn, Zea mays, respectively. Many mutants of these and other plants with defects in flavonoid biosynthesis have been investigated (37, 41). The gene encoding the key enzyme of this pathway, chalcone synthase, has been isolated, cloned, and partially sequenced from parsley, Petroselinum crispum, snapdragon, Antirrhinum majus, Petunia hybrida, and French bean, Phaseolus vulgaris. Its transcriptional control by transposable elements is being studied. The number of chalcone synthase genes per haploid genome seems to vary from one or two in parsley, snapdragon, and Petunia to somewhere around five to ten in bean (unpublished results by the groups of K. Hahlbrock, H. Saedler, and C.J. Lamb). It is unknown in all of these plants whether duplicate or multiple copies of the chalcone synthase gene, when present, are expressed differentially during development, with cell specialization, or according to the subclasses of flavonoids.

On the other hand, for the majority of secondary plant products the genetics are virtually unknown. This applies even to the small, otherwise extensively studied group of enzymes of "general phenylpropanoid metabolism" which is common to most phenylpropanoid derivatives (Fig. 3).

PROSPECTS OF GENE TRANSFER IN VITRO

The paper by Peacock and Dennis (this volume) has outlined the state of the art of plant gene engineering and its expected impact in plant breeding. Basic considerations will therefore not be repeated here. A rapid further development of the now established principles of gene transfer techniques and of the regeneration of fertile plants from single transformed cells may be taken for granted, even though further research will be required, notably with monocotyledonous plants, to make these methods routinely applicable to any given plant species.

These new techniques will very probably have a large impact in secondary product formation, with cultured cells as well as with intact plants. Although this paper has its major emphasis on practical applications, it should be realized that the expected impact will be as important in the advancement of basic scientific knowledge (if such a distinction can be made at all). To give only one example, secondary products are often conspicuous compounds (e.g., colored anthocyanins in genetic analyses) and therefore useful phenotypic markers for studies involving in vitro mutagenesis, recombination, and transfer of the related genes to elucidate the mechanisms of their expression.

In principle, the basic methods to achieve such goals through gene engineering are available. As an example in the area of secondary product formation, parsley cell suspension cultures have been used for the isolation and cloning of genes involved in flavonoid or furanocoumarin biosynthesis (Fig. 3) via the conventional steps of enzyme purification, production of antisera, and generation of cDNA-containing plasmids as probes for the identification of "genomic clones" (24). The clones are potential sources of a) the complete genes, b) signal-specific promoters, and c) promoter-less structural genes, thus enabling various in vitro recombinations. Recent studies using the light-dependent promoter of the gene encoding the small subunit of ribulose-1,5-bisphosphate carboxylase/oxygenase in combination with a structural gene from bacteria have demonstrated that chimeric genes can be expressed and inherited in transformed plants (see Peacock and Dennis, this volume).

Using these or analogous materials, various strategies with many thematic variations are conceivable. For instance, regulatory genes or genes coding for new enzymes, new pathways, or pathway modifications could be introduced into recipient plant cells; copy numbers of more or less unaltered genes, from homologous or heterologous sources, could be increased; structural genes could be combined with more efficient or otherwise different promoters responding either to the same or to different regulatory signals; genes coding for "anti-sense" mRNA could be constructed and possibly be used to inactivate undesired transcripts. Other possibilities could be listed, but this area of research will naturally develop such that each individual case will require its specific adaptations.

Thus, although many questions concerning the physiology, biochemistry, and genetics of secondary product formation remain to be answered, the basic methods of gene engineering have proved applicable to plants and have therefore opened up new prospects of producing larger amounts, smaller amounts, or structural variants of a given compound. In all likelihood, this field will fully take part in the beginning biotechnological revolution, even though the first case of commercial application might take some more time.

Finally, the potential of transforming microorganisms with plant genes should be mentioned. A detailed discussion would be outside the scope of this paper.

CONCLUSION

Plants and plant cell cultures are extremely valuable, medically indispensable, and economically important sources of secondary products. Recent advances in cell culture and gene transfer techniques have a foreseeable, large impact in the field of secondary plant product formation and its commercial applications. Many technical problems have been solved in principle. Future research efforts will have to concentrate largely on the physiology, biochemistry, and genetics of secondary product formation, particularly on the mechanisms of gene expression, in order to replace trial and error by more directed, widely applicable approaches.

Acknowledgement. I thank W. Barz and M.H. Zenk for valuable contributions in the preparation of this manuscript.

REFERENCES

(1) Alfermann, A.W.; Schuller, I.; and Reinhard, E. 1980. Planta Med. **40:** 218-223.

(2) Amrhein, N.; Deus, B.; Gehrke, G.; and Steinrücken, H.C. 1980. The site of inhibition of the shikimate pathway by glyphosate. Plant Physiol. **66:** 830-834.

(3) Bailey, A.J., and Mansfield, J.W. 1982. Phytoalexins. Glasgow and London: Blackie.

(4) Barz, W., and Hüsemann, W. 1982. Aspects of photoautotrophic cell suspension cultures. In Plant Tissue Culture 1982, ed. A. Fujiwara, pp. 245-248. Tokyo: Japanese Association for Plant Tissue Culture.

(5) Barz, W., and Köster, J. 1981. Turnover and degradation of secondary (natural) products. In Secondary Plant Products, eds. P.K. Stumpf and E.E. Conn, vol. 7, pp. 35-80. New York: Academic Press.

(6) Bell, E.A. 1981. The physiological role(s) of secondary (natural) products. In Secondary Plant Products, eds. P.K. Stumpf and E.E. Conn, vol. 7, pp. 1-18. New York: Academic Press.

(7) Berlin, J. 1984. Plant cell cultures - a future source of natural products? Endeavour **8:** 5-8.

(8) Böhm, H. 1982. The inability of plant cell cultures to produce secondary substances. In Plant Tissue Culture 1982, ed. A. Fujiwara, pp. 325-328. Tokyo: Japanese Association for Plant Tissue Culture.

(9) Britsch, L.; Heller, W.; and Grisebach, H. 1981. Conversion of flavanone to flavone, dihydroflavonol and flavonol with an enzyme system from cell cultures of parsley. Z. Naturforsch. **36c:** 742-750.

(10) Brodelius, P., and Mosbach, K. 1982. Immobilised plant cells. Adv. Appl. Microbiol. **28:** 1-26.

(11) Chappell, J.; Hahlbrock, K.; and Boller, T. 1984. Rapid induction of ethylene biosynthesis in cultured parsley cells by fungal elicitor and its relationship to the induction of phenylalanine ammonia-lyase. Planta **161:** 475-480.

(12) Conn, E.E. 1980. Cyanogenic compounds. Ann. Rev. Plant Physiol. **31:** 433-452.

(13) Conn, E.E. 1981. Secondary plant products. In The Biochemistry of Plants, eds. P.K. Stumpf and E.E. Conn, vol. 7. New York: Academic Press.

(14) Deus-Neumann, B., and Zenk, M.H. 1984. Instability of indole alkaloid production in Catharanthus roseus cell suspension cultures. Planta Med. **50:** 427-431.

(15) Ebel, J., and Hahlbrock, K. 1982. Biosynthesis. In The Flavonoids, eds. J.B. Harborne and T.J. Mabry, pp. 641-675. London: Chapman and Hall.

(16) Farnsworth, N.R. 1984. The role of medicinal plants in drug development. In Natural Products and Drug Development, eds. P. Krogsgaard-Larsen, S.B. Christensen, and H. Kofod, p. 18. Copenhagen: Munksgaard.

(17) Farnsworth, N.R., and Morris, R.W. 1976. Higher plants - the sleeping giant of drug development. Am. J. Pharm. **148:** 46-52.

(18) Friedman, J., and Waller, G.R. 1985. Allelopathy and autotoxicity. Trends Biochem. Sci. **10:** 47-50.

(19) Fujita, Y.; Hara, Y.; Ogino, T.; and Suga, C. 1981a. Production of shikonin derivatives by cell suspension cultures of Lithospermum erythrorhizon. I. Effects of nitrogen sources on the production of shikonin derivatives. Plant Cell Rep. **1:** 59-60.

(20) Fujita, Y.; Hara, Y.; Suga, C.; and Morimoto, T. 1981b. Production of shikonin derivatives by cell suspension cultures of Lithospermum erythrorhizon. II. A new medium for the production of shikonin derivatives. Plant Cell Rep. **1:** 61-63.

(21) Fujiwara, A. 1982. Plant Tissue Culture 1982. Tokyo: Japanese Association for Plant Tissue Culture.

(22) Hagmann, M.-L.; Heller, W.; and Grisebach, H. 1983. Induction and characterization of a microsomal flavonoid 3'-hydroxylase from parsley cell cultures. Eur. J. Biochem. **134:** 547-554.

(23) Hahlbrock, K. 1981. Flavonoids. In Secondary Plant Products, eds. P.K. Stumpf and E.E. Conn, vol. 7, pp. 425-455. New York: Academic Press.

(24) Hahlbrock, K.; Chappell, J.; and Scheel, D. 1984. Genes involved in resistance reactions in higher plants: possible candidates for gene transfer? In The Impact of Gene Transfer Techniques in Eukaryotic Cell Biology, eds. P. Starlinger and J. Schell, pp. 155-166. Heidelberg: Springer-Verlag.

(25) Harborne, J.B.; Mabry, T.J.; and Mabry, H. 1975. The Flavonoids. London: Chapman and Hall.

(26) Hinz, H., and Zenk, M.H. 1981. Production of protoberberine alkaloids by cell suspension cultures of Berberis species. Naturwiss. **68:** 620-621.

(27) Hornberg, C., and Weiler, E.W. 1984. High-affinity binding sites for abscisic acid on the plasmalemma of Vicia faba guard cells. Nature **310:** 321-324.

(28) Kasha, K.J.; Kott, L.; and Seguin-Swartz, G. 1982. Genetic stability of haploid cell cultures. In Plant Tissue Culture 1982, ed. A. Fujiwara, pp. 405-408. Tokyo: Japanese Association for Plant Tissue Culture.

(29) Kistler, H.C., and VanEtten, H.D. 1984. Regulation of pisatin demethylation in Nectria haematococca and its influence on pisatin tolerance and virulence. J. Gen. Microbiol. **130:** 2605-2613.

(30) Kistler, H.C., and VanEtten, H.D. 1984. Three non-allelic genes for pisatin demethylation in the fungus Nectria haematococca. J. Gen. Microbiol. **130:** 2595-2603.

(31) Lindsey, K., and Yeoman, M.M. 1986. The synthetic potential of immobilised cells of Capsicum frutescens Mill cv. annum. Planta, in press.

(32) Linsefors, L., and Brodelius, P. 1985. Immobilization of plant protoplasts; viability studies. Plant Cell Rep. **4:** 23-27.

(33) Luckner, M. 1984. Secondary Metabolism in Microorganisms, Plants and Animals. Jena: Veb Gustav Fischer.

(34) Melchers, G. 1982. The first decennium of somatic hybridization. In Plant Tissue Culture 1982, ed. A. Fujiwara, pp. 13-18. Tokyo: Japanese Association for Plant Tissue Culture.

(35) Moritz, S.; Schuller, I.; Figur, C.; Alfermann, W.; and Reinhard, E. 1982. Biotransformation of cardenolides by immobilized digitalis cells. In Plant Tissue Culture 1982, ed. A. Fujiwara, pp. 401-402. Tokyo: Japanese Association for Plant Tissue Culture.

(36) Nakagawa, K.; Konagai, A.; Fukui, H.; and Tabata, M. 1984. Release and crystallization of berberine in the liquid medium of Thalictrum minus cell suspension cultures. Plant Cell Rep. **3:** 254-257.

(37) Nevers, P.; Shepherd, N.; and Saedler, H. 1986. Plant transposable elements. Adv. Bot. Res., in press.

(38) Paxton, J.D. 1981. Phytoalexins - a working redefinition. Phytopathol. Z. **101:** 106-109.

(39) Schopfer, P. 1977. Phytochrome control of enzymes. Ann. Rev. Plant Physiol. **28:** 223-252.

(40) Seigler, D.S. 1981. Secondary metabolites and plant systematics. In Secondary Plant Products, eds. P.K. Stumpf and E.E. Conn, vol. 7, pp. 139-171. New York: Academic Press.

(41) Seyffert, W. 1982. Beiträge zur Genetik und Enzymologie der Flavonoide. Biol. Zbl. **101**: 465-483.

(42) Shields, S.E.; Wingate, V.P.; and Lamb, C.J. 1982. Dual control of phenylalanine ammonia-lyase production and removal by its product cinnamic acid. Eur. J. Biochem. **123**: 389-395.

(43) Staba, E.J. 1982. Production of useful compounds from plant tissue cultures. In Plant Tissue Culture 1982, ed. A. Fujiwara, pp. 25-30. Tokyo: Japanese Association for Plant Tissue Culture.

(44) Stöckigt, J. 1980. In Indole and Biogenetically Related Alkaloids, eds. J.D. Phillipson and M.H. Zenk, p. 113. London: Academic Press.

(45) Stöckigt, J.; Pawelka, K.-H.; Tanahashi, T.; Danieli, B.; and Hull, W.E. 1984. Voafrine A and voafrine B, new dimeric indole alkaloids from cell suspension cultures of Voacanga africana. Stapf. Helv. Chim. Acta **66**: 2525-2533.

(46) Ulbrich, B.; Wiesner, W.; and Arens, H. 1985. Large scale production of rosmarinic acid from plant cell cultures of Coleus blumei Benth. In Proceedings of the Symposium on Primary and Secondary Metabolites of Plant Cell Cultures, ed. K.-H. Neumann, pp. 293-303. Heidelberg: Springer-Verlag.

(47) Walter, M., and Hahlbrock, K. 1984. Cinnamic acid induces PAL mRNA and inhibits mRNA translation in cultured parsley cells. Plant Physiol. (Suppl.) **75**: 155.

(48) Weiler, E. 1983. Immunoassay of plant constituents. Biochem. Soc. Trans. **11**: 485-495.

(49) Wiermann, R. 1981. Secondary plant products and cell and tissue differentiation. In Secondary Plant Products, eds. P.K. Stumpf and E.E. Conn, vol. 7, pp. 86-112. New York: Academic Press.

(50) Wink, M., and Witte, L. 1983. Evidence for a wide-spread occurrence of the genes of quinolizidine alkaloid biosynthesis. FEBS Lett. **159**: 196-200.

(51) Zenk, M.H. 1978. The impact of plant cell culture on industry. In Frontiers of Plant Tissue Culture 1978, ed. T.A. Thorpe. Calgary: University Press.

(52) Zenk, M.H. 1980. Enzymatik synthesis of ajmalicine and related indole alkaloids. J. Nat. Prod. **43**: 438-451.

(53) Zenk, M.H. 1982. Pflanzliche Zellkulturen in der Arzneimittelforschung. 112. Versammlung der Gesellschaft Deutscher Naturforscher und Ärzte, pp. 145-162. Stugart: Wissenschaftliche Verlagsgesellschaft.

(54) Zenk, M.H., and Deus, B. 1982. Natural product synthesis by plant cell cultures. In Plant Tissue Culture 1982, ed. A. Fujiwara, pp. 391-394. Tokyo: Japanese Association for Plant Tissue Culture.

(55) Zenk, M.H.; El-Shagi, H.; and Schulte, U. 1975. Anthraquinone production by cell suspension cultures of Morinda citrifolia L. Plant Med. Suppl. **75:** 79-101.

(56) Zenk, M.H.; El-Shagi, H.; and Ulbrich, B. 1977. Production of rosmarinic acid by cell-suspension cultures of Coleus blumei. Naturwiss. **64:** 585-586.

Biotechnology: Potentials and Limitations, ed. S. Silver, pp. 259-273. Dahlem Konferenzen 1986. Berlin, Heidelberg, New York, Tokyo: Springer-Verlag.

Plant-Microbe Interactions

J.E. Beringer
Unit of Molecular Genetics
University of Bristol
Bristol BS8 1UG, England

Abstract. All plants are exposed to extremely large numbers of microorganisms; some are pathogenic and some beneficial. The potential damage that can be caused by pathogens appears to be held in check by beneficial microorganisms, many of which are probably unknown to us. Future exploitation of such interactions will be as dependent on a better understanding of the biology of plant-microbe interaction as on developments in biotechnology. Plant nutrition is influenced by nitrogen-fixing microorganisms, mycorrhizal fungi, and possibly other microorganisms. These existing symbioses can be exploited to improve the activities that we understand sufficiently well and to introduce novel functions into the symbionts. Whether the production of growth-promoting substances by microorganisms can be exploited for crop production or not remains to be established.

HARMFUL INTERACTIONS

The most obvious and best-known example of these are diseases caused by pathogenic microorganisms because they reduce yields or even occasionally destroy crops completely. However, it should be recognized that microorganisms can affect plant growth adversely in other ways, for example, by competing for nutrients, producing acids or toxic compounds that can interfere with root growth, and by acting as foci for ice crystal formation. How important each of these is on a global scale is impossible to assess because only extreme effects tend to be recognized. Indeed, the primary source of information on two of these effects comes from work on incorporating straw where nitrogen uptake by microorganisms decomposing the straw is significant and depletes available soil reserves. Also, acid production during straw decomposition is often a serious problem to germinating seeds (14). Indeed, it is the production of undesirable metabolic products that is the primary cause of undesirable effects from ploughing straw into the soil rather than burning or otherwise removing it.

The most obvious role for biotechnology in reducing the effect of harmful organisms is to find ways for avoiding the problem (e.g., burning straw), to develop existing forms of biological control, and to invent new ones. It is clear that in nature there are a series of complex and effective interactions between plants and microorganisms that usually prevent massive outbreaks of disease. Indeed, it may be more realistic to think of crop diseases as an artificial problem created by man's desire to optimize yields and grow large areas of almost identical crops.

Plant Manipulations

Probably the best long-term solution to the control of harmful microorganisms is to breed plants which are no longer affected. This has been done for many years with major crop species and has led to significant improvements in yields. There are, however, a number of problems that limit the extent to which traditional breeding can be used to generate totally new forms of resistant plant lines. These are discussed by Peacock and Dennis (this volume), as are the extremely important interactions between the bacterium Agrobacterium tumefaciens and some viruses with plants which are being used to introduce cloned genes into plants. The great advantage that "molecular genetics procedures" for introducing genes into organisms have over traditional methods is that sources of resistance need not be limited to variations of genes within the existing gene pool of a species. Theoretically any gene, or group of genes, whose functioning will make plants less attractive to pests but still safe for human and animal consumption can be introduced. While the long-term future prospects for producing new types of resistant crop plants look very good, it is unclear how much crop yields will be improved by such breeding if a significant effort is not put into identifying other ways in which microorganisms influence crop yields - particularly at the level of nutrient uptake by roots.

Modification of Harmful Microorganisms

An attractive approach towards reducing the severity of plant pathogens is to inoculate plants with strains that are not pathogenic but which are still able to colonize plants effectively. Theoretically, and sometimes in practice, these will interfere with the establishment of the pathogen and reduce or prevent disease. A very good example of this concept is the use of Agrobacterium radiobacter to control crown gall disease caused by Agrobacterium tumefaciens. Kerr and colleagues (19) have shown that A. radiobacter strain 84 protects roses and stonefruit crops if cuttings are dipped into cultures of it. Strain 84 produces a bacteriocin which inhibits the growth of pathogenic strains. This is an unusually effective form of biological control which works well in soils in Australia, New Zealand, South Africa, and parts of the USA, where the local A. tumefaciens strains are sensitive (nopaline, but not octopine strains). Interestingly, even though this type of competitive inhibition would be extremely useful for producing strains of Rhizobium (which is very closely related to Agrobacterium) with increased competitive ability, there have been no encouraging reports that the "natural" production of antagonistic compounds can be exploited for Rhizobium inoculants.

Another potentially very exciting example of the use of a modified pathogen in biological control is the use of a Rhizoctonia solani strain carrying three "killer" double-stranded (ds) RNA elements (10). When such strains are inoculated into soil containing a healthy pathogen which is sensitive to the killer dsRNA, control of the pathogens can be achieved. This is thought to be due to transfer of dsRNA following anastomosis. Virus particules are not known to be produced and it is assumed that the dsRNA must be transferred for the sensitive strain to be killed. It is clear that this procedure is of limited value because its general success in agriculture would depend upon most R. solani strains being sensitive, which is not a general phenomenon. However, it is a beautiful example of a system that could possibly be exploited through the use of molecular genetic techniques to make modified dsRNA elements that are much more virulent. Whether or not dsRNA elements and fungal viruses can be exploited for biological purposes remains to be determined. Their inability to infect, like "normal" viruses, appears to be a major limitation (6).

This last example points to two major problems in developing good biological methods for controlling plant pathogens. One is that we know of a number of potentially very attractive interactions but, with the clear exception of Agrobacterium, which has been studied in great detail because of its role in tumor formation, very little genetic work has been done with the relevant organisms and one is so often left thinking, "Has no one thought of the potential for manipulating the system?" The other problem is that the environment (soil) is extremely complex biologically, usually containing about 10^9 - 10^{10} microorganisms per gram. It is very difficult to introduce sufficient control organisms to have a major effect in the short term, let alone to expect chosen organisms that have been cultured in the laboratory to become established and compete successfully over a long period. Perhaps the best argument that can be made for using attenuated derivatives of pathogens for control measures is that these organisms should have the best chance of colonizing the same sites and competing directly with their wild-type relations. Also, it may be less difficult than anticipated to introduce control microorganisms into particular niches (for example, leaf surfaces) where populations of other microorganisms are low.

A topical example of the modification of harmful microorganisms for biological control is that of modifying ice-nucleating bacteria (21). The temperature at which cooling water forms ice crystals and thus causes frost damage to the leaves of plants is related to the presence of foci for crystallization. It has been shown that some saprophytic pseudomonas species produce proteins which act as foci. Deletion of the genes involved produced strains that no longer induced ice crystal formation. When large numbers of these mutant strains were inoculated on test plants, these interfered with crystal formation by wild-type strains by reducing their potential to colonize leaves (21). It is a pity that this simple but fine example of a practical use of molecular genetics has not yet been exploited commercially because of the fears of environmental groups in the USA. However, it is perhaps a blessing that the first major battle to be fought over the release of manipulated organisms should be over one in which DNA had been deleted, rather than about organisms in which novel genes had been introduced.

BENEFICIAL INTERACTIONS

Surprisingly there are only two well-known types of beneficial plant-microbe interaction: symbiotic associations involving nitrogen-fixing microorganisms and mycorrhizas. This is presumably because these are of major importance to the growth of participating plant species and research on them has had a good chance of leading directly to improvements in plant growth through inoculation with modified microorganisms. It is important to recognize that other beneficial interactions are possible and that the identification and exploitation of these could lead to major improvements in crop production (3, 8).

Examples of two other relationships that are of major importance for the plant species involved are the microbially induced production of proteoid roots and the removal of H_2S from the rhizosphere of rice. Proteoid roots are dense clusters of rootlets which are induced under low nutrient conditions in members of the Proteaceae. They are only produced in the presence of microorganisms which apparently do not need to invade the roots to induce rootlet proliferation (22, 23). These are of undoubted benefit to the host as growth is enhanced in their presence (23). Perhaps the best known example of an organism which induces root multiplication after invasion is Agrobacterium rhizogenes, which is not generally thought to be beneficial, though Strobel and Nachmias (34) have reported that treatment of almond cuttings with A. rhizogenes promotes initial root development. One might perhaps speculate on the potential that exists for using interactions such as these to manipulate root development on agricultural plants to enable them to exploit soil reserves more efficiently and, indeed, to ask whether root development for many species is solely determined by the particular plant species.

The other example of a locally very important interaction is that between Beggiatoa and rice plants (16, 18) in which the bacteria oxidize H_2S generated by the anaerobic reduction of sulfate by other bacteria. In this symbiosis the plant provides oxygen and catalase to foster the growth of Beggiatoa, and presumably other H_2S-oxidizing bacteria, which in turn facilitate plant growth through the oxidation of H_2S which can reach inhibitory levels in paddy soils. It seems likely that the interaction is of general importance for wetland plants (16, 20) and may be representative of an important role for microorganisms facilitating plant growth.

Biological Nitrogen Fixation

This process, the conversion of atmospheric N_2 gas to ammonia, is extremely important in nature because "fixed nitrogen" is often a major factor limiting plant growth. The enzyme system, nitrogenase, which reduces atmospheric N_2 to ammonia, is only known to be present in prokaryotic microorganisms. Nitrogenase genes are found in species of many different genera of bacteria and cyanobacteria. Nitrogen is fixed by free-living, associative, and symbiotic microorganisms. Free-living fixation is of very limited value in agriculture (usually less than 50-150 kg N per ha per annum) (2). Because in general absolute amounts of N_2 fixation are low, and the potential for exploiting biological nitrogen fixation by free-living microorganisms is remote, relatively little research is being done with these interesting organisms.

An exception to this generalization is the upsurge of interest in Azotobacter, which is an N_2-fixing obligate aerobe. Nitrogenase is irreversibly inactivated by oxygen and any attempts to introduce nitrogen fixation genes into eukaryotic organisms, and plants in particular, must be based on procedures that will reduce or avoid damage. Azotobacter produces proteins which bind to nitrogenase and protect it from oxygen (31). Unfortunately, this protective mechanism does not allow the enzyme to function and N_2 fixation can only restart once the oxygen tension is reduced and the proteins are released. While this protective system is not the solution to obtaining N_2 fixation at fairly high oxygen tensions, it does provide a valuable method for protecting nitrogenase and reducing delays and energy costs in synthesizing new enzyme, a particularly important consideration with nitrogenase, which can represent up to 20% of the protein in a nitrogen-fixing microorganism (30). Another major limitation to the manipulation of nitrogen fixation genes is that biological nitrogen fixation is expensive in terms of its requirement for ATP and produces a toxic product, ammonia. Unless plants are able to regulate the amount of N_2 fixation to their needs and assimilate all the N_2 that is fixed, they are unlikely to benefit from it.

Associative biological nitrogen fixation has aroused a lot of interest because early workers claimed that as much as 150 kg per ha per annum could be fixed by bacteria associating with maize roots (37). Unfortunately, much of the interest and enthusiasm has waned because later experiments have shown that fixation is much lower and almost unmeasurable in temperate soils (28). Significant yield increases of some crops have been measured, but recent evidence tends to suggest that these are due more to the production of growth-promoting compounds than to biological nitrogen fixation, a situation which reflects the rise and fall of interest in Azotobacter as an important nitrogen-fixing microorganism (5). At present there seems to be little point in expending much effort on manipulating the nitrogen fixation genes of associative nitrogen-fixing bacteria, such as Azospirillum, though there seems to be every reason to study their ability to modify plant growth and improve nutrient uptake. It will be a pity if work on Azospirillum falters because of its limited value in N_2 fixation. The same happened with Azotobacter, which may also have a valuable and exploitable role in modifying plant growth through the production of growth-promoting substances (5).

Biological nitrogen fixation arising from symbiotic associations between plants and Rhizobium, Frankia, or cyanobacteria can be extremely important agriculturally and appears to have the potential to provide all the plant's requirements for fixed N. The rhizobia are the only members of this group which affect crops of direct commercial value (legumes) and probably also have the potential to fix the largest amounts of N (good estimates of up to 600 kg N per ha per annum are available) (2). When one considers that a 5-10 tonne yield of wheat per hectare requires only about 150 kg added N per hectare, the potential significance of N_2 fixation by legumes is obvious, particularly as "average" levels of N_2 fixation are in the region of 50-200 kg by N per ha per annum, essentially all of which is available to the host, whereas it is unusual for a standing crop to assimilate as much as 70% of fertilizer nitrogen.

While it may appear heretical to say, it is important to note that there is little real evidence to show that when sufficient rhizobia are present in soil the growth of legume crops is limited by their ability to fix nitrogen. There is a lot of enthusiasm for improving nitrogen fixation and reducing the energy cost (estimated to be about 10 to 30 moles ATP per mole N_2 fixed) which is considered to be very high and potentially growth-limiting. Whether this is true or not remains to be demonstrated as we do not know the true cost and cannot calculate the expected cost accurately until we know more about N_2 reduction in root nodules. Much more work is needed to measure the various parameters involved in carbon and electron allocation to nitrogenase, and to establish sensible methods for measuring what is happening to allow for the selection and generation of improved rhizobia and legumes (33, 38). Furthermore, the cost of nitrate assimilation is not known and may even be the same as that for N_2 fixation. However, it is clear that if the efficiency of biological nitrogen fixation could be improved significantly, extra photosynthate would be available for plant growth, and if photosynthate is growth-limiting, yields should increase.

The fast-growing species of Rhizobium are relatively easy to manipulate and when the functions of genes that have a major role in the efficiency of nitrogen fixation are understood, it should be possible to isolate and manipulate them without serious difficulty. Sadly, the only example that we have of the large-scale manipulation of such genes is the work that is being done with hydrogen uptake (Hup) genes (11). At least a quarter of the electrons flowing through nitrogenase are being used (wasted) in the evolution of H_2 in a reaction which seems to be an inescapable part of N_2 reduction. Some nitrogen-fixing microorganisms have an uptake hydrogenase which can recover energy from the oxidation of the released H_2. Many rhizobia do not and there is tremendous interest in producing Hup^+ strains. This has been done on a limited scale and growth responses way in excess of those expected from the energy savings have been reported, together with responses that suggest that hydrogen uptake is of only very marginal value (11). Energetically the latter might have been expected, which suggests that hydrogen uptake may be important to remove H_2 or to mop up excess oxygen.

Possibilities obviously exist to modify nitrogenase to make it more efficient. At present we know so little about the molecular structure of this enzyme that it is difficult to see how changes in amino acid sequence can be productive until active sites can be related to specific amino acids in the enzyme. Once this is known, the excellent data available for the DNA sequences of the nitrogenase structural genes in different microorganisms can be exploited to generate site-directed mutations of the DNA. Initially random mutagenesis could be tested to determine whether enzyme activity could be altered in a beneficial way and later, when more is known about the function of the enzyme, specified amino acids could be changed. Whether or not such procedures have potential is difficult to determine because nitrogenase has been so highly conserved, and functions such as oxygen sensitivity and hydrogen evolution may be an inescapable part of the chemistry of N_2 reduction by this enzyme.

A lot of work needs to be done to identify characteristics that are important for determining the ability of Rhizobium strains to colonize soil and compete with each other for nodule formation. No matter how clever molecular geneticists are in producing novel Rhizobium strains with improved nitrogen-fixing properties, these strains will have little impact on agriculture unless methods are available to ensure that they form a significant proportion of nodules on host plants. Unfortunately, research on microbial ecology is extremely difficult, and at present there are few good ideas as to how rhizobia might be manipulated to increase their opportunity to colonize plants. Indeed, the main emphasis of most work on legume inoculation is on ways to increase the number of rhizobia that are added to seeds to help to circumvent competition by giving the introduced strain a major numerical advantage. Perhaps the random approach of Brewin et al. (4) of transferring plasmids between strains and looking for alterations in competitiveness may hold the answer in the short term, though surely in the future predictive methods will become available. Because the plant is very important in determining how nodules develop, it is likely that in the future host plant preference can be exploited as a way of selecting out inoculant strains. In this respect it is encouraging that there are a number of reports that preference for different Rhizobium strains can be influenced by the genotype of the host (12, 25).

Symbiotic associations between the actinomycete Frankia and a range of mostly shrubby dicotyledenous plants are important ecologically because similar amounts of nitrogen are fixed in the nodules that are formed on the roots of these plants as are fixed in leguminous plants by Rhizobium. Commercially, however, they are of little real value though inoculation of alder and Casuarina could be important locally for timber production. The real interest in these symbioses is that the actinomycete induces the formation of nitrogen-fixing root nodules on a range of unrelated plants, suggesting that the ability of plants to form root nodules is widespread and not restricted to legumes. Thus, it would seem that a genetic analysis of the interaction between Frankia and its host could yield valuable information about how to extend the range of nitrogen-fixing plants in the future. It is arguable that this is potentially a much more attractive symbiosis to study than that between Rhizobium and legumes because Frankia already has found a way of nodulating a wide host range and thus might be easier to modify to extend this. However, it is important to remember that the plant is probably the most important partner in the symbiosis in terms of establishing the nodule and providing energy. Any attempt to make a new symbiosis must, therefore, involve plant genetic studies. In this respect it is encouraging that we are beginning to learn something about the genetics of nodulation by Rhizobium (36).

Unfortunately, Frankia isolates have been difficult to handle and only recently have a number of strains been grown in pure culture (7, 9). However, once cultural problems have been resolved, this organism should be susceptible to genetic analysis, probably utilizing the range of excellent techniques that are available for a number of other actinomycetes.

As with Frankia, the associations between cyanobacteria and a number of different plant species are interesting because of their diversity (2). With the exception of the water fern Azolla, which is an important source of nitrogen for rice cultivation in some parts of the Far East, the associations are of no agricultural value. Unfortunately, though many cyanobacteria are simple to grow, developments in our understanding of their genetics have been limited. Interestingly, root nodules are formed on at least one host plant (Cycas), suggesting that root nodule formation may be very largely determined by the plant. Basic studies of the genetics of these symbioses should shed light on how it may be possible to extend the range of nitrogen-fixing plants.

Much stress has been made in this section of the paper on exploiting existing symbioses to produce new types of nitrogen-fixing plants. The reason for this is that the symbiotic associations are an excellent example of the way in which the interactions have evolved to regulate the amount of nitrogen-fixing tissue to the plant's requirements for fixed nitrogen. In all the symbiotic associations there is a very close relationship between the total volume of nitrogen-fixing tissue and size of the host. It appears that the host is able to regulate the number of infections that become established to form nodules and to control the eventual size and number of nodules. Indeed, in most systems there is a further important interaction in that the nitrogen-fixing microorganisms lose their ability to assimilate ammonia and therefore excrete ammonia directly into the plant (27, 29). I believe that the successful exploitation of biological nitrogen fixation in crop plants in the future will be based on the exploitation of symbiotic associations. Alternatively, one might ask whether breeding legumes to produce varieties which could replace cereals as a source of starch-based food products (such as flour) in developing countries might be a simpler alternative. We already have a much better understanding of how to breed plants for optimal yields of a particular storage product than we do of the gene functions that would be needed to enable a plant to be nodulated and to integrate nodulation and nitrogen fixation into the physiology of the host. This last point is of great importance when considering the introduction of *nif* genes directly into plants. It is essential that the amount of energy provided for N_2 fixation is related to the need for fixed N. If N_2 fixation is not properly regulated the plant may not have sufficient fixed N at some stages of growth and may have too much at other stages, wasting energy and possibly leading to ammonium toxicity.

Mycorrhizas

Symbiotic associations between the roots of most plant species and a range of different fungi are almost ubiquitous in nature. These have been called mycorrhizas (literally, fungus roots) because of the extensive nature of the interactions. Two main types are recognized: ectomycorrhizas in which the fungus grows largely outside the plant, forming a sheath round the roots and also infecting the root itself, and endomycorrhizas in which the fungus is predominantly within the root (13). In both cases mycellium grows out from the root, taking up nutrients from regions of the soil that are beyond the range of root hairs. One very valuable aspect of these associations is that the mycorrhizal fungi assimilate and transport phosphate to the roots of the host plant. This is particularly important because phosphate ions are relatively immobile in the soil and depletion zones form rapidly around roots. Because the mycelium explores soil beyond these depletion zones a much more efficient extraction of phosphate is possible (15).

Ectomycorrhizal fungi, which are largely confined to woody perennials, also have an important role in water uptake and the assimilation of fixed nitrogen and other compounds from the soil. A number of different fungal species are able to form ectomycorrhizas, and some of these have the great advantage that they can be cultured readily in the laboratory. Unfortunately for biotechnologists, most crop plants are not ectomycorrhizal but are endomycorrhizal and the fungi involved are not culturable on laboratory media.

Endomycorrhizal fungi are the most interesting mycorrhizal fungi for biotechnological studies because they form symbioses with most crop plants, the only exceptions being Brassicas and members of the Chenopodiaceae. The uptake of phosphate is the most important result of symbiosis with these fungi. However, they also have a role in the uptake of copper, zinc, and some trace elements and are generally thought to be responsible for reducing the susceptibility of infected plants to water stress (26). Another potentially important commercial role is that there is some evidence that mycorrhizal plants may be less susceptible to certain root pathogens (32).

Unfortunately, it is not yet possible to grow these fungi in pure culture. Thus, material for research purposes is dependent on the production of mycelium or spores on growing plant roots. Much work has gone into trying to obtain pure cultures, but as yet there is little indication that they will be available in the near future. Fortunately, the host range of mycorrhizal fungi is extremely broad and therefore a culturable isolate from lettuce or carrot could be used on almost any other crop plant species. Indeed, the commercial use of mycorrhizal fungi in the citrus industry in California (24) requires that the fungi are not produced on the same plant species to reduce potential cross-infection with pathogens.

There are a number of different ways in which the manipulation of mycorrhizal associations could be exploited to improve crop growth. The most obvious is to improve the existing properties of isolates to take up and transport phosphates and to modify other beneficial functions. As with any other biological system, it is difficult to predict how easy it will be to improve an existing function. Perhaps one of the most attractive aspects that might be exploited in the future is the possible role of these fungi in reducing the susceptibility of plants to pathogens. In some cases this may simply be a physical problem in which the mycorrhizal fungus displaces or impedes the development of a pathogen. We have no idea whether specific inhibitory substances are produced. Certainly the construction of such strains could be of great value for reducing the invasion of roots by pathogens. However, as with Rhizobium, it is important to remember that the fungi are ubiquitous and that it may be extremely difficult to introduce inoculant strains.

It is hard to predict how rapidly our understanding of and ability to manipulate mycorrhizal symbioses will develop. It seems likely, especially for the endomycorrhizal fungi, that this will be determined largely by our ability to culture the fungi.

Plant Growth-promoting Rhizobacteria

Recently there has been considerable interest in the role of rhizosphere microorganisms in assisting plants to grow well. They have been implicated in reducing the effect of minor pathogens, producing plant growth hormones, and generally being of value. The best-known example of this group of microorganisms are the bacteria that have been studied by Schroth and colleagues (35) which appear to have an important role in reducing the susceptibility of plants to minor root pathogens. Schroth and colleagues have shown that their cultures of rhizosphere bacteria are only growth-promoting when added to soils to which the host plant has been grown recently. It appears to be essential for there to be a buildup of minor pathogens because no effect is seen when plants are inoculated in sterile soil or in soil in which the host has not been cultured previously. Unfortunately, as yet there has been no realistic commercial exploitation of this observation because the extensive field trials that have been undertaken have been disappointing. This is probably because the types of minor pathogens that might be controlled by a given rhizosphere isolate are limited, and thus an isolate may only be expected to be beneficial in a very similar environment to the one from which it had been isolated.

The potential for exploiting plant growth-promoting rhizobacteria through gene manipulation appears to be limited at present because of our lack of understanding of the interactions which are occurring in the rhizosphere, and our almost complete lack of understanding of the mode of action. There has been much excitement about the role of microbially produced siderophores which chelate iron, because they might provide an environment which is less suitable for the growth of pathogens. However, it is unclear whether siderophore production in the rhizosphere is of any consequence at all. An obvious reservation about the role of siderophores is that if they were sufficiently effective at chelating iron to prevent the growth of pathogenic microorganisms, they would also affect plant growth. Recently, Becker et al. (1) have shown that this could potentially be a problem in a study of the effect of siderophores produced by different soil microorganisms. Some plants were able to strip iron from the siderophores, but others could not and growth was adversely affected.

To be able to exploit such organisms we also need to have a much clearer definition of what minor pathogens are and how they affect plant growth, so that realistic methods for measuring the effect of potential control agents can be developed. Workable assay procedures are also needed to select for and test modifications to potential controlling microorganisms. At present the selection of growth-promoting rhizosphere microorganisms is largely on the basis of antagonism of known pathogens and may, therefore, have little relevance to the control of untested organisms and will almost certainly exclude other ways in which microorganisms may stimulate plant growth.

It has been realized for many years that some soil microorganisms produce plant growth-promoting substances, such as cytokininis, auxins, and probably gibberellins. The best example of the recognition of this has been the unravelling of the role of Azotobacter in plant growth promotion (5). For many years there was a lot of excitement about inoculation trials in which growth responses were obtained. This was thought to be due to nitrogen fixation, but the amount of nitrogen fixed by Azotobacter species is so low that some other method must have been involved. Auxin production was recognized and is generally thought to be the main beneficial response. However, the response obtained is extremely limited and probably never commercially worthwhile.

Presumably all plants are exposed to microorganisms which produce growth-promoting substances, and therefore the development of root systems must to a certain extent be influenced by the population of microbes in the rhizosphere. Whether or not rooting patterns can be altered by manipulating the populations of microorganisms, or more particularly by altering their ability to produce growth promoters, remains to be determined. There would appear to be considerable potential for modifying the growth of plant roots in this way.

There also appears to be considerable potential for developing communities of microorganisms which will colonize root systems and make the host plants less susceptible to root pathogens (17). Whether this should be done by isolating new species of rhizosphere microorganisms or by the genetic exploitation of mycorrhizal fungi is debatable. Plant root systems are extremely important for the uptake of nutrients and water but have been little studied. This is not surprising because of the complexity of looking at such a large network of fine organic material in a large bulk of soil. However, realistically it is probably essential that we pay considerably more attention to root development if we are to obtain truly optimal yields of crop plants.

PROBLEMS OF COMMERCIAL EXPLOITATION

In discussing the possible role of exploiting microorganisms to improve plant growth it is easy to state that inoculation with various types of microorganisms is likely to be beneficial. This is indeed true for plant growth, but is only likely to be of agricultural benefit if there is a sensible return on investment. A classic example of this is the exploitation of biological nitrogen fixation. In the United Kingdom, average yields of cereals are in the region of 5 tonnes per hectare and this yield is driven largely by an addition of nitrogen fertilizer which costs the farmer approximately the equivalent of his income from half a ton of yield. This is extremely cost-effective and does not encourage the exploitation of biological nitrogen fixation if it cannot promise equivalent economic returns without introducing complications in agronomy. On the other end of the scale in developing countries, yields are on average about 1 tonne per hectare and are often limited to this level by a shortage of nitrogen fertilizer. In this case, investment in inoculation may well be worthwhile. For Rhizobium, production is relatively cheap and inoculum can be sold at a cost as low as about 50p per hectare, though usually the cost is nearer 5 per hectare. Conversely, it seems most unlikely that inoculum of the non—culturable endomycorrhizas could be produced at present for less than 25 per hectare. At this sort of cost, benefits to plant growth must be obvious and significant financially as they cannot be marketed solely on the basis of an insurance of adequate yields.

I believe that the future large-scale exploitation of microorganisms in agriculture will depend upon an ability to produce large amounts of cheap inoculum with long shelf life. Desirable characteristics such as a resistance to seed dressings and a packaging formulation that makes it simple to add to seed at the time of sowing, or to be distributed with fertilizer granules, will also be needed. Such inoculants must be reliable in use and give a realistic return on investment. Perhaps the best way that one can ensure a realistic return on inoculation costs in the future will be to use our understanding of molecular genetics to construct inoculant strains which have the ability to confer a number of desirable attributes on the host plant in question. Whether or not we shall see such inoculants in the near future would appear to depend upon the investment of much time and facilities on developing a better understanding of the biology and ecology of plant-microbe interactions.

REFERENCES

(1) Becker, J.; Hedges, R.W.; and Messen. 1985. Inhibitory effects of pseudobactins on uptake of iron by higher plants. Applied Envir. Microbiol. **49:** 1090-1093.

(2) Beringer, J.E. 1984. The significance of symbiotic nitrogen fixation in plant production. CRC Crit. Rev. Plant Sci. **1:** 269-286.

(3) Beringer, J.E., and Tinker, P.B. 1983. The role of microorganisms in plant nutrition. **In** Proceedings of the British Sulphur Corporation's 7th International Conference, pp. 113-118. Dulwich: Purley Press.

(4) Brewin, N.J.; Wood, G.A.; and Young, J.P.W. 1983. Contribution of the symbiotic plasmid to the competitiveness of Rhizobium leguminosarum. J. Gen. Microbiol. **129:** 2973-2977.

(5) Brown, M.E. 1982. Nitrogen fixation by free-living bacteria associated with plants - fact or fiction? **In** Bacteria and Plants, eds. M.E. Rhodes-Roberts and F.A. Skinner, pp. 25-41. Society for Applied Bacteriology Symposium Series No. 10. London, New York: Academic Press.

(6) Brown, M.E., and Beringer, J.E. 1983. Potential of antagonists for fungal control. Agr. Ecosyst. Envir. **10:** 127-141.

(7) Burggraaf, A.J.P.; Quispel, A.; Tak, T.; and Valstar, J. 1981. Methods of isolation and cultivation of Frankia species from actinorhizas. Plant Soil **61:** 157-168.

(8) Burr, T.J., and Caesar, A. 1984. Beneficial plant bacteria. CRC Crit. Rev. Plant Sci. **2:** 1-20.

(9) Callaham, D.; Tredici, P. del; and Torrey, J.G. 1978. Isolation and cultivation in vitro of the actinomycete causing root nodulation in Comptonia. Science **199:** 899-902.

(10) Castanho, B.; Butler, E.E.; and Shepherd, R.J. 1978. The association of double-stranded RNA with Rhizoctonia decline. Phytopathology **68:** 1515-1519.

(11) Eisbrenner, G., and Evans, H.J. 1983. Aspects of hydrogenase metabolism in nitrogen fixing legumes and other plant microbe interactions. Ann. Rev. Plant Physiol. **34:** 105-136.

(12) Hardarson, G., and Jones, D.G. 1979. The inheritance of preference for stains of Rhizobium trifolii by white clover (Trifolium repens). Ann. Appl. Biol. **92:** 329-333.

(13) Harley, J.L., and Smith, S.E. 1983. Mycorrhizal Symbioses. London, New York: Academic Press.

(14) Harper, S.H.T., and Lynch, J.M. 1981. The kinetics of straw decomposition in relation to its potential to produce the phytotoxin acetic acid. J. Soil Sci. **32:** 627-637.

(15) Hayman, D.S. 1983. The physiology of VA mycorrhizal symbiosis. Can. J. Bot. **61:** 944-963.

(16) Heritage, A.D., and Foster, R.C. 1984. Catalase and sulfur in the rice rhizosphere: an ultrastructural histochemical demonstration of a symbiotic relationship. Microb. Ecol. **10:** 115-121.

(17) Hornby, D. 1983. Suppressive soils. Ann. Rev. Phytopathol. **21:** 65- 85.

(18) Joshi, M.M., and Hollis, J.P. 1977. Interaction of Beggiatoa and rice plant: detoxification of hydrogen sulphide in the rice rhizosphere. Science **197:** 179-180.

(19) Kerr, A. 1982. Biological control of soil-borne microbial pathogens and nematodes. **In** Advances in Agricultural Microbiology, ed. N.S. Subba Rao, pp. 429-463. New Delhi: Oxford and IBP.

(20) King, G.M.; Klug, M.J.; Wiegert, R.G.; and Chalmers, A.G. 1982. Relation of soil water movement and sulfide concentration to Spartina alterniflora production in a Georgia salt marsh. Science **218:** 61-64.

(21) Kozloff, L.M.; Schofield, M.A.; and Lute, M. 1983. Ice nucleating activity of Pseudomonas syringae and Erwinia herbicola. J. Bacteriol. **153:** 222-231.

(22) Lamont, B.B., and McComb, A.J. 1974. Soil microorganisms and the formation of proteoid roots. Austr. J. Bot. **22:** 681-688.

(23) Malajczuk, N., and Bowen, G.D. 1974. Proteoid roots are microbially induced. Nature **251:** 316-317.

(24) Menge, J.A. 1983. Utilization of vesicular-arbuscular mycorrhizal fungi in agriculture. Can. J. Bot. **61:** 1015-1024.

(25) Mytton, L.R. 1975. Plant genotype x rhizobium strain interactions in white clover. Ann. Appl. Biol. **80:** 103-107.

(26) Nelsen, C.E., and Safir, G.R. 1982. Increased drought tolerance of mycorrhizal onion plant caused by improved phosphorus nutrition. Planta **154:** 407-413.

(27) O'Gara, F., and Shanmugam, K.T. 1976. Regulation of nitrogen fixation by rhizobia. Export of fixed N2 as NH4+. Biochim. Biophys. Acta **437:** 313-321.

(28) O'Hara, G.W.; Davey, M.R.; and Lucas, J.A. Effect of inoculation of Zea mays with Azospirillum brasilense strains under temperate conditions. Can. J. Microbiol. **27:** 871-877.

(29) Peters, G.A.; Calvert, H.E.; Kaplan, D.; Ito, O.; and Toia, R.E. 1982. The Azolla-Anabaena symbiosis: morphology, physiology and use. Isr. J. Bot. **31:** 305-323.

(30) Postgate, J.R. 1985. Nitrogenase. Biologist **32:** 43-48.

(31) Robson, R.L., and Postgate, J.R. 1980. Oxygen and hydrogen in biological nitrogen fixation. Ann. Rev. Microbiol. **34:** 183-207.

(32) Schonbeck, F. 1979. Endomycorrhiza in relation to plant diseases. In Soil-borne Plant Pathogens, eds. B. Schippers and W. Gams, pp. 271-280. London: Academic Press.

(33) Schubert, K.R., and Wolk, C.P., eds. 1982. The Energetics of Biological Nitrogen Fixation, pp. 30. Rockville: American Society of Plant Physiologists.

(34) Strobel, G.A., and Nachmias, A. 1985. Agrobacterium rhizogenes promotes the initial growth of bare root stock almond. J. Microbiol. **131:** 1245-1249.

(35) Suslow, T.V., and Schroth, M.N. 1982. Rhizobacteria of sugar beets: effects of seed application and root colonization on yield. Phytopathology **72:** 199-206.

(36) Veeger, C., and Newton, W.E., eds. 1984. Advances in Nitrogen Fixation Research. Proceedings of the 5th International Symposium on Nitrogen Fixation, pp. 760, Noordwijkerhout, The Netherlands, August 28-September 3, 1983. The Hague: Martinus Nijhoff/Dr W. Junk.

(37) Von Bulow, J.F.W., and Dobereiner, J. 1975. Potential for nitrogen fixation in maize genotypes in Brazil. Proc. Nat. Acad. Sci. USA **72:** 2389-2393.

(38) Witty, J.F.; Minchin, F.R.; and Sheehy, J.E. 1985. Carbon costs of nitrogenase activity in legume root nodules determined using acetylene and oxygen. J. Exp. Bot. **34:** 951-963.

Biotechnology: Potentials and Limitations, ed. S. Silver, pp. 275-282. Dahlem Konferenzen 1986. Berlin, Heidelberg, New York, Tokyo: Springer-Verlag.

Importance of the Rhizosphere in Plant-Microbe Interactions

J. Sobieszczański
Academy of Agriculture, Dept. of Technical Microbiology
50-375 Wroclaw, Poland

INTRODUCTION

Harmful interactions, beneficial interactions, and problems of commercial exploitation all occur. Future exploitation of such interactions will be as dependent on better understanding of the biology of plant-microbe interactions as on developments in biotechnology. In my opinion, more attention should be placed on the details of the rhizosphere of microorganisms and their role in plant life. In addition, more stress should be placed on methods which should be used for better recognition of the biology and ecology of plant-microbe interactions. We should come to a better understanding of this problem and try to utilize this information in practice in order to direct the growth and development of plants.

More attention should be placed the rhizosphere microflora, which are composed of epiphytic seed microflora and soil microflora. Soil and plants play essential roles in the composition of the rhizosphere microorganisms, too. There are various numbers and kinds of microorganisms on the seed surface. For example, on clover seeds there are twenty times more bacteria and fungi than on timothy grass seeds. Similarly, in the clover rhizosphere there are six times more bacteria than in the timothy grass rhizosphere (1).

In general, the number of rhizosphere microorganisms is 100-200 times higher than the number of microorganisms in root-free soil. Krasilnikov (11) showed that the number of bacteria in the plant rhizosphere was from 10^5 to 5×10^5 per g of soil. The number and the variety were dependent on the kind of plants and the developmental phase (31). Such a great number of active microbial cells living in the plant rhizosphere must play a great role in the plant life. It should be mentioned here that on the surface of the upper parts of plants there are many epiphytic microorganisms (phyllosphere); these microorganisms have their role in plant life as well (25, 26, 44, 50).

INTERACTIONS IN THE ROOT-SYSTEM AREA

The rhizosphere microorganisms purify the root-system area from excessive concentrations of various organic and inorganic substances, root exudates, inorganic salts, pesticides, and other substances. Plant roots exude various kinds of organic compounds such as amino acids, sugars, organic acids, and other compounds (19, 51, 55) which are excellent nutrients for microorganisms. These facilitate their development and help to form various kinds of bacterial and fungal complexes, which may be beneficial or harmful; for example, a chemotactic response of zoospores of a soil-inhabiting plant pathogenic fungus, Phytophtora cinnamonii, for roots of avocado seedlings has been observed.

The chemotaxis of the zoospores and chemotropy of their germ tubes were directly related to infection and disease production. Induction was obtained with specific pathogen-attracting root exudates; the interesting implication is evident with regard to mechanisms of invasion and pathogenicity and to disease resistance (56). On the other hand, root exudates can play a role in the colonization of the rhizosphere by various kinds of beneficial microorganisms such as Azotobacter and Rhizobium and antagonists of pathogens (19, 52, 53). These results show that plant exudates are of great importance to microorganisms in the rhizosphere. The problems mentioned above are little understood and require wider research work.

THE ROLE OF RHIZOSPHERE MICROORGANISMS IN PLANT NUTRITION PROCESSES

There are two possible roles for rhizosphere microorganisms in plant nutrition.

First, rhizosphere microorganisms facilitate or participate in the uptake of nutrients from the soil, especially those that are growth-limiting. For example, Barber (3) found that the roots of mycorrhized (sterile) plants contained almost twice as much phosphate as those of non-mycorrhized (non-sterile) plants and the amount transferred to the shoot was nearly twenty times greater. A much larger fraction of the absorbed phosphate was therefore moved to the shoots of the sterile plants. At higher concentrations of phosphate there was no significant difference between the contents of roots or shoots of sterile and non-sterile plants. It has also been shown (30, 47) that bacteria can facilitate the uptake of phosphate from various insoluble forms of phosphate.

Similarly, microorganisms can facilitate the uptake of nitrogen, calcium, sodium, magnesium, and other macro- and micro-elements from mineral salts and various organic compounds. Plants which have been grown in artificial soil and inoculated with microorganisms from barley or mustard rhizosphere have taken up much more nitrogen, phosphorus, potassium, sodium, calcium, and magnesium than control plants growing without microorganisms. The fresh and dry matter production of the plants was higher in inoculated than in non-inoculated plants (31). More vitamin B_{12}

has been found in artificially inoculated soil (31). Similar results have been obtained by Saric (27).

These results show that inoculation with microorganisms may be very useful, but more research must be done in this direction (49).

Second, various microbial metabolites may act as plant growth regulators. In the pot experiment mentioned above we have observed that the root system in inoculated plants was richer than in non-inoculated ones and that experimental plants grew faster and were greener than in those not inoculated with microorganisms (31). A closer investigation of some isolated strains of bacteria shows that filtrates from liquid culture influence the growth and development of lettuce, stimulate its hypocotyl growth and influence formation of side-roots; the presence of more roothairs than in controls accelerates blooming and causes earlier formation of seed (7, 9, 10, 31, 36-38, 45, 46). There is very little known about these problems although there have been some trials to apply the metabolites in plant cultivation (28, 48); therefore, further investigation is necessary.

PLANT PROTECTION AGAINST PATHOGENIC MICROORGANISMS

The rhizosphere can also be important in protecting plants against phytopathogenic microorganisms. As mentioned above, root exudates play an important role in the development of phytopathogenic microorganisms whose development is limited by other, saprophytic microorganisms: Streptomyces and many other bacteria are antagonists of phytopathogenic microorganisms (14). For example, in the rhizosphere of various kinds of plants, antagonists of Xantomonas phaseoli, Pseudomonas lachrimans, Erwinia carotovora, Phoma beta, and Pseudomonas phaseoli have been found (17, 18). Some results of the present investigations show the possibility of biotechnological usage of antagonistic strains for production of preparations (antibiotics) against phytopathogenic microflora. It is interesting that the antibiotic substances are easily taken up by plants and are deposited in their tissues (8, 18).

BIOLOGICAL NITROGEN FIXATION

The problem of nitrogen increase in soil facilitated by microorganisms has been studied for over 100 years. Through genetic engineering it should be possible to transfer nitrogen-fixing genes to rhizosphere bacteria (23). It is also important to exploit Rhizobium strains and use them to inoculate leguminous crop plants (4-6, 19, 52-54).

Evidence indicates that Azotobacter plays a greater role as a producer of biologically active substances than as a nitrogen fixer, as Beringer (this volume) rightly emphasizes. A few years ago Azotobacter was used to inoculate seeds of various kinds of plants. Cruciferous plants yielded the best results but these results varied due to the lack of knowledge about the ecological conditions necessary for the development of the bacteria. In our experiments the filtrates of Azotobacter liquid cultures affected plant growth in the same way as auxines or giberellic acid (43). Further research of ecological conditions, which may give good results for Rhizobium and Azotobacter, as well as other less known nitrogen-fixing bacteria, should be done.

THE INFLUENCE OF PESTICIDES ON PLANT-MICROBE INTERACTIONS

Soil has continually been treated with more and more chemicals such as herbicides, insecticides, fungicides, and other synthetic compounds, all of which affect the whole biocenosis. For example, the effect of herbicides (urea and carbamate herbicides) in doses of 10, 50, and 100 ppm had inhibitory effects on the antibiotic activity of the tested bacterial strains. The effect of lower doses of herbicide was different in that the stimulation of antibiotic production by bacteria was more frequent. Strains of Erwinia carotovora and Pseudomonas lachrimans were resistant to the herbicide in the administered doses (2, 15, 16, 39, 40). Generally, herbicides disturbed the metabolic processes of bacteria that produced gibberelin-like substances. The herbicide reduced the amount of these substances produced in liquid media by the bacteria; only tenoran in the culture of Pseudomonas sp. brought about an increase. On the other hand, when seeds were dressed with crude preparations of Pseudomonas sp., the toxic effect of the herbicide chloropropham on the growth of young clover, seradella, lucerne, and carrot was clearly diminished (40).

The analysis of the results obtained leads to the conclusion that significant interactions between herbicides and microorganisms exist. These problems are little known and require further research.

METHODS OF STUDYING PLANT-MICROBE INTERACTIONS

In order to examine these very complicated processes, one can use two kinds of methods, direct and indirect. Direct methods include the collection of soil samples in various phases of plant development and qualitative identification of soil and rhizosphere microorganisms (35-38).

Another direct method is plant cultivation in non-sterile (pot experiment) and sterile conditions: a) open system in which the roots grow and develop in sterile conditions but upper parts of the plants grow in non-sterile conditions (22, 47); b) closed system in which the whole plants grow and develop in sterile conditions. For this purpose the various types of apparatus and appliance for plant cultivation must be sterile (9, 21, 29, 32, 33, 42) and methods and apparatus for plant seed sterilization are required (9, 21, 34).

Indirect methods would involve the examination of some microorganisms (from direct methods) and their improvement in biotechnological ways (35-38).

CONCLUSION

The evidence presented encourages further research on plant-microbe interactions which should be a basis for the practical usage of microorganisms in plant cultivation. It is possible to use inexpensive preparations for plant inoculation (Beringer, this volume), but on the condition that the proper microorganisms are chosen for the particular kinds of plants inoculated, and provided that they are applied at the appropriate time in well recognized ecological conditions. It seems that the best results of inoculating can be obtained in greenhouse cultivations (vegetables and flowers).

Another opportunity has been created through biotechnology and genetic engineering. Appropriately prepared antagonistic strains of Streptomyces, other bacteria, or fungi can be used for producing compounds (antibiotics) that are antagonistic to phytopathogenic microorganisms. On the other hand, there is a possibility of using microorganisms for the production of plant growth regulators. There are strains of bacteria or fungi which can produce metabolites which have the ability to stimulate plant growth by promoting the formation of a rich root system, accelerating the development of plants (blooming and seed formation).

For this purpose, strains of higher productivity might be obtained by means of mutation or genetic engineering. The use of microorganisms will make it possible to eliminate some chemical preparations which are harmful to the natural environment by causing serious, unforeseen deleterious effects, such as, for example, contamination of soil, water, food, and fodder, disturbing not only the metabolism of microorganisms but the interaction between microorganisms and plants as well.

REFERENCES

(1) Balicka, N. 1952. Influence of mixed-sowing of clover with timothy grass on rhizosphere-microflora. Acta Microbiol. Polon. **1:** 59-65.

(2) Balicka, N., and Krezel, Z. 1977. The effect of microorganisms on phytotoxicity of herbicides. Part V. Interaction of some microorganisms with betanal. Acta Phytopath. Acad. Sci. Hung. **12:** 61-65.

(3) Barber, D.A. 1966. Effect of microorganisms on nutrient absorbtion by plants. Nature **212:** 638-640.

(4) Bessems, E.P.M. 1973. Nitrogen fixation in the phyllosphere of Graminea. Doctoral Thesis, Center for Agricultural Publication and Documentation, Wageningen, Netherlands.

(5) De Bont, J.A.M. 1976. Nitrogen fixing methane-utilizing bacteria. Doctoral Thesis, Agricultural University, Wageningen, Netherlands.

(6) De Bont, J.A.M. 1976. Nitrogen fixation by hydrogen-utilizing bacteria. Arch. Microbiol. **107:** 235-240.

(7) Hirota, A.; Isogai, A.; and Sakai, H. 1982. New plant growth regulators, Cladospolide A and B, produced by a strain of Cladosporium cladospor ides. Ibidem **8/IIIb-4.**

(8) Ichihara, A., and Sakamura, S. 1982. Phytotoxins produced by plant pathogenic microorganisms. Fifth International Congress of Pesticide Chemistry (IUPAC) on Human Welfare - Pesticides, August 29 - September 4, 1982. Abstracts Kyoto-Japan **IIIb-1.**

(9) Kandler, O. 1951. Über den Einfluss von Bodenbakterien und deren Filtraten auf das Wachstum in vitro kultivierter Wurzeln. Archiv. Mikrobiol. **15:** 430-438.

(10) Kimura, Y.; Hamasaki, T. and Nakajima. 1982. Plant growth regulators produced by fungi. Ibidem **8/III-5.**

(11) Krasilnikov, N.A. 1956. Soil microorganisms and higher plants. Izd. Akad. Nauk SSR, Moscow.

(12) Krasilnikov, N.A. 1957. The role of soil microorganisms in plants' nutrition. Mikrobiologija **22:** 659-667.

(13) Krasilnikov, N.A. 1966. Research methods of soil microorganisms and their metabolites. Izdatielstvo Moskowskogo Universiteta.

(14) Krezel, Z. 1966. Rhizosphere microorganisms as a factor influencing resistance of plants to infections. Acta Microbiol. Polon. **15:** 163-172.

(15) Krezel, Z., and Kosinkiewicz, B. 1972. The effect of herbicides on morphology of colonies and antibiotic activity of some Streptomycetes. Acta Microbiol. Polon. **(Ser.B) 4:** 3-8.

(16) Krezel, Z., and Leszcynska, D. 1977. The effect of herbicides on the antibiotic activity of Bacillus subtilis 160. Acta Phytopath. Acad. Sci. Hung. **12:** 13-17.

(17) Krezel, Z., and Leszcynska, D. 1978. Antibiotic activity of Bacillus subtilis 93 to some phytopathogenic microorganisms. Med. Fac. Landbouw. Rijksuniv. Gent **43/2:** 850-865.

(18) Krezel, Z., and Stankiewicz, M. 1984. Effect of Bacillus subtilis 93 on the infection of sugar beet seedlings with Phoma betae. Med. Fac. Landbouw. Rijksuniv. Gent **49:** 517-523.

(19) Lie, T.A. 1964. Modulation of leguminous plants as affected by root secretion and red light. H. Veenman et Zonen N.V. Wageningen.

(20) Lochhead, A.G. 1957. Soil bacteria and growth-promoting substances. Bact. Rev. **22(3):** 1-10.

(21) Nilsson, P.E. 1957. Asceptic cultivation of higher plants. Archiv. Mikrobiol. **26:** 285-301.

(22) Prazmowski, A. 1981. Die Wurzelknöllchen der Erbse, Teil II. Die biologische Bedeutung der Wurzelknöllchen. Landwiss. Vers. Stat. **38:** 5-62.

(23) Pühler, A. 1984. Potentials and Limits of Applied Genetic Engineering. Forum Mikrobiologie, Supplement to the 3rd European Congress on Biotechnology, September 10-14, 1984.

(24) Rovira, A.D. 1963. Microbial inoculation of plants. Establishment of free-living nitrogen-fixing bacteria in the rhizosphere and their effects on maize, tomato, and wheat. Plant Soil **3:** 304-313.

(25) Ruinen, J. 1956. Occurrence of Beijerinckia species in the phyllosphere. Nature **117:** 220-221.

(26) Ruinen, J. 1961. The phyllosphere. Plant Soil **15:** 81-109.

(27) Saric, Z. 1969. Vitamin B_{12} synthesis by bacteria in the rhizosphere of wheat, maize and sugar beet. Matica Srpska - Zbornik za Prirodne Nauke **37:** 64-73.

(28) Schutting, and Fritsche, W. 1983. Mikrobielle Stoffwechselprodukte als potentielle Herbizide. Potsdamer Forschungen Wissenschaftlicher

Schriftreihe der Pädagogischen Hochschule Potsdam, Series B, vol. 35, pp. 92-109.

(29) Shemakhanova, N.M. 1959. A study of mycotrophy of arboreal plants by the pure culture method. Izwestija Akad Nauk SSR/Seria biologiczeskaja **5:** 674-686.

(30) Sobieszczański, J. 1961. The part of rhizosphere microorganisms in concentration and translocation of insoluble phosphate. Cz. 1. Zeszyty Naukowe WSR we Wroclawiu **40:** 109-118.

(31) Sobieszczański, J. 1961. The research methods of influence of microorganisms on higher plants. Zeszyty Naukowe WSR, Wroclaw, Rolnictwo **38:** 3-42.

(32) Sobieszczański, J. 1963. An appliance for plant cultivation under sterile conditions. Inform. Tech. Microbiol. Sol **4:** 45-48.

(33) Sobieszczański, J. 1963a. New modification of the apparatus for sterile plant cultivation. Acta Microbiol. Polon. **12:** 301-306.

(34) Sobieszczański, J. 1963b. Methods of seed sterilisation. Acta Microbiol. Polon **12:** 307-309.

(35) Sobieszczański, J. 1965a. Role of microorganisms in life of cultivated plants. I. Quantitative and qualitative changes in the microflora of the rhizosphere of rye and winter vetch during the vegetation process. Acta Microbiol. Polon. **14:** 161-182.

(36) Sobieszczański, J. 1965b. Role of microorganisms in life of cultivated plants. II. Effect of microorganisms from rye and vetch rhizosphere and from root-free soil on the growth of plants. Acta Microbiol. Polon. **14:** 183-202.

(37) Sobieszczański, J. 1966. Biologically Active Substances Produced by Bacteria of Some Cultivated Plants' Rhizosphere. Proceedings of the International Symposium on Plant Stimulation, Sofia, October 25-26, 1966, pp. 835-843.

(38) Sobieszczański, J. 1966. Studies on the role of microorganisms in the life of cultivated plants. III. Origin of the bacterial substances stimulating the growth of plants. Acta Microbiol. Polon. **15:** 67-84.

(39) Sobieszczański, J. 1970. Influence of herbicides on the production of gibberelin-like substances by Pseudomonas sp. and Arthrobacter sp. Meddelingen Faculteit Landbouwwetenschappen, Gent **35(3):** 681-688.

(40) Sobieszczański, J. 1975. Microorganisms as a factor changing the action of herbicides on plants. Roczniki Gleboznawcze **26(2):** Warsaw, 80-89.

(41) Sobieszczański, J. 1977. Fungi as producers of plant growth regulators. Part II. Plant growth inhibitors. Proceedings of the Vienna Symposium of the Federation of European Microbiological Society, March 28 - April 1, 1977, pp. 48-50.

(42) Sobieszczański, J. 1981. Apparatus for plant cultivation under sterile conditions (germfree). Zeszyty Naukowe Akademii Rolniczej we Wroclawiu, Technologia Zywnosci II- No. 136, p. 143 (Patent PRL No. 223143).

(43) Sobieszczański, J., and Niewiadoma, T. 1966. The influence of metabolites of Azotobacter upon the growth and development of plants. Proceedings of the IX International Congress for Microbiology, Moscow, p. 285.

(44) Stempniewicz, R. 1984. Studies on corn-seed microflora. Zeszyty Naukowe Akademii Rolniczejwe Wroclawiu Technologia Zywnosci **3(149):** 46-63.

(45) Stenz, E. 1962. Über den Einfluss von Bakterienfiltraten und Wuchsstoffen auf Wurzelhaare. Wiss. Z. Universität Leipzig, Math.-Natur.-Wiss. **4:** 642-646.

(46) Stolp, H. 1952. Beiträge zur Frage der Beziehungen zwischen Mikroorganismen und höheren Pflanzen. Archiv. Mikrobiol. **17(28):** 1-28.

(47) Szember, A. 1959. Influence on plant growth of the breakdown of organic phosphorus compounds by microorganisms. Plant Soil **2:** 147-158.

(48) Tamura, S.; Sakurai, A.; Kainuma, K.; and Takai, M. 1965. Isolation of Helmintosporol as a natural plant growth-regulator and its chemical structure. Agr. Biol. Chem. **29(3):** 216-221.

(49) Welte, E., and Trolldenier, G. 1962. Der Einfluss der Bodenmikroorganismen auf Trockensubstanzbildung und Aschengehalt in Nährlösungen wachsender Pflanzen. Archiv. Mikrobiol. **43:** 138-147.

(50) Wichner, S., and Libbert, E. 1968. Interaction between plants and epiphytic bacteria - regarding their auxin metabolism detection of IAA-producing epiphytic bacteria and their role in long duration experiments on tryptophan metabolism in plant homogenates., and Li Physiol. Plant. **21:** 227-241.

(51) Vancura, V., and Hovadik, A. 1965. Root exudates of plants II. Composition of root exudates of some vegetables. Plant Soil **22(1):**21-32.

(52) Van Egeraat, A.W.S.M. 1975. Changes in free ninhydrin-positive compounds of young pea plant as affected by different nutritional and environmental conditions. Plant Soil **42(1):** 16-36.

(53) Van Egeraat, A.W.S.M. 1975. The possible role of homoserine in the development of Thizobium leguminosarum in the rhizosphere of pea seedlings. Plant Soil **42:** 381-386.

(54) Van Egeraat, A.W.S.M. 1976. Response of Rhizobium sp. and pea seedlings to glycine. Plant Soil **45:** 191-199.

(55) Zak, J. 1959. Beitrag zu den Methodiken der Gewinnung von Wurzelsekreten steril angebauter Pflanzen. Sbornik Ceskoslovenske Akademie Zemedelskich Ved, Rostlinna Vyroba, Rocnik **5(32)9:** 1337-1346.

(56) Zentmyer, G.A. 1961. Chemotaxis of zoospores for root exudates. Science **133(3464):** 1595-1596.

Biotechnology: Potentials and Limitations, ed. S. Silver, pp. 283-291. Dahlem Konferenzen 1986. Berlin, Heidelberg, New York, Tokyo: Springer-Verlag.

The Production and Utilization of Lignocellulose

P. Broda
Dept. of Biochemistry and Applied Molecular Biology
University of Manchester Institute of Science and Technology
Manchester M60 1QD, England

Abstract. Many poorer countries depend for energy and chemicals for development upon renewable resources. Biotechnology has a part to play in improving production of biomass and in possible processes for its conversion to fuels and chemical feedstocks. Economic, political, social, and environmental considerations as well as scientific and technological advances will determine how much progress is made. Although there is a diversity of materials, ranging from wood and sugarcane bagasse to municipal solid waste and animal waste, there are some unifying features. There is the need to understand the biological attack of lignin and of crystalline cellulose and the design of processes such as solid-substrate fermentations that would be applicable to developing countries.

INTRODUCTION

Much of the world's energy is moved around as oil. As well as being used for generating electricity and for motor vehicles, it has also become the basis of much chemical industry. Between 1940 and 1978 the proportion of the world's total organic chemical industry that was based on coal dropped from 95% to 3%. Such a major technological revolution may be followed by others, as oil becomes less available. There are large reserves of coal (perhaps over 200 years of proven reserves), and unconventional sources of oil such as tar sands and oil shale may be used. This will require a solution of the problems of sulfur and nitrogen, which are the probable causes of acid rain.

However, such fossil fuels are inadequate for the world's needs since they will eventually become depleted and because they are distributed very unevenly. There is also the anxiety that if the CO_2 that is liberated is not recycled there will be long-term climatic and environmental consequences. For many developing countries no fossil fuel sources exist and there is no money to import them. In such countries total energy derived from biomass (for instance, wood, charcoal, crop residues, and animal dung) is usually over 50%. However, these resources are becoming depleted through deforestation, erosion, deterioration of soil, and the need to use land for crops. This will result in a second energy crisis. Without more available energy such countries are unlikely to improve their economic status (4).

For the world as a whole, a number of major options must all contribute in the future. These include the development of additional fossil fuel sources and of nuclear energy. Much effort is being spent to develop controlled nuclear fusion processes. Within the scientific community, there is uncertainty about how to capture and store solar energy. But one obvious way is to do it how nature has always done it - by growing plants. There is a pressing need to develop indigenous biomass-based energy sources for developing countries. The amount of energy from the sun that could be trapped in plant biomass is a couple of orders of magnitude greater than is presently generated as electricity. Present economics inhibit investment in this field, for reasons such as the presently depressed oil market and the fact that in the United States, the engine of such investment, biomass could only make a relatively modest contribution given their present extraordinary level of energy usage (1). Interest in biomass in the U.S. is more likely to come from the need to dispose of urban and farm wastes. However, from a global point of view, it is essential that applied molecular biologists and biotechnologists push forward their own research on the production of biomass and its utilization by biological processes. There is a need for supranational long-term investment; it is an area of biotechnology which lends itself particularly to open science, conducted in universities and other nonprofit organizations, and in the end the single most important contribution of biotechnology may be in this area.

ETHANOL FROM SUGARCANE

The biotechnology of production and utilization of biomass is in fact already with us. Brazil has accumulated the world's largest foreign debt, primarily by importing petrochemicals to fuel its industrial development over the last decade, because Brazil has relatively little indigenous oil. A decision was made in 1975 to push forward a program for the production of fuel alcohol (16). Currently there is spending of $1 billion per annum on the technology that would provide 10 billion liters of alcohol per year in 1987. This program is based on the fermentation of sucrose to ethanol. Presently about 120 million tons of sugarcane are produced, yielding 8 million tons of sugar. The scale of the program can also be judged from the fact that a million cars already run on alcohol and all others run on alcohol-petroleum blends. In addition, ethanol is used as a feedstock for chemical industry, for instance, for the production of ethylene.

In some ways Brazil provides a paradigm: transport depends almost entirely on cars and trucks; there is a rapidly increasing population; there is the foreign debt referred to; underemployment in the countryside; intense sunlight; land that can be used (or at present grossly misused); and compelling strategic/economic reasons. When such a major program is being pushed through the opportunities for improvement are immense. One aspect is the utilization of the millions of tons of bagasse, which is already brought to the sugar mill. At present it is mainly burned to fuel the distillation process needed to recover ethanol. However, if some of it (there is more than is needed for this purpose) were to be broken down into its constituent carbohydrates and lignin, there would be the possibility of saccharification of the carbohydrates and the recovery of the lignin. Also, the economics of distillation would be radically improved with higher yields of ethanol (e.g., by using ethanol-tolerant strains) or if the fermentation were conducted at higher temperatures with thermophiles (5, 13).

The Brazilian process depends primarily on the sugar component of sugarcane. Yet it is likely that its success as an economic process depends on progress on the utilization of lignocellulose. Moreover, there are other crops that are attractive as converters of solar energy, e.g., elephant grass (120 tonnes/hectare dry weight compared with sugarcane's 70 tons wet weight) and eucalyptus wood, where lignocellulose is the primary product.

THE COMPOSITION OF LIGNOCELLULOSE

Lignocellulose is a complex of three classes of polymer. Cellulose, which consists of repeating glucose units largely in crystalline fibers, is what is sought in the pulp and paper process. The hemicelluloses are a more mixed group of polysaccharides, in which the monomers are mainly pentoses. In the case of lignin the monomers are three cinnamyl alcohols. There are two limiting steps in the disruption and biological utilization of lignocellulose: one is the crystallinity of the cellulose, which makes it substantially more difficult to attack than, for instance, starch or the hemicelluloses. The second is the nature of the intermonomeric bonds in lignin. Lignin is assembled by an oxidative polymerization of free-radical species of its monomers to yield a highly irregular polymer with a number of intermonomeric bond types. The predominant class is the β-aryl ether bond, but others include direct linkages and ethers between aromatic rings. Any system that frees lignin from lignocellulose must be able to attack at least a high proportion of the bonds.

PRODUCTION OF BIOMASS

Much of what can be done for its production is not in the biological domain. The technologies of planting, fertilizing, harvesting, collection, handling, transportation, and storage could all be greatly improved. There is a need to minimize energy use (in developed countries agriculture uses more energy than it yields), make use of local labor, and halt deforestation and erosion. There may be conflicts of interest between fuel production and the utilization of good land for food for local populations. The interests of the car-owning middle classes are not always the same as those of the peasants whose land is being used, and large-scale biomass production may result in the concentration of land ownership in fewer hands.

Perhaps the industry that stands to be improved in efficiency more than any other is that of forestry. Forestry management with planned rotations and whole tree farming, and the reduction of the amount of waste from the present 50%, would be coupled with breeding programs based on tissue culture and somatic embryogenesis, together with effective screening programs (14). One can seek, both with forest crops and with others, disease-resistant and herbicide-resistant lines, and ones where fractions of the product have maximum potential as animal feed. Another need is to minimize nitrogen depletion. This implies major investment on studies on biological nitrogen fixation both with Rhizobium-like systems and with associative nitrogen fixers. In addition, the merits of C_3 and C_4-photosynthesizers must be considered in different situations, i.e., when availability of sunlight or CO_2 is the limiting factor.

UTILIZATION OF WHOLE LIGNOCELLULOSE

The range of biomass materials includes wastes from forests, agriculture, stock-raising, and towns. Much lignocellulose is and will be used in an unfractionated form. Crude burning is likely to be supplanted by chemical processes that yield various mixtures of gas, liquid, and solid that all have caloric value. However, the materials obtained from lignocellulose tend to be corrosive and viscous (17). Pilot schemes for utilization of municipal solid waste suggest a potential return for investment comparable to that for oil; this is a resource that is estimated at 360 Mt per annum in the U.S. alone. It is an area where there will be changes in the appropriateness of different technologies depending on economic and resource-availability considerations. Nevertheless, these are processes that are being developed and it is for biotechnology to find a competing or complementary role in such processes (15).

It should be recognized that at least two biological systems processes already exist that utilize lignocellulose. One is the production of biogas, which has been practiced for centuries in China. There are substantial efforts to develop such technology in India, and it has a future in the utilization of wastes in landfills and on farms in developed countries. The other is in the production of mushrooms in solid-substrate fermentation systems (20). The value of this industry in the United States is about $500 M per annum. The major problems for the process are in the physiology of the organisms and in protection against disease. The preparation and management of the starting compost and the casing layer substrates are critical, but a very wide range of starting materials might be used. There is much work to be done on the analysis of the parameters and of conditions that could be employed for different fungal strains and substrates. It would be absurd to suggest that the world's biomass should be utilized for mushroom production, but this is a model that is already operating on a large scale.

FRACTIONATION OF LIGNOCELLULOSE

By far the largest process currently in operation is pulping for both the paper industry and the production of cellulose fiber. There are a number of pulping methods, but in each what is obtained is cellulose, with a proportion of the hemicelluloses and lignin in a modified form. The bulk of the hemicelluloses are broken down, and lignin is converted into products (Kraft lignins or lignosulfonates) that have little value other than for burning. Such industrial lignin have been considered as a source of aromatic chemicals such as phenols and catechols, for fiberboard, as an emulsion stabilizer, as an additive in oil drilling, and for cement processing, but none of these are high value products (6). Novel uses must be sought for the derivatives of lignin that match the enormous scale in which they are produced in nature.

Possible processes for utilization of the different components of lignocellulose, whether chemical or biological, typically fall into three stages: pretreatment to disrupt the gross structure, depolymerization, and utilization of the monomers. In the pretreatment, mechanical chipping is followed in a number of schemes that are in the pilot stage by steam explosion, and in others by solvent extractions. Such methods render the mixtures of lignin, cellulose, and hemicellulose amenable to further treatment. Another process that has been reinvestigated recently is treatment with anhydrous hydrogen fluoride. This requires minimal pretreatment and yields high glucose values at low chemical cost, but the need to recycle the hydrogen fluoride is critical. The demands on the materials of the equipment are supposed not to be severe, and this may be an attractive possibility since the lignin obtained is relatively undamaged. The methods used to fractionate the carbohydrates include treatment with dilute acid, which removes the hemicelluloses from the celluloses, or with concentrated acids, which have high material costs. Both steam explosion and solvent extraction yield hemicellulose that can serve as a substrate for enzymatic hydrolysis (18). Relatively mild treatments with ammonia are used to upgrade the digestibility of straw for ruminants.

It would seem that some rapid mechanical process to obtain particulate material, e.g., chipping, is likely to be common to all processes, but that biological methods can then be considered in competition with chemical and physical methods (or as complements to them). In general, biological methods will operate at lower temperatures and pressures, with cheaper starting materials and the possibility of operating on a smaller scale; on the other hand, the throughput time may be much slower.

There are at least four areas in which scientific progress might lead to biologically based processes. Two of these are concerned with the polymers, namely, disruption of the crystallinity of the cellulose allowing greater access to cellulases, and the solubilization of lignin. The other two are concerned with the utilization of monomers, i.e., the fermentation of pentose sugars and the production of ethanol in high yield together with its separation from the aqueous medium. I believe that the solubilization of the polymers is the key to biological processes and will therefore discuss these.

CELLULOSE BIODEGRADATION

Cellulose in lignocellulose is a much larger resource than starch for breakdown to sugars. But it is also a much more complex problem because of its crystallinity and its admixture with hemicellulose and lignin. Most work on its biodegradation has been done with the filamentous fungus Trichoderma reesei, which, given optimum culture conditions, can produce yields of 20 g extracellular cellulose protein from 60 g cellulose. It is now evident that cellulase is a complex of at least eight proteins that act synergistically and are subject to complex controls (10, 12). Trichoderma produces five or more endo β-glucanases which convert the crystalline cellulose to oligocellodextrins and amount to 20-35% of the protein by weight. In addition, there are at least two cellobiohydrolase components (65-80% of total protein), which convert these to cellodextrins and cellobiose. Both classes of enzymes are subject to end product inhibition by cellobiose. Finally there is a cellobiase that converts the cellodextrins and cellobiose to glucose; this enzyme constitutes a mere 1% of the protein, and its action is inhibited by glucose. The whole complex is also subject to catabolite repression by glucose.

The pilot studies show that with preparations of cellulase that contain these various activities it is possible in stirred tanks, under realistic conditions, to hydrolyze various types of cellulose to soluble sugars with substrate concentrations of 10-30% and yields of 30-60%. One could convert 1000 kg of lignocellulose of substrate (wood chips, straw, corn stover, or fiber from urban waste) to 350 kg of glucose in 24 h. But the economic barrier is the very large amounts of enzyme complex needed to hydrolyze the cellulose (perhaps a hundred times more than the amylase needed to break down starch).

The reasons have to do with the recalcitrant and impure nature of cellulose, with the requirement for a synergistic complex, with product inhibition, and with the difficulty of desorbing the enzyme complex from the substrate. It seems premature to dismiss the biological attack of cellulose on grounds of present economics. The appropriate view to take is that the methods of molecular genetics, allowing cloning of individual enyzmes, will result in a far better understanding of how, in one organism, the system works. There will then be opportunities for site-directed mutagenesis to remove product inhibition, to improve the rate at which the enzymes are produced (very little is secreted while the fungus has sufficient cellulose for growth; it occurs as cellulose and nitrogen are approaching depletion), and perhaps to improve the activities of the enzymes themselves.

Other fungi have also been studied, but none is qualitatively more attractive than Trichoderma, and this organism is certainly an excellent model for this difficult system. However, other organisms should also be studied further.

These include two prokaryotes, the thermophilic actinomycete Thermomonospora (3) and a species that converts cellulose to ethanol, Zymomonas mobilis (19).

LIGNIN BIODEGRADATION

Discussions on lignin breakdown and utilization are still primarily based on chemical engineering approaches. However, recent reviews recognize the potential for biological lignin degradation (6). Breakdown of lignin is the true limiting step in lignocellulose utilization, as lignin is the most resistant component to solubilization. The achievement of this step by biological methods would have advantages of releasing the cellulose and hemicelluloses efficiently, as well as lignin-derived monomers that could be converted into useful chemicals.

In nature the function of lignin solubilization may be primarily to allow the degrading organism to obtain access to the carbohydrates since the energetic investment in lignin solubilization is rather substantial; thus the hymenomycete fungus Phanerochaete chrysosporium will only degrade lignin if it is depleted for carbon and nitrogen sources. Phanerochaete produces apparent ligninases extracellularly (21). The type enzyme is a peroxidase that has an associated heme group and requires H_2O_2 as a co-substrate. It is possible that these novel enzymes, which have a wide substrate specificity, could have applications in oxidations beyond the solubilization of lignin.

Recently molecular genetical methods have started to be applied to this problem. We have shown that the control of ligninolytic attack is likely to be at the level of gene expression, mediated through cyclic AMP (9). We have also isolated both non-degrading and hyper-degrading mutant strains (8), and are seeking to relate mutations to specific DNA sequences in genomic and cDNA gene libraries. A current objective is to understand more of the control of expression of the ligninolytic functions. It may then be possible to obtain strains in which lignin degradation is uncoupled from the cessation of growth. This is necessary for solid substrate processes if lignin degradation is to take place efficiently. However, this requires the development of appropriate cloning vectors and transformation methods.

As with cellulose degradation, organisms other than fungi may be important; it has recently been shown that some actinomycete bacteria can solubilize lignin (2). The mode of attack is different from that by Phanerochaete, since here a high-molecular-weight soluble product is formed from lignin. In addition, the control of this attack is different since it can occur even in rich medium, in normal growth (11). These organisms therefore have considerable promise for biological processes for lignocellulose breakdown.

CONCLUSION

The prospects for a major contribution from biotechnology to these very large-scale problems of energy and chemicals for the next century are excellent. The solution of basic scientific questions, in particular on the degradation of lignocellulose, would pay handsome dividends. The amount of work that has been done, particularly that using molecular genetical methods, is very small indeed compared with the effort extended to health care topics. There is some interest in more developed countries in specific problems of lignocellulose treatment. One example is in the pulp and paper industry, where partial delignification can reduce energy requirements and improve pulp properties in mechanical pulping; and pulps might themselves be improved or bleached. Another example is in upgrading agricultural wastes for ruminant digestibility (7). These may lead to more general applications. With proper support, we can realize the potential for developing processes that could occur on a modest scale, using intermediate technology, without any extremes of temperature or pressure, and without over-sophisticated equipment, to produce locally useful products.

Acknowledgement. Work done in the author's laboratory was part of a program supported jointly by British Petroleum's Venture Research Unit and the Agricultural and Food Research Council.

REFERENCES

(1) Biomass Panel of the Energy Research Advisory Board. 1983. Biomass energy. Solar Energy **30:** 1-31.

(2) Crawford, D.L. 1978. Lignocellulose decomposition by selected Streptomyces strains. Appl. Envir. Microbiol. **35:** 1041-1045.

(3) Hagerdal, B.; Ferchak, J.D.; and Pye, K.E. 1978. Cellulolytic enzyme system of Thermoactinomyces sp. grown on microcrystalline cellulose. Appli. Envir. Microbiol. **36:** 606-612.

(4) Hall, D.O.; Bernard, G.W.; and Moss, P.A. 1983. Biomass for Energy in the Developing Countries. Oxford: Pergamon Press.

(5) Hartley, B.S., and Payton, M.A. 1983. Industrial prospects for thermophiles and thermophilic enzymes. In Biochemical Society Symposium on Biotechnology, eds. C.F. Phelps and P.H. Clarke, vol. 48, pp. 133-146. London: The Biochemical Society.

(6) Janshekar, H., and Fiechter, A. 1978. Lignin: biosynthesis, application and breakdown. Adv. Biochem. Eng. **27:** 119-178.

(7) Kerr, T.J.; Windham, W.R.; and Benner, R. 1984. Chemical and bacterial pretreatment of lignocellulosic materials to increase ruminant digestibility. Biotechnology **2:** 805-807.

(8) Liwicki, R.; Paterson, A.; MacDonald, M.J.; and Broda, P. 1986. Phenotypic classes of phenol oxidase-negative mutants of the lignin-degrading fungus Phanerochaete chrysosporium. J. Bacteriol. **162,** in press.

(9) MacDonald, M.J.; Paterson, A.; and Broda, P. 1984. A possible relationship between cyclic AMP and idiophasic metabolism in the white-rot fungus Phanerochaete chrysosporium. J. Bacteriol. **160:** 470-472.

(10) Mandels, M. 1981. Enzymatic conversion of cellulosic biomass: a viable alternative to ethanol from grain? Am. Soc. Microbiol. News **47:** 174-178.

(11) McCarthy, A.J., and Broda, P. 1984. Screening for lignin-degrading actinomycetes and characterisation of their activity against [^{14}C]-lignin-labelled wheat lignocellulose. J. Gen. Microbiol. **130**: 2905-2913.

(12) Montencourt, B.S. 1983. Trichoderma reesei cellulases. Trends Biotechnol. **1**: 156-161.

(13) Oliver, S.G. 1984. Economic possibilities for fuel alcohol. Chem. Ind. **18th June**: 425-427.

(14) Powledge, T.M. 1984. Biotechnology touches the forest. Biotechnol. **2**: 763-772.

(15) Proceedings of the Fifth Symposium on Biotechnology for Fuels and Chemicals held in Gatlinburg, Tennessee, May 10-13, 1983, ed. C.D. Scott. Biotechnol. Bioeng. Supplement **13**.

(16) Rothman, H.; Greenshields, R.; and Rosillo Calle, F. 1983. The Alcohol Economy: Fuel Ethanol and the Brazilian Experience. London: Frances Pinter.

(17) Russell, J.A.; Molton, P.M.; and Landsman, S.D. 1983. Chemical comparisons of liquid fuel produced by thermochemical liquefaction of various biomass materials. Altern. Energy Sour. **3**: 307-322.

(18) Saddler, J.N.; Yu, E.K.C.; Mes-Howtree, M.; Levitin, N.; and Brownell, H.H. 1983. Utilisation of enzymatically hydrolysed wood hemicelluloses by microorganisms for production of liquid fuels. Appl. Envir. Microbiol. **45**: 153-160.

(19) Skotnicki, M.L.; Warr, R.G.; Goodman, A.E.; Lee, K.J.; and Rogers, P.L. 1983. High-productivity alcohol fermentations using Zymomonas mobilis. In Biochemical Society Symposium on Biotechnology, eds. C.F. Phelps and P.H. Clarke, vol. 48, pp. 53-86.

(20) Tautorus, T.E., and Townsley, P.M. 1984. Biotechnology in commercial mushroom fermentation. Biotechnology **2**: 696-701.

(21) Tien, M., and Kirk, T.K. 1983. Lignin-degrading enzyme from the hymenomycete Phanerochaete chrysosporium Burds. Science **222**: 661-663.

Biotechnology: Potentials and Limitations, ed. S. Silver, pp. 293-294. Dahlem Konferenzen 1986. Berlin, Heidelberg, New York, Tokyo: Springer-Verlag.

Social and Ethical Considerations for Biotechnology

J.E. Davies
Dept. of Biotechnology
Institut Pasteur
75724 Paris 15
France

Modern biotechnology has provided techniques that allow the planned and directed modification of living organisms: the realization of the concept of biological factories, and of changing certain forms of plant and animal life to suit man's needs, is about to occur. However, when a technical revolution of some substance occurs, there are always aspects other than the direct benefit to mankind that have to be considered. In all cases a risk/benefit ratio should be evaluated.

It was not the brief of the group to present their conclusions on possible "misuses" of the biotechnological revolution and certainly not to pontificate on this subject! Nonetheless, as responsible scientists we cannot avoid making some comment on the use of genetic engineering, and the following is a summary (not necessarily a consensus) of our thoughts on this matter.

Our sentiments are directed primarily towards the concern that the products of this new revolution should be made available to all who need them. There should be no discrimination. It is obvious that many of the new genetically engineered pharmaceuticals will be expensive and therefore restricted to the more affluent societies. A new generation of vaccines is in the offing; we earnestly hope that the research and development necessary to provide such vaccines should be based on man's needs worldwide, regardless of cost. More than one million African children die of malaria every year; a malaria vaccine should be a major goal of genetic engineering technology, even if it is not a disease common to the industrialized nations of the world.

To give another medical example, diagnostic methodology is becoming more sophisticated and sensitive with every year. This technology should be generally available, but should one diagnose disease states where there is no hope of treatment? Should diagnostic methods be used to deny or limit insurance coverage or the right to accept employment? Such ethical questions need careful consideration and their many aspects have to be evaluated to guarantee appropriate and effective applications.

There is considerable public suspicion of new developments such as genetic engineering which can often be elevated to hysteria by unreasonable public presentation; in some circumstances this type of response could lead to delayed acceptance of a new product even when the need is great. The "compassionate" or "orphan" drug release approval is an important route to providing new treatments; regulatory authorities must meet the challenge of evaluating new and unusual products by accepting new methods and standards for release so that novel products can be used quickly and effectively. One can state, without fear of contradiction, that recombinant proteins are being produced at higher standards of purity than any protein products previously used in medicine.

We are obviously concerned about misuses of science in biological warfare but believe that public opinion - hopefully worldwide - will be opposed to any extension of the existing military armamentarium.

The potential and promise of genetic engineering is enormous and the dissemination of the products of this technology must not be limited by anything other than technical considerations. This applies to short-term benefits such as recombinant pharmaceuticals and to medium- and long-term benefits that will be developed in agriculture, energy, etc. These products are being developed for the benefit of mankind as a whole and not for just a segment of society. The World Bank approach to technology sharing provides a good model, and we hope that other international organizations (both profit and nonprofit) will adopt similar approaches to guarantee optimal apportionment of the advantages of recombinant DNA methodology.

List of Participants with Fields of Research

ABELSON, P.H.
SCIENCE Magazine
1515 Massachusetts Avenue NW
Washington, DC 20005
USA

Editor, SCIENCE Magazine

AHARONOWITZ, Y.
Tel Aviv University
George S. Wise Faculty of
Life Sciences
Dept. of Microbiology
Ramat-Aviv 69 978
Israel

β-lactam antibiotics in streptomyces; metabolic and genetic regulation of biosynthesis; microbial metabolites for agricultural use; biotransformations

BARNES, W.M.
Dept. of Biological Chemistry 8094
Washington University
School of Medicine
660 South Euclid
St. Louis, MO 63110
USA

Plant genetic engineering; DNA analysis technology; bacterial genetic control

BARZ, W.
Lehrstuhl für Biochemie der Pflanzen
Westfälische Wilhelms-Universität
Hindenburgplatz 55
4400 Münster
F.R. Germany

Plant biochemistry; microbial degradation of phyto-alexins; aromatic metabolism in plant cell cultures

BERINGER, J.E.
Unit of Molecular Genetics
University of Bristol
Woodland Road
Bristol BS8 1UG
England

Genetic studies of the Rhizobium-legume symbiosis; beneficial plant-microbe interactions

BETZ, H.
ZMBH
Universität Heidelberg
Im Neuenheimer Feld 364
6900 Heidelberg
F.R. Germany

Neurobiology

BLÖCKER, H.
Gesellschaft für Biotechnologische
Forschung mbH
Mascheroder Weg 1
3300 Braunschweig
F.R. Germany

DNA synthesis, protein design

BRODA, P.
Dept. of Biochemistry and
Applied Molecular Biology
University of Manchester
Institute of Science and
Technology
Manchester M6O 1QD
England

Molecular biology of lignin breakdown

BU'LOCK, J.D.
Dept. of Chemistry
University of Manchester
Manchester M13 9PL
England

General biotechnology

CHAKRABARTY, A.M. Dept. of
Microbiology and Immunology
University of Illinois
Medical Center
835 South Wolcott Ave.
Chicago, IL 60612
USA

Biodegradation of toxic chemicals; microbial recovery of oil; pseudomonas infection in cystic fibrosis

COLLINS, J.
Abt. Genetik
Gesellschaft für Biotechnologische
Forschung mbH
Mascheroder Weg 1
3300 Braunschweig
F.R. Germany

Molecular genetics as applied to biotechnology

DAVIES, J.E.
Dept. of Biotechnology
Institut Pasteur
75724 Paris 15
France

Biotechnology: production of pharmaceutically useful human and animal proteins in microorganisms

EVELEIGH, D.E.
Dept. of Biochemistry and Microbiology
Cook College, Lipman Hall
Rutgers - The State University of
New Jersey
New Brunswick, NJ 08903
USA

Microbial conversion of biomass; cellulase cloning and gasohol production

FIERS, W.
Laboratory of Molecular Biology
State University of Ghent
Ledeganckstraat 35
9000 Ghent
Belgium

Molecular biology: expression of heterologous genes in microorganisms and in animal cells; influenza; lymphokines; molecular biology of cancer

FLAVELL, R.B.
Plant Breeding Institute
Trumpington
Cambridge CB2 2LQ
England

Plant molecular biology: gene structure, control of expression

HAHLBROCK, K.
Max-Planck-Institut für
Züchtungsforschung
Abt. Biochemie
5000 Köln 30
F.R. Germany

Plant biochemistry/molecular biology, phytopathology

HOPWOOD, D.A.
John Innes Institute
Norwich NR4 7UH
England

Genetics and molecular biology of streptomyces

KAHMANN, R.
Max-Planck-Institut für
Molekulare Genetik
Otto-Warburg-Laboratorium
Ihnestraße 63
1000 Berlin (West) 33

Gene regulatory mechanisms in prokaryotes and eukaryotes

KRAWINKEL, U.
Institut für Genetik
der Universität Köln
Weyertal 121
5000 Köln 41
F.R. Germany

Antibody-variable region; differentiation of lymphocytes

KULA, M.-R.
Gesellschaft für Biotechnologische
Forschung mbH
Mascheroder Weg 1
3300 Braunschweig
F.R. Germany

Enzyme technology: coenzyme-dependent processes, technology of protein isolation and purification

LEHRACH, H.R.
European Molecular Biology
Laboratory
Postfach 10.2209
6900 Heidelberg
F.R. Germany

Mammalian genetics at the molecular level, development of cloning techniques

MARQUARDT, R.
Hoechst AG
Abt. Biotechnik D528
6230 Frankfurt 80
F.R. Germany

Cloning in E. coli and methylotrophic bacteria; vector development

MATTIASSON, B.
Pure and Applied Biochemistry
Chemical Center
University of Lund
P.O. Box 124
221 00 Lund
Sweden

Process intensification in biotechnology; biochemical analysis

MORRIS, J.G.
Dept. of Botany and Microbiology
University College of Wales
Aberystwyth SY23 3DA
Wales

Microbial physiology, especially anaerobic fermentations

NEIJSSEL, O.M.
Laboratory for Microbiology
University of Amsterdam
Nieuwe Achtergracht 127
1018 WS Amsterdam
Netherlands

Physiology of metabolite overproduction; energetics of microbial growth

PAOLETTI, E.
Laboratory of Immunology
Wadsworth Center for Laboratories
and Research
New York State Dept. of Health
Albany, NY 12201
USA

Live recombinant vaccines

PEACOCK, W.J.
CSIRO Division of Plant Industry
GPO Box 1600
Canberra, ACT 2601
Australia

Alcohol dehydrogenase genes and controlling elements of maize

PFAFF, E.
Institut für Mikrobiologie
Universität Heidelberg
Im Neuenheimer Feld 230
6900 Heidelberg
F.R. Germany

Structure of antigenic determinants, analysis of known and unknown gene products of hepatitis B virus

ROWLANDS, D.J.
Wellcome Biotechnology Ltd.
F.M.D. Division
Pirbright, Woking
Surrey GU24 0NQ
England

Foot and mouth disease virus structure and replication and the mimicking of immunogenic determinants

SAEDLER, H.
Max-Planck-Institut für
Züchtungsforschung
Egelspfad
5000 Köln 30
F.R. Germany

Plant molecular biology

SAHM, H.
Institut für Biotechnologie
Kernforschungsanlage Jülich
Postfach 1913
5170 Jülich 1
F.R. Germany

Microbial amino acid production; ethanol fermentation with bacteria; anaerobic wastewater treatment

SALAMINI, F.
Istituto Sperimentale per la
Cerealicoltura
Sezione di Bergamo
Via Stezzano 24
P.O. Box 164
24100 Bergamo
Italy

Genetics and breeding of maize

SCHALLER, H.
Mikrobiologie & ZMBH
Universität Heidelberg
Im Neuenheimer Feld 230
6900 Heidelberg
F.R. Germany

Molecular biology: control of gene expression; pathogenic microorganisms and host defense mechanism

SIEWERT, G.
Schering AG
Mikrobiol. Biochemie
Müllerstraße 170-178
1000 Berlin (West) 65

Application of recombinant DNA techniques for the production of biologically active peptides and proteins

SILVER, S.
Biology Dept.
Washington University
St. Louis, MO 63130
USA

Molecular genetics; plasmid-determined resistance systems; bacterial metabolism toward inorganic cations and anions

SIPPEL, A.E.
ZMBH
Universität Heidelberg
Im Neuenheimer Feld 364
6900 Heidelberg 1
F.R. Germany

Molecular mechanisms of eukaryotic gene expression and differentiation

SMITH, M.
Dept. of Biochemistry
Faculty of Medicine
University of British Columbia
2146 Health Sciences Mall
Vancouver, BC V6T 1W5
Canada

Synthesis and use of oligodeoxyribonucleotides as tools in molecular biology and genetics: as probes in gene isolation, as primers in RNA and DNA sequence determination and in the production and isolation of mutants

SOBIESZCZANSKI, J.
Akademia Rolnicza
Zaklad Mikrobiologii Technicznej i Zywnosci
ul. Norwida 25
50-375 Wroclaw
Poland

Microbiology

STAHL, U.
FG Mikrobiologie
Technische Universität Berlin
Seestraße 13
1000 Berlin (West) 65

Strain improvement of microorganisms by recombination, vector/host systems in fungi and bacteria, formal and molecular genetics of fungi

STARLINGER, P.
Institut für Genetik
Universität Köln
Weyertal 121
5000 Köln 41
F.R. Germany

Plant molecular biology

TIMMIS, K.N.
Dept. of Medical Biochemistry
The Central Medical University
University of Geneva
1211 Geneva 4
Switzerland

Design of complex metabolic pathways for industrial processes; pathogenicity of Streptococcus and Shitella

VAN MONTAGU, M.
Laboratory of Genetics
State University of Ghent
Ledeganckstraat 35
9000 Ghent
Belgium

Plant molecular biology; gene transfer

WAGNER, E.F.
European Molecular Biology Laboratory
Postfach 10.2209
6900 Heidelberg
F.R. Germany

Gene regulation in mammalian development and differentiation

WEGLENSKI, P.
Dept. of Genetics
Warsaw University
Al. Ujadowskie 4
00-478 Warsaw
Poland

Molecular genetics, gene transfer in filamentous fungi

WEILER, E.
Ruhr-Universität Bochum
Lehrstuhl für Pflanzenphysiologie
Universitätsstraße 150
4630 Bochum 1
F.R. Germany

Plant hormone function, hormone receptors, regulation of plant development

WINTER, G.
MRC Laboratory of Molecular Biology
Hills Road
Cambridge CB2 2QH
England

Site-directed mutagenesis and protein engineering of enzymes and immunoglobulins

WODAK, S.J.
Unité Conformation de Macromolécules Biologiques
Université Libre de Bruxelles
Avenue P. Héger, P2, CP 160
1050 Brussels
Belgium

Conformation of biological macromolecules

ZENK, M.
Pharmazeutische Biologie
Universität München
Karlstraße 29
8000 München 2
F.R. Germany

Plant cell culture for the production of pharmaceutically important metabolites; biosynthesis

Subject Index

Acid, citric, 85
-, clavulanic, 74, 103, 108, 109
-, giberellic, 277
-, gluconic, 85, 88, 89
-, lactic, 85
- probes, nucleic, 134, 226
-, rosmarinic, 243, 244
ACTH-precursor, 127
Actinomycetes, 76, 95, 106, 208, 266, 289
Active proteins, biologically, 131
ACV tripeptide, 77, 109, 110
Adenovirus major late promoter, 24
Adjuvants, 133, 137, 152, 171, 172
Adsorbents, solid, 118, 119
Affinity, substrate, 9, 56, 61-65
Agents, antineoplastic, 104, 105
Agriculture, 1, 2, 5, 73, 175, 199-221, 223-237, 242, 248-250, 261, 263, 285
Agrobacterium, 201, 202, 220, 228, 229, 260-262
AIDS virus, 129, 135, 171
Alkaline phosphatase, 168
Alkaloid, 243-246, 251
Allograft rejection, 173
α-amylase, 116, 169
-/tumor growth factor, 130
Alzheimer's disease, 130, 134
Amplification, DNA, 35, 36
Amylases, 92, 116, 169
Analogues, growth factor, 130
-, - - receptor, 130
Animal disease, screening, 135
- models for human disease, 130, 136, 189-192
Animals, transgenic, 135, 136, 186-189
Antibacterial vaccines, 170
Antibiotic synthetases, 77, 107-109
Antibiotics, 74-77, 83, 100-109, 128, 169, 175, 242, 277, 279
-, resistance to, 102
Antibodies, chemotherapy - targeting toxins with, 134
-, monoclonal, 11, 28, 127, 131, 133, 146-149, 179-182
-, - antitumor, 180
Antibody tools, 11, 134, 159
Antigenic variation, 143-147
Antigens, cell surface, 16, 130, 159
-, class II HLA, 172
-, hepatitis B virus surface, 33, 133, 144, 148, 149, 155, 159, 160, 169
-, synthetic, 139-152
Anti-idiotype interactions 134, 174
Antineoplastic agents, 104, 105
Anti-sense mRNA, 252
Antitumor antibodies, monoclonal, 180
Approach, segmental support, 43, 44
Approaches, recombinant DNA, 128
Aqueous two-phase systems, 118, 119

Area, root-system, 276
Arginase, 174
Associative biological nitrogen fixation, 263
Attenuated vaccine, 132, 141-145
Attenuation, 144, 145
AUG, Shine-initiating, 168
Auxines, 277
Azospirillum, 221, 263, 264
Azotobacter, 221, 263, 264, 269, 276-278

B cell differentiation factor (BCDF), 172
- - growth factor (BCGF), 172
Bacillus amyloliquefacien, 116
- subtilis, 15, 131, 169
- thuringiensis, 73, 206, 221, 232
Bacteria, ice-nucleating, 261
-, rhizosphere, 277
Bacterial expression systems, 29-31
Bagasse, 283, 285
Beggiatoa, 262
β-galactosidase, 13, 28, 91
-/lactams, 77, 102-104, 109, 110
Bioconversion process, 77, 78, 105, 243, 244
Bioconversions, extractive, 117-120
Biogas, 286
Biological control, 260
- nitrogen fixation, 263-267, 270, 277, 278, 286
- - -, associative, 263
Biologically active proteins, 131
Biology, molecular, 199, 200, 227
-, plant molecular, 227
Biomass, 72, 87, 201, 208, 283-286
Bioreactors, 241, 243
Biosensor, 115, 121, 122
Biosynthesis, mutational, 77
Biotransformation, 243, 246
Blood homeostasis, 127, 130
Bond formation, disulfide, 9, 29, 31, 34
Bonds, hydrogen, 55, 61-67
Bone calcification, 130
Bone marrow transplantation, 179, 182
Bovine papilloma virus (BPV), 35
Brazil, 284, 285
Bypass, methylglyoxal, 87

Calcification, bone, 130
Callus, 201-203, 212-219, 229
Cardiovascular disease, 130, 175
Carotenoids, 246, 247
Cassette mutagenesis, 52
Catabolic pathway, 78
Catalytic rate, 61-63
cDNA libraries, 13, 20-22, 26, 166, 228, 289
Cell cultures, 242-246, 251, 253
- growth and differentiation, 130, 173
- - factor, nerve, 173
- lines, immortalized differentiated, 136
- recycling, 115, 117
- sorting, 16
- surface antigens, 16, 130, 159
Cells, Chinese hamster ovary, 11-14, 36, 132, 169
-, embryonic stem, 136, 185, 189-192
-, endothelial, 172
-, haploid, 205, 246
-, hemopoetic, 185, 192, 193
-, immobilized, 72, 78, 80, 115, 116
-, natural killer, 172
-, stem, 136, 185-193
Cellulase, 92, 288
Cellulose, 208, 227, 241, 283-289
Central nervous system, 130
Chalcone synthase, 249, 251
Channeling, metabolic, 251
Chemical synthesis of oligodeoxy-ribonucleotides, 42, 50
Chemotaxis, 276
Chemotherapy - targeting toxins with antibodies, 134
Chickens, Marek's disease in, 135, 136, 139

Chinese hamster ovary (CHO) cells, 11-14, 36, 132, 169
Chloramphenicol acetyltransferase gene, 159
CHO cells, 11-14, 36, 132, 139
Cholera, 129, 170
Cholesterol, 175
Citric acid, 85
Class II HLA antigens, 172
Claviceps purpurea, 116
Clavulanic acid, 74, 103, 108, 109
Cloning, expression, 26-28
-, gene, 14, 19-36, 76, 78, 108-111, 128, 165-170, 252
- techniques, 1, 20-25, 79, 108
Clostridium acetobutylicum, 116, 119
Coagulation Factor VIII, 30, 128, 131, 166, 167, 171
- - IX, 19, 128, 171
Coal and oil, desulfurization of, 73
Codon usage, 30-33, 165
Colitis, 170
Commune, nostoc, 28
Computer graphic model, 10, 52, 62
Conditions, hybridization, 23
Consequences, social, 135, 137
Construction, diagnostic gene, 236
-, gene, 46-48, 236
Continuous culture, 71, 78
Control, biological, 260
-, pollution, 73
-, process, 114, 120-122
Cornea, 130
Corynebacterium renale, 119
Co-transformation, 34-36
CO_2, supercritical, 118
Counselling, genetic, 128, 134, 137
Culture, continuous, 71, 78
Cultures, cell, 242-246, 251, 253
Current, leakage, 84, 88
Cyanobacteria, 263-266
Cyanogenic glycoside, 251
Cytomegalovirus, 133
Cytotoxic T lymphocytes, 161

Degradation, 78, 87, 93
Design, protein, 8, 41, 52, 61
Desulfurization of coal and oil, 73
Detoxification, 78
DHFR, 11, 35, 36, 61, 169, 170
Diagnostic gene construction, 236
Diagnostics, 133-135, 140, 179, 203, 204, 225, 226
Differentiated cell lines, immortalized, 136
Differentiation, cell growth and, 130, 173
- factor, B cell, 172
Dihydrofolate reductase (DHFR), 11, 35, 36, 61, 169, 170
Disease, Alzheimer's, 130, 134
-, animal models for human, 130, 136, 189-192
-, cardiovascular, 130, 175
-, foot and mouth, 129, 140-152
-, graft-versus-host, 173
- in chickens, Marek's, 135, 136, 139
- resistance, 201-207, 224, 231, 241, 247-249, 276
-, screening animal, 135
Disulfide bond formation, 9, 29, 31, 34
DNA amplification, 35, 36
- approaches, recombinant, 128
-, double-stranded, 5-7, 21, 41, 42, 46-52, 229
- libraries, genomic, 5, 20, 21, 166, 228, 289
- probes, 203, 204
- recombination, site-specific, 136
- synthesis, 21, 22, 41-43, 49, 50, 165, 166
- technology, recombinant, 1, 36, 41, 52, 77, 83, 111, 130, 137, 155, 158, 161, 223, 235, 294
Double-stranded DNA, 5-7, 21, 41,

Double-stranded DNA (cont.), 42, 46-52, 229
-/- RNA, 24, 236, 261
Downstream processing, 81, 114, 120
Dwarfism, 171

Ectomycorrhizal fungi, 267
Egg injection, 185-189, 191
Elements, transposable, 201, 213-218, 230, 251
Embryonic stem cells, 136, 185, 189-192
Endogenous pyrogen, 172
Endomycorrhizal fungi, 267, 268
Endothelial cells, 172
Energy requirement, maintenance, 84, 86
-/spilling reactions, 85, 87
Engineering, genetic (see Genetic engineering)
-, plant genetic, 223-237
-, protein, 5, 8-10, 55-67, 131, 174
Environment, growth, 71, 72, 83-85
Enzyme inhibitors, 74, 75, 105, 106
- production, 78
Epidermal growth factors, 128, 173
Epiphytic microorganisms, 275
Epitopes, immunogenic, 174
Erwinia carotovora, 277, 278
Erythropoietin, 166, 171
Ethanol, 6, 79, 84-87, 93, 116, 119, 284-289
Explosion, steam, 287
Expression cloning, 26-28
-, gene, 5, 11-13, 19-36, 75, 165-170, 185-193
- systems, 11, 12, 15, 24-36, 161, 230
- -, bacterial, 29-31
- -, mammalian, 12, 29, 33, 34
- -, yeast, 11, 15, 31-33
-, transient, 24-26, 230
- vectors, 25, 32, 57, 79, 157

Extractive bioconversions, 117-120
- fermentations, 114, 117, 118
Exudates, root, 276

Factor, α-tumor growth, 130
- analogues, growth, 130
-, B cell differentiation, 172
-, - - growth, 172
- VIII, coagulation, 30, 128, 131, 166, 167, 171
-, macrophage activating, 173
-, nerve cell growth, 173
- IX, coagulation, 19, 128, 171
-, platelet-derived growth, 173
- receptor analogues, growth, 130
-, skeletal growth, 128
-, tumor growth, 173
-, - necrosis, 128, 129, 167, 173
Factors, epidermal growth, 128, 173
-, social, 210, 283, 293, 294
Fermentation, extractive, 114, 117, 118
- products, 72, 79, 83-95, 107
-, solid-substrate, 283, 286
-, vacuum, 118-120
Fever, Lassa, 133
Fibroblasts, 172
Fixation, associative biological nitrogen, 263
-, biological nitrogen, 263-267, 270, 277, 278, 286
Flavonoid, 248-252
Flower pigments, 242, 246
FMDV, 129, 140-152
Folding, 8-11, 20, 31, 67, 131
Foot and mouth disease (FMDV), 129, 140-152
Forestry, 285
Formation, disulfide bond, 9, 29, 31, 34
4-coumarate:CoA ligase (4CL), 249
Fowlpox, 162
Frame shifting, 168
Frankia, 209, 264-266
Fungi, ectomycorrhizal, 267
-, endomycorrhizal, 267, 268

Fungicides, 278
Furanocoumarin, 252

Gene, chloramphenicol acetyltransferase, 159
- cloning, 14, 19-36, 76, 78, 108-111, 128, 165-170, 252
- construction, 46-48, 236
- -, diagnostic, 236
- expression, 5, 11-13, 19-36, 75, 165-170, 185-193
- insertion, 79, 159, 201, 202, 205-207, 210, 233, 234
- machines, 42
- regulation, 136, 189
-, silent, 76
- therapy, somatic, 136, 185
-, thymidine kinase, 159, 169, 186-190
- transfer, 78, 108, 127, 135, 136, 185, 189-193, 201, 223, 228, 230, 236, 241, 242, 246, 252, 253
Genes, synthetic, 41-53
Genetic counselling, 128, 134, 137
- engineering, 1, 16, 36, 42, 75-81, 111-114, 144, 156, 165-175, 180, 199, 200, 207, 223-237, 252, 253, 279, 293, 294
- -, plant, 223-237
Genomic DNA libraries, 5, 20, 21, 166, 228, 289
Giberellic acid, 277
Gluconic acid, 85, 88, 89
Glycoprotein D, 155, 159, 160
-, rabies virus, 155, 159, 161
Glycoside, cyanogenic, 251
Graft-versus-host disease, 173
Gramineae, 212
Graphic model, computer, 10, 52, 62
Growth and differentiation, cell, 130, 173
- environment, 71, 72, 83-85
- factor, -tumor, 130
- - analogues, 130
- -, B cell, 172
Growth factor, nerve cell, 173
- -, platelet-derived, 173
- - receptor analogues, 130
- -, skeletal, 128
- -, tumor, 173
- factors, epidermal, 128, 173
- hormone, 19, 30, 34, 35, 128, 130, 135, 168-173, 186-190, 268
- -, human, 34, 35, 168, 171, 173
- -, skeletal, 130
-, hypocotyl, 277
- promoting substances, plant, 269
- regulators, plant, 277, 279

Hairs, root, 267, 277
Haploid cells, 205, 246
Heavy metal toxicity, 207
Hemagglutination inhibition, 159
Hemagglutinin, influenza virus, 149, 150, 155, 159
Hemicellulose, 208, 285-289
Hemophilia, 171
Hemopoetic cells, 185, 192, 193
Hepatitis B vaccine, 170
- - virus, 33, 129, 133, 141, 144, 148, 149, 155, 159, 160, 169, 170, 180
- - - surface antigens, 33, 133, 144, 148, 149, 155, 159, 160, 169
Herbicides, 175, 206, 224, 231, 232, 278
Herpes simplex virus, 129, 155, 159, 160, 170, 187
- - - thymidine kinase, 159
HLA antigens, class II, 172
Homeostasis, blood, 127, 130
Hormone, growth, 19, 30, 34, 35, 128, 130, 135, 168-173, 186-190, 268
-, human growth, 34, 35, 168, 171, 173
-, skeletal growth, 130
Hormones, parathyroid, 128, 130
-, peptide, 2, 127
Human disease, animal models for, 130, 136, 189-192
- growth hormone, 34, 35, 168, 171, 173

Hybrid molecules, 76
Hybrid selection screening, 27
- -/translation, 166, 167
Hybridization conditions, 23
-, somatic, 246
-, subtractive, 28
Hydrogen bonds, 55, 61-67
Hypocotyl growth, 277

Ice-nucleating bacteria, 261
IL-1, 128, 172
IL-2, 2, 19, 27, 61, 128, 133, 169, 172
Immobilization, 115, 116, 120, 210, 244, 245
Immobilized cells, 72, 78, 80, 115, 116
Immortalized differentiated cell lines, 136
Immune system, 2, 130-133, 137, 142, 146, 149, 150, 162, 171-173, 179
Immunization, passive, 133, 180
Immunoassays, 173
Immunogenic epitopes, 174
Immunoglobulin, 8, 9, 172, 180, 186, 189
Immunomodulators, 105-107, 133, 172
Immunosurveillance, 172
Improvements of second generation proteins, 130, 131
In vitro transcription system, 27, 28
Industry, 1-5, 16, 34, 41, 71, 72, 100, 137, 165, 175, 201, 241, 242
-, pharmaceutical, 1, 3, 34, 41, 137, 165, 175, 241, 242
Inert vaccines, 133
Infection, 139, 225, 227, 276
-, virus, 139, 225, 227
Influenza virus, 129, 140, 143, 147-150, 155, 159
- - hemagglutinin, 149, 150, 155, 159
Inhibition, hemagglutination, 159
Inhibitors, enzyme, 74, 75, 105, 106
Injection, egg, 185-189, 191
Inoculation, 155, 161, 209, 260-262, 265-271, 277, 278
Insecticides, 242, 278
Insertion, gene, 79, 159, 201, 202, 205-207, 210, 233, 234
Insulin, 19, 29, 30, 79, 128, 134, 171
Intensification, process, 114-116, 123
Interactions, anti-idiotype, 134, 174
-, plant-microbe, 259-271, 275-279
Interferon, 19, 27, 30-33, 44, 61, 127-130, 165-174
Interleukin-1 (IL-1), 128, 172
Interleukin-2 (IL-2), 19, 27, 61, 128, 133, 169, 172
Isolation and characterization of oligonucleotides, 44

Lactic acid, 85
Lassa, 129, 133
- fever, 133
Leaching, metal, 73
Leakage current, 84, 88
Leguminosae, 214
Lettuce, 268, 277
Libraries, cDNA, 13, 20-22, 26, 166, 228, 289
-, genomic DNA, 5, 20, 21, 166, 228, 289
Library screening method, 22-28
Ligase, 4-coumarate:CoA, 249
Ligation strategy, 6
Lignin, 208, 241, 249, 285-289
Ligninase, 208, 289
Lignocellulose, 208, 283-290
Lines, immortalized differentiated cell, 136
Linkage map, 203, 237
Lipases, 92
Live vaccines, recombinant, 129, 132, 133, 137, 155, 159, 161
Lupus, 173
Lymphocytes, cytotoxic T, 161
Lymphokines, 130, 172, 173
Lymphotoxin, 173
Lytic viral vectors, 34

Machines, gene, 42
Macrophage Activating Factor (MAF), 173
Maintenance energy requirement, 84, 86
Major late promoter, adenovirus, 24
Malaria, 129, 133, 170, 293
Mammalian expression systems, 12, 29, 33, 34
Map, linkage, 203, 237
Marek's disease in chickens, 135, 136, 139
Marrow transplantation, bone, 179, 182
Mass spectrometry, 45, 51
MDP, 133, 151
Media, production, 245
Medicine, 2, 5, 16, 127-137, 165-175
Metabolic channeling, 251
Metabolite overproduction, 92-94, 100
Metabolites, secondary, 74, 75, 78, 88, 95, 99-111, 209, 210, 234
Metal leaching, 73
- toxicity, heavy, 207
Method, library screening, 22-28
Methylglyoxal bypass, 87
Mice, transgenic, 186-189
Microbial physiology, 95, 123
Microflora, rhizosphere, 275
-, soil, 275
Microorganisms, epiphytic, 275
-, phytopathogenic, 277, 279
-, rhizosphere, 268-270, 275, 276
-, saprophytic, 277
Micropropagation, 202, 203, 206, 216-219
Mismatch primer mutagenesis, 52
Model, computer graphic, 10, 52, 62
Modeling, protein, 5, 9, 10
Models for human disease, animal, 130, 136, 189-192
Modification, post-transcriptional, 15, 16, 19, 131, 132, 136
Molecular biology, 199, 200, 227
- -, plant, 227
Molecules, hybrid, 76
Monasus sp., 119
Monoclonal antibodies, 11, 28, 127, 131, 133, 146-149, 179-182
- antitumor antibodies, 180
Monokines, 173
Morinda citrifolia, 116
mRNA, anti-sense, 252
Multiple sclerosis, 173
Muramyl dipeptide (MDP), 133, 151
Mushrooms, 286
Mutagenesis, 5-9, 14, 31, 41, 52, 55-61, 130, 205, 208, 227, 230, 233, 252, 265, 288
-, cassette, 52
-, mismatch primer, 52
-, site-directed, 31, 41, 52, 55-61, 130, 265, 288
-, transposon, 14, 205, 227, 230, 233
Mutants, recA, 79
Mutational biosynthesis, 77
Mutations, 11, 33, 55, 58-60, 65, 77, 108, 204, 205, 233, 279
Mycorrhizas, 73, 202, 207, 209, 221, 259, 262, 267, 268
Mycorrhized plants, 276

Natural killer cells, 172
Necrosis factor, tumor, 128, 129, 167, 173
Nematode, 75, 207, 221
Nerve cell growth factor, 173
Nervous system, central, 130
N-fixing, 209, 259, 262-266, 277, 278
Nitrogen fixation, associative biological, 263
- -, biological, 263-267, 270, 277, 278, 286
Nitrogenase, 116, 263-265
Non-lytic viral vectors, 34, 35
Non-mycorrhized plants, 276
Nostoc commune, 28
Nucleic acid probes, 134, 226
Nutrient uptake, 260, 264, 267, 270, 276

Oil, desulfurization of coal and, 73
- recovery, 73
Oligodeoxyribonucleotides, chemical synthesis of, 42, 50
Oligonucleotide probes, 22, 23, 167
Oligonucleotides, 5-7, 13, 22, 23, 42-51, 56, 57, 60, 134, 167, 228
-, isolation and characterization of, 44
-, synthetic, 7, 13, 167
Oncogenes, 130, 135, 136, 173, 174, 188
Organic solvent, 115, 118
Ovalbumin, 167
Overproduction, metabolite, 92-94, 100

PAL, 249, 250
Papilloma virus, bovine, 35
Parathyroid hormone, 128, 130
Passive immunization, 133, 180
Pathway, catabolic, 78
PDGF, 173
Penicillium chrysogenum, 116
Peptide hormones, 2, 127
Peptides, pharmaceutical, 128
-, synthetic, 127, 133, 139, 145-152
Pesticides, 175, 276
Phage, SP6, 27
Phanerochaete chrysosporium, 208, 289
Pharmaceutical industry, 1, 3, 34, 41, 137, 165, 175, 241, 242
- peptides, 128
Phenylalanine ammonia-lyase (PAL), 249, 250
Phosphatase, alkaline, 168
Physiology, microbial, 95, 123
Phytoalexins, 242, 247-250
Phytopathogenic microorganisms, 277, 279
Pigments, flower, 242, 246
Plant genetic engineering, 223-237
- growth promoting substances, 269
- - regulators, 277, 279
Plant-microbe interactions, 259-271, 275-279
- molecular biology, 227
- transformation, 229, 230, 237
Plants, mycorrhized, 276
-, non-mycorrhized, 276
Plasmodium knowlesi, 159, 160
Platelet-derived growth factor (PDGF), 173
Poliovirus, 129, 140-150
Pollution control, 73
Polymers, storage, 85
Polyvalent vaccines, 161, 162
- vaccinia virus, 161
Post-translational modification, 15, 16, 19, 131, 132, 136
Potential, water, 116
Poxvirus, 155-162
- family, 157, 162
Primer mutagenesis, mismatch, 52
Probes, DNA, 203, 204
-, nucleic acid, 134, 226
-, oligonucleotide, 22, 23, 167
Process, bioconversion, 77, 78, 105, 243, 244
- control, 114, 120-122
- intensification, 114-116, 123
Processing, downstream, 81, 114, 120
Production, enzyme, 78
- media, 245
-, protein, 131, 165-175
Products, fermentation, 72, 79, 83-95, 107
-, secondary, 200, 206, 241-253
Proinsulin, 165, 169
Promoter, adenovirus major late, 24
Promoters, signal-specific, 252
-, transcriptional, 12, 24, 28, 32
Proteases, 79, 80, 92, 119, 130, 169, 171, 237
Protein design, 8, 41, 52, 61
- engineering, 5, 8-10, 55-67, 131, 174
- modeling, 5, 9, 10
- production, 131, 165-175
-, rabies surface, 169

Proteins, biologically active, 131
-, improvements of second generation, 130, 131
Proteoid roots, 262
Protoplast, 201, 205, 220, 228-230, 245
Pseudomonas aeruginosa, 119
- lachrimans, 277, 278
Pyrogen, endogenous, 172

Rabies, 129, 155, 159, 161, 169, 170
- surface protein, 169
- virus glycoprotein, 155, 159, 161
Rate, catalytic, 61-63
Reactions, energy-spilling, 85, 87
recA mutants, 79
Receptor analogues, growth factor, 130
Recombinant DNA approaches, 128
- -technology, 1, 36, 41, 52, 77, 83, 111, 130, 137, 155, 158, 161, 223, 235, 294
- live vaccines, 129, 132, 133, 137, 155, 159, 161
- pharmaceutical peptides, 128
- vaccinia virus, 155-162
Recombination, site-specific DNA, 136
Recovery, 79-81
Recycling, cell, 115, 117
Reductase, dihydrofolate, 11, 35, 36, 61, 169, 170
Regulation, gene, 136, 189
Regulators, plant growth, 277, 279
Rejection, allograft, 173
Requirement, maintenance energy, 84, 86
Resistance, disease, 201-207, 224, 231, 241, 247-249, 276
- to antibiotics, 102
Responsibilities, social, 41, 52, 53
Retroviral vectors, 35, 135, 136, 185, 190-192
Rhizobium, 73, 202, 207, 209, 221, 227, 260, 261, 264-270, 276-278
Rhizoctonia solani, 261
Rhizosphere bacteria, 277
- microflora, 275
- microorganisms, 268-270, 275, 276
RNA, double-stranded, 24, 236, 261
Root exudates, 276
-/free soil, 275
- hairs, 267, 277
-/system area, 276
Roots, proteoid, 262
Rosmarinic acid, 243, 244

Saccharomyces cerevisiae, 116, 119, 169
Saprophytic microorganisms, 277
Sclerosis, multiple, 173
Screening animal disease, 135
-, hybrid selection, 27
- method, library, 22-28
- systems, 74, 75, 105
Second generation proteins, improvements of, 130, 131
Secondary metabolites, 74, 75, 78, 88, 95, 99-111, 209, 210, 234
- products, 200, 206, 241-253
Secretion, selective, 80
Secretory leader, 31, 33
Segmental support approach, 43, 44
Selection screening, hybrid, 27
-/translation, hybrid, 166, 167
Selective secretion, 80
Sequential sublining, 181
Serine deaminase, 91
Serum albumin, 128, 173
Shifting, frame, 168
Shikonin, 243, 244
Shine-Dalgarno sequence, 168
-/initiating AUG, 168
Side-roots, 277
Siderophores, 269
Signal-specific promoters, 252
Silent gene, 76
Site-directed mutagenesis, 31, 41, 52, 55-61, 130, 265, 288
-/specific DNA recombination, 136

Skeletal growth factor, 128
- - hormone, 130
Smallpox, 137, 139, 155-157
Smallpox vaccine virus, 156, 157
Social consequences, 135, 137
- factors, 210, 283, 293, 294
- responsibilities, 41, 52, 53
Soil microflora, 275
-, root-free, 275
Solid adsorbents, 118, 119
-/substrate fermentation, 283, 286
Solvent, organic, 115, 118
Somatic gene therapy, 136, 185
- hybridization, 246
Somatomedins, 128, 173
Somatostatin, 165
Sorting, cell, 16
Soya, 175
Specificity, substrate, 61, 66, 67
Spectrometry, mass, 45, 51
SP6 phage, 27
Stability, strain, 78, 79
Steam explosion, 287
Stem cells, 136, 185-193
- -, embryonic, 136, 185, 189-192
Steroid, 242, 244, 247
Storage polymers, 85
Strain stability, 78, 79
Strategy, ligation, 6
Streptomyces, 11, 15, 76, 77, 103, 104, 108-110, 116, 169, 277, 279
- aurefaciens, 116
- clavuligerus, 103, 108
- griseus, 119
Stress, water, 267
Sublining, sequential, 181
Substances, plant growth promoting, 269
Substrate affinity, 9, 56, 61-65
- specificity, 61, 66, 67
Subtractive hybridization, 28
Sugarcane, 283-285
Supercritical CO_2, 118
Support approach, segmental, 43, 44
Surface antigens, cell, 16, 130, 159
Surface antigens, hepatitis B virus, 33, 133, 144, 148, 149, 155, 159, 160, 169
- protein, rabies, 169
SV40 virus, 12, 24, 25, 34, 142, 169, 186-190
Symbiosis, 209, 227, 262-267
Synthase, chalcone, 249, 251
Synthesis, DNA, 21, 22, 41-43, 49, 50, 165, 166
- of oligodeoxyribonucleotides, chemical, 42, 50
Synthetases, antibiotic, 77, 107-109
Synthetic antigens, 139-152
- genes, 41-53
- oligonucleotides, 7, 13, 167
- peptides, 127, 133, 139, 145-152
System, central nervous, 130
-, immune, 2, 130-133, 137, 142, 146, 149, 150, 162, 171-173, 179
-, in vitro transcription, 27, 28
Systems, aqueous two-phase, 118, 119
-, bacterial expression, 29-31
-, expression, 11, 12, 15, 24-36, 161, 230
-, mammalian expression, 12, 29, 33, 34
-, screening, 74, 75, 105
-, yeast expression, 11, 15, 31-33

T lymphocytes, cytotoxic, 161
Targeting toxins with antibodies, chemotherapy, 134
T-DNA, 228-234, 237
Techniques, cloning, 1, 20-25, 79, 108
Technology, recombinant DNA, 1, 36, 41, 52, 77, 83, 111, 130, 137, 155, 158, 161, 223, 235, 294
Terpenoids, 246, 247
TGF, 173
Therapy, somatic gene, 136, 185
Thermomonospora, 289
Thermostability, 67
Thrombosis, 171
Thymidine kinase gene, 159, 169,

Thymidine kinase gene (cont.), 186-190
- -, herpes simplex virus, 159
Tissue plasminogen activator (TPA), 29, 128, 129, 171
- transplantation, 133
TNF, 128, 129, 167, 173
Tools, antibody, 11, 134, 159
Totipotency, 246
Toxicity, heavy metal, 207
TPA, 29, 128, 129, 171
Transcription system, in vitro, 27, 28
Transcriptional promoters, 12, 24, 28, 32
Transfer, gene, 78, 108, 127, 135, 136, 185, 189-193, 201, 223, 228, 230, 236, 241, 242, 246, 252, 253
Transformation, plant, 229, 230, 237
Transgenic animals, 135, 136, 186-189
- mice, 186-189
Transient expression, 24-26, 230
Transplantation, bone marrow, 179, 182
-, tissue, 133
Transposable elements, 201, 213-218, 230, 251
Transposon mutagenesis, 14, 205, 227, 230, 233
Trichoderma reesei, 288
Tripeptide, ACV, 77, 109, 110
Tumor growth factor (TGF), 173
- - -, α-, 130
- necrosis factor (TNF), 128, 129, 167, 173
Two-phase systems, aqueous, 118, 119

Uptake, nutrient, 260, 264, 267, 270, 276
Usage, codon, 30-33, 165

Vaccination, 139, 140, 143, 157, 172
Vaccine, attenuated, 132, 141-145
-, hepatitis B, 170
- virus, smallpox, 156, 157
Vaccines, 127, 129, 130-137, 139-152, 155-162, 169, 170, 189, 293
-, antibacterial, 170
-, inert, 133
-, polyvalent, 161, 162
-, recombinant live, 129, 132, 133, 137, 155, 159, 161
Vaccinia, 132, 139, 142-145, 155-162
- virus, polyvalent, 161
- -, recombinant, 155-162
Vacuum fermentation, 118-120
Variation, antigenic, 143-147
Variola, 155
Vectors, expression, 25, 32, 57, 79, 157
-, lytic viral, 34
-, non-lytic viral, 34, 35
-, retroviral, 35, 135, 136, 185, 190-192
-, virus, 12, 34, 144
Virulence, 139-145
Virus, AIDS, 129, 135, 171
-, bovine papilloma, 35
- glycoprotein, rabies, 155, 159, 161
- hemagglutinin, influenza, 149, 150, 155, 159
-, hepatitis B, 33, 129, 133, 141, 144, 148, 149, 155, 159, 160, 169, 170 180
-, herpes simplex, 129, 155, 159, 160, 170, 187
- infection, 139, 225, 227
-, influenza, 129, 140, 143, 147-149, 150, 155, 159
-, polyvalent vaccinia, 161
- recombinant, vaccinia, 155-162
-, smallpox vaccine, 156, 157
- surface antigen, hepatitis B, 33, 133, 144, 148, 149, 155, 159, 160, 169
-, SV40, 12, 24, 25, 34, 142, 169, 186-190
- thymidine kinase, herpes simplex, 159

Virus vectors, 12, 34, 144

Water potential, 116
- stress, 267

Xenopus oocyte, 24, 27, 166

Yeast expression systems, 11, 15, 31-33

Zymomonas 86, 87, 208, 289

Author Index

Abelson, P.H.; 199-221
Aharonowitz, Y.; 71-81, 99-112
Barnes, W.M.; 5-18
Barz, W.; 199-221
Bedouelle, H.; 55-69
Beringer, J.E.; 199-221, 259-273
Betz, H.; 127-138
Blöcker, H.; 5-18, 41-54
Broda, P.; 199-221, 283-291
Bu'lock, J.D.; 71-81
Carter, P.; 55-69
Chakrabarty, A.M.; 71-81
Cohen, G.; 99-112
Collins, J.; 127-138
Davies, J.E.; 127-138, 293-294
Dennis, E.S.; 223-239
Eveleigh, D.E.; 199-221
Fersht, A.R.; 55-69
Fiers, W.; 127-138, 165-177
Flavell, R.B.; 199-221
Frank, R.; 41-54
Hahlbrock, K.; 199-221, 241-257
Hopwood, D.A.; 71-81
Kahmann, R.; 199-221
Kamen, R.; 19-40
Kaufman, R.; 19-40
Kay, R.M.; 19-40
Krawinkel, U.; 5-18, 179-183
Kula, M.-R.; 71-81
Leatherbarrow, R.J.; 55-69
Lehrach, H.R.; 5-18
Lipinskas, B.R.; 155-164
Lowe, D.; 55-69
Marquardt, R.; 5-18
Mattiasson, B.; 71-81, 113-125
Mercer, S.R.; 155-164
Morris, J.G.; 71-81
Neijssel, O.M.; 71-81, 83-97
Paoletti, E.; 127-138, 155-164
Peacock, W.J.; 199-221, 223-239
Perkus, M.E.; 155-164
Pfaff, E.; 127-138
Piccini, A.; 155-164
Rowlands, D.J.; 127-138, 139-154
Ruether, U.; 185-196
Saedler, H.; 199-221
Sahm, H.; 71-81
Salamini, F.; 199-221
Schaller, H.; 127-138
Schendel, P.; 19-40
Siewert, G.; 127-138
Silver, S.; 1-3, 71-81
Sippel, A.E.; 127-138
Smith, M.; 5-18
Sobieszczański, J.; 71-81, 275-282
Stahl, U.; 71-81
Starlinger, P.; 199-221
Stewart, C.L.; 185-196
Tempest, D.W.; 83-97
Timmis, K.N.; 71-81
Turner, K.; 19-40
van Montagu, M.; 5-18
Wagner, E.F.; 127-138, 185-196
Weglenski, P.; 5-18
Weiler, E.; 199-221
Winter, G.; 5-18, 55-69
Wodak, S.J.; 5-18
Zenk, M.; 199-221

MIX
Papier aus verantwortungsvollen Quellen
Paper from responsible sources
FSC® C105338

If you have any concerns about our products,
you can contact us on
ProductSafety@springernature.com

In case Publisher is established outside the EU,
the EU authorized representative is:
Springer Nature Customer Service Center GmbH
Europaplatz 3, 69115 Heidelberg, Germany

Printed by Libri Plureos GmbH
in Hamburg, Germany